*Instructor's Resource Guide to accompany*

# Chemistry

# *& Chemical Reactivity*

SECOND EDITION

**Kotz & Purcell**

**John C. Kotz**
State University of New York at Oneonta

*In collaboration with,*

**William Vining**
Hartwick College

**David Koster**
Southern Illinois University

**Keith Purcell**
Kansas State University

**Saunders Golden Sunburst Series**
**SAUNDERS COLLEGE PUBLISHING**
Philadelphia Fort Worth Chicago
San Francisco Montreal Toronto
London Sydney Tokyo

Printed in the United States of America.

Kotz: Instructor's Resource Guide to accompany: CHEMISTRY AND CHEMICAL REACTIVITY, 2/e.

ISBN # 0-03-053484-4

123 082 987654321

# TABLE OF CONTENTS

# PREFACE

In the Preface to the *Instructor's Resource Manual* for the first edition of *Chemistry and Chemical Reactivity*, we stated that "teaching chemistry today is an increasingly important but difficult and time-consuming task. We seem to deal with a broader spectrum of student abilities and with classes having students that range from the well-prepared to those with ability but with poor preparation." Now, four years later, this is more true than ever, and we are continuing to see a decline in enrollment in chemistry courses in many universities. Therefore, it is increasingly important that we make chemistry even more interesting and important while retaining the intellectual rigor of the subject. To this end, our courses must be well-organized with clear, understandable, and reasonable goals. Thus, in addition to providing answers to many of the Study Questions in *Chemistry and Chemical Reactivity*, this *Instructor's Resource Manual* addresses the organization of the course and has useful information on a most important aspect of our task of teaching chemistry, *lecture demonstrations.*

## Answers to Study Questions

One purpose of this *Instructor's Resource Manual* is to provide answers to the "non-designated" Study Questions in *Chemistry and Chemical Reactivity.* Many Study Questions in the text are numbered with a blue number. The answers to these questions are given in the *Student Solutions Manual* prepared by Alton Banks. This *Instructor's Resource Manual* provides answers to questions not answered in the *Student Solutions Manual.* However, unlike the very complete answers in the *Student Solutions Manual,* relatively brief answers are provided in the present manual. Further, in most chapters the first ten to twenty questions are "review questions" whose answers are readily found in the text. Thus, in most cases answers to these questions are not given in this *Manual.*

## Lecture Demonstrations

We hope that the textbook conveys a clear and organized view of chemistry as an interesting and valuable subject. Of course, these are also the goals of the lectures we all present. To help in this, we believe lecture demonstrations are useful and important, and we attempt to incorporate at least one demonstration into every lecture. (Some of these demonstrations were photographed for the textbook, as were reactions too dangerous to be done in the classroom.)

Therefore, in each chapter of this *Manual* we have provided information on the demonstrations we have used in our lectures. Further notes on demonstrations are given in the next section of this *Manual.*

You will notice that many of the photos in the book and the demonstrations we do are taken from the excellent series of books by Professor Bassam Shakhashiri, *Chemical Demonstrations.* Therefore, we are very pleased that an additional feature of the second edition of our text is the **Chemical Demonstrations Videotape** of Professor Shakhashiri performing a number of demonstrations. This tape is available to adopters of the text.

Also Available with this text are videotapes based on the television series *World of Chemistry,* with Roald Hoffmann as the host. These tapes are each about 15 minutes long and provide introductory material on such subjects as the mole, the atom, the periodic table, bonding, water, atmospheric chemistry, kinetics, catalysis, acids and bases, oxidation-reduction, and other subjects. They are suitable for classroom use as an introduction to a topic. Most importantly, they serve to place a particular area of chemistry in context.

## Overhead Transparencies that Accompany the Text

Accompanying the textbook are 150 overhead transparencies of figures and tables in the book. The transparencies available for each chapter are listed in this *Manual.*

## Computer Software and Videodisc

A computer program called **KC? Discoverer** from *JCE: Software,* is available in both MS-DOS and Macintosh format. The program is available at no charge to adopters of the text or it may be purchased directly from *JCE:Software:* Department of Chemistry, University of Wisconsin-Madison, 1101 University Avenue, Madison, WI 53706. This program is an extensive database of information on the chemical elements. As the database was carefully assembled by a team of chemists, *it has been chosen as the "official" database for the text and all the tables in the text that refer to atomic properties reflect the information in* ***KC? Discoverer.*** Further information on the program is given in the next section of this *Manual,* and we shall call attention to places where the program can be used throughout.

The **Periodic Table Videodisc: Reactions of the Elements**, also published by *JCE: Software,* is available to adopters of the text. This videodisc, which can be driven by **KC? Discoverer** or can stand alone, shows still and moving images of the elements, their uses, and their reactions with air, water, acids, and bases. It is particularly useful as a way to demonstrate chemical reactions in a large lecture room.

## Instructor's Notes

In each chapter we have described the number of lectures roughly allocated to each topic and have noted where the topics in the text may be placed as an alternate to the order in which they are presented in the text.

## Acknowledgements

This *Manual* was put together as a collaborative effort. Alton Banks worked out answers for many of the problems in the last few chapters of the first edition of the text. I thank him for his meticulous work on these as well as his outstanding work on the *Student Solutions Manual* that accompanies the text. Susan Hughson checked the answers for all of the Exercises and Study Questions in the first edition of the text (a number of which were repeated in the second edition), and Professor William Vining of Hartwick College and Professor David Koster of Southern Illinois University worked all of the Study Questions in the second edition. Their help and cooperation are greatly appreciated. Any errors that remain in this *Manual* are mine, and you should contact me if you have corrections or comments.

This *Manual* was prepared using a Macintosh computer and printed on an Apple LaserWriter IISC. Text was prepared using Wordperfect™, and the figures were drawn with ChemDraw™ and Canvas 2.1™. In addition, plots were prepared with "Drexel Plot" from Kinko's Academic Courseware Exchange. Molar masses were checked with ChemIntosh™ MM Calc, and some calculations were done using Excel 2.2™. Finally, I plan to make available the computer files of some of the sketches made for the text as "chemistry clip art." These will be available from Project SERAPHIM, Department of Chemistry, University of Wisconsin-Madison, Madison, WI 53706.

As a final note, I want to again thank my wife Katie. None of this would have been possible without her patience and love.

John C. Kotz
January 1991

# LECTURE DEMONSTRATIONS COMPUTER SOFTWARE

## LECTURE DEMONSTRATIONS

We believe lecture demonstrations should serve to motivate student interest and to illustrate the point at hand. Further, because most of us to do not have a technician to prepare these demonstrations, and because we are all busy with many other tasks, lecture demonstrations should be simple to prepare. And finally, because most lecture rooms do not have hoods, these demonstrations must not generate noxious fumes.

Each chapter in this *Manual* gives some information on the demonstrations we have found helpful. Most need relatively little preparation, and all are safe for the classroom. (Where there is doubt in this regard, there is mention of this fact.) Many of the demonstrations are in fact quite simple and can be prepared in the hour before class. To help in this, however, I keep on a separate shelf in a stockroom all of the items of equipment I need from year to year, and many of the required solutions are stored in the stockroom and have been prepared in amounts large enough to last for several years.

Some of the ideas for demonstrations are my own, but many have been obtained from colleagues, from other books, and from the literature. You will see many of these ideas in the photographs in the textbook. One of the best sources of demonstrations is the following:

Bassam Z. Shakhashiri, *Chemical Demonstrations,* The University of Wisconsin Press, Madison. Volume 1, 1983; Volume 2, 1985; and Volume 3, 1989.

I have used many of the demonstrations from these books in the classroom, and you will notice in the textbook that a number of the photographs are based on Shakhashiri's experiments. In addition, I have noted in this *Manual* where demonstrations from Shakhashiri can be incorporated. Finally, we are pleased that Professor Shakhashiri has prepared a **Chemical Demonstrations Videotape**, which are available to adopters of this text.

The *Journal of Chemical Education* is an excellent source of demonstrations and experiments. To facilitate looking for material, Stanley Bunce of Renssaeler Polytechnic Institute has compiled a database of over 2000 experiments and demonstrations published in the *Journal* from 1957-1982. The disks, readable on an IBM or compatible computer, can be obtained for $25 by writing to Project SERAPHIM, Department of Chemistry, University of Wisconsin-Madison, Madison, WI 53706. In various places in this *Manual* I have given literature references to demonstrations. In many cases, these references were found through this database.

A problem that many of us face in the lecture room is that there are no facilities for gas and water. Enough water can be brought into the lecture room in a large beaker to do many of the demonstrations outlined in this *Manual.* However, a source of gas is a problem. We have solved this by using a small, butane-fueled backpacking stove.

Many of the suggested demonstrations appear in the text as **color photographs.** As noted in the text, these photos were done in collaboration with Charles Winters of SUNY-Oneonta. A great deal of effort went into this project, and we believe we have developed some useful techniques and have come to understand better the limitations of the medium. If you wish to share information on chemical photography, Charlie and I would be pleased to hear from you.

Finally, I am especially interested in expanding my own efforts in the use of lecture demonstrations. I would welcome the opportunity to exchange information and ideas with colleagues at other colleges and universities.

## COMPUTER SOFTWARE: USING KC? DISCOVERER AND THE PERIODIC TABLE VIDEODISC

As noted in the Preface, the Project SERAPHIM program KC? Discoverer has been chosen as the official database for the book.[1] For further information on the program and its use in a classroom setting, see J. Kotz, *J. Chem. Educ.*, **1989**, *66*, 750. What follows are excerpts from that article.

The database for KC? Discoverer contains 49 different pieces of information for each of the first 103 elements. The computer presents the information for each element in a series of three screens. The screens for boron from the Macintosh Hypercard version of the program are given on pages 6-8. The first of these, as illustrated for boron on page 6, consists of macroscopic properties such as melting point, hardness, and conductivity. Furthermore, this screen lists some useful information on reactivity: the results of reacting the element with air, water, or common acids and bases.

The second screenful of information for each element (page 7) consists of atomic-level properties, and the formulas of the hydrides, oxides, and chlorides formed by the element are given. Finally, the third information screen for an element (page 8) illustrates the solid state structure of the element.

The program is menu-driven so that students can learn to use KC? Discoverer with a minimum of instruction. For example, after starting the program the first screen of instructions asks if the user wishes to:

- FIND the element(s) that has (have) a specified range of values for a selected property.
- GRAPH one numeric property against another.
- LIST the name, symbol, atomic number, and up to four other properties for all elements or for a selected group of elements, arranged in order of increasing atomic number.

---

1 The MS-DOS version of the program is called *KC? Discoverer* and the Macintosh version is a Hypercard stack called *The Periodic Table Stack*. The KC portion of the name stands for "knowledgeable counselor." The original conception of the program was that it could be used to *assist* the user in finding a solution to a problem. This variation on the program is under active development.

- SORT elements in increasing alphabetic or numeric order for one property.
- use the periodic TABLE to graph or sort information regarding a property within a group or period.
- ask for HELP with any of the functions of the programs. Help is always available from within the program simply by typing "?".

The HELP function is useful for two reasons. First, the user can learn about each of the functions of the program and how they can be used. And second, one can obtain background information on each type of data in the database. The user is first given a definition of the term, some discussion concerning the meaning of the information and, in some cases, how such values are obtained experimentally. Following this the program indicates the source of the data in the chemical literature. Our students find this function to be quite useful, not only because it reminds them of the correct definition of terms, but also because it is important for them to know the source of the data in order to assess its accuracy.

The FIND function allows you to search for all the elements having a particular property or attribute. If you choose a numerical property the program allows you to specify a range of values within which to search. When searching for non-numerical information, the user is prompted to choose from a list of characteristics for a given property. For example, one may find all the elements having a certain color or a particular type of solid state structure Alternatively, the user can search for all the elements having a given reaction with air, water, 6 M HCl, 15 M $HNO_3$, or 6 M NaOH. In the case of reactions, furthermore, it is also possible to find elements that have no reaction, a mild reaction, or a vigorous reaction. Finally, using the version of the program that is tied to the **Periodic Table Videodisc**, you can then see video of these reactions occurring in real time or in slow motion. It is simple, for example, to compare the relative reactivities of the alkali or alkaline earth metals and use this to discuss periodic trends. (When watching these reactions, you also notice many other examples of interesting chemistry, such as the colors of transition metal ions in acid media and the solubilities of common salts. The **Videodisc** is a particularly useful way to explore oxidation-reduction reactions in general.)

GRAPH is one of the most useful options available in KC? Discoverer. Although plots of some numerical quantity against atomic number are perhaps most useful, it is possible that a plot using some parameter other than atomic number on the ordinate would be informative. For example, plotting the heat of vaporization against boiling point illustrates the essential validity of Trouton's rule (that a wide range of liquids have an entropy of vaporization of about 85 J/K·mol). Another important way to use GRAPH is to plot some numerical parameter on the x-axis against a *combination* of parameters on the y-axis. For example, a plot of the quotient of the second and first ionization energies of the elements versus atomic number can be used as experimental evidence for the shell/subshell electronic structure of the atom.

The GRAPH function is especially useful in the classroom. Graphs may be viewed in class on a large-screen monitor or by means of a computer projector device. (Alternatively, a printed copy can be made into a transparency on a copy machine.) After showing the class a series of graphs illustrating various periodic trends, the students are asked if they wish to try new relationships or to ask "what-if" questions using the computer. For example, what do we expect to see if electron affinity is plotted against the first ionization energy for the first 36 elements, or what is the relation between the expected cost of the element and its abundance in the earth's crust?

The LIST and SORT commands are similar in that both provide a list of values of properties for a selected list of elements. With LIST the user can, for example, list all the elements, in order of increasing atomic number, that have melting points somewhere between room temperature, say 20 °C, and 100 °C (Na, K, Rb, Cs, Fr, P, and Ga). At the same time, up to three other properties of each element can be listed (such as boiling point, ionization energy, and electron affinity). With SORT, on the other hand, the program will list all elements having a value within a range of values, the list being arranged in order of ascending value for the property of interest. In addition, the user may ask to have one other property of the element listed.

The periodic TABLE function is as useful as GRAPH, but it offers other possibilities. With TABLE you are shown a periodic table on the screen and given the option to SORT or GRAPH numerical information. These functions work as described above, except that one has the option of sorting or graphing by group or by period. The particular advantage of the GRAPH function here, however, is that the graph for one group or period may be superimposed on another, thus making direct comparisons possible.

After several years of using KC? Discoverer in the classroom, we believe the program has proven its worth. The advantage of the program is that is truly open-ended. That is, the student or instructor asks the question and then the database is used to *explore* possible answers; the program does not prompt the question, tutor the user on finding the answer, nor provide the answer directly. (Indeed, the program often invokes the most desirable possible response from students and instructors alike: "Gee, whiz! I didn't realize that that periodic correlation was possible or that element had those properties.") And finally, as instructors, we find that it is often much more convenient to use KC? Discoverer on a computer disk instead of searching through printed databases.

## Table of Macroscopic Properties from The Periodic Table Stack

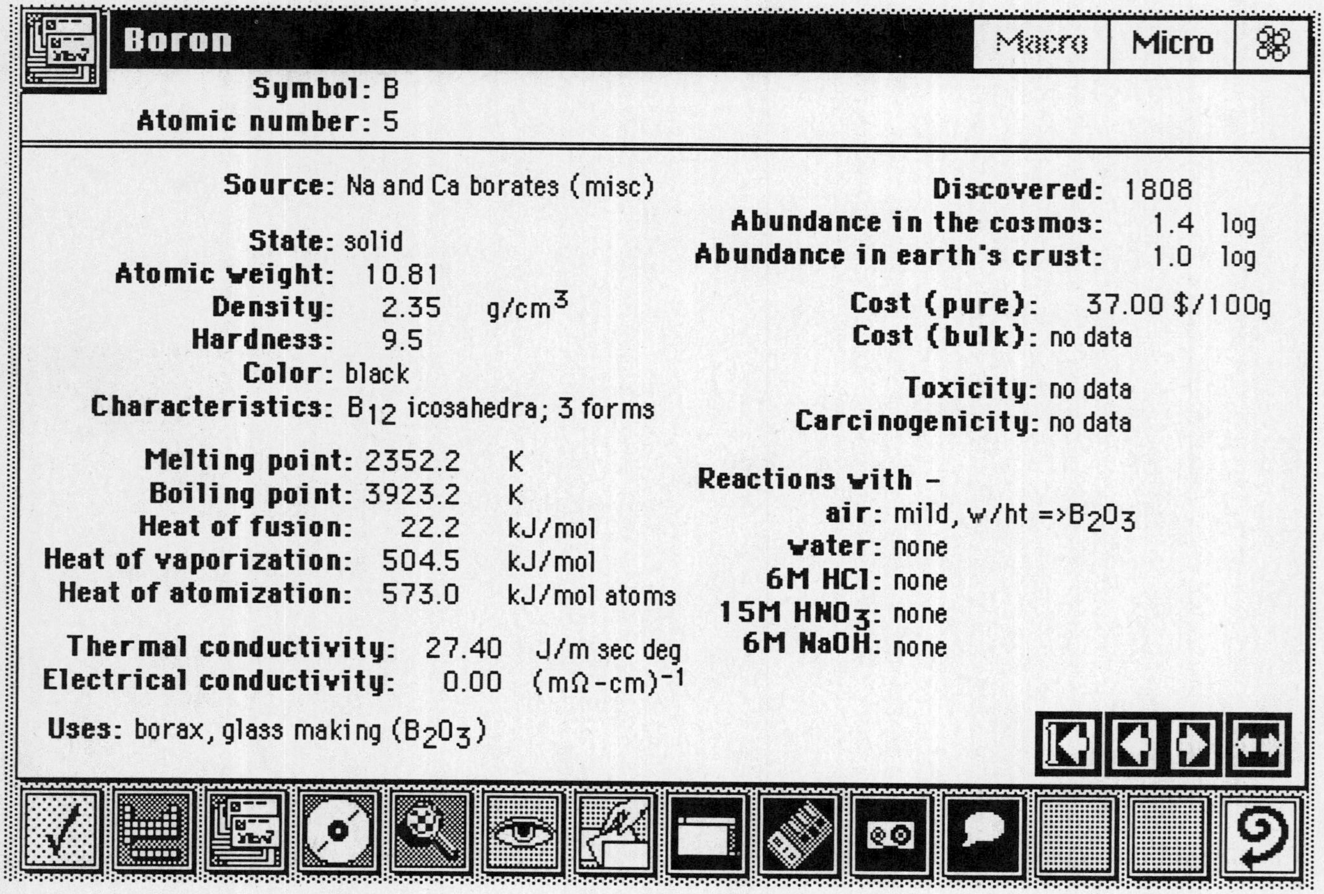

# Table of Microscopic Properties from The Periodic Table Stack

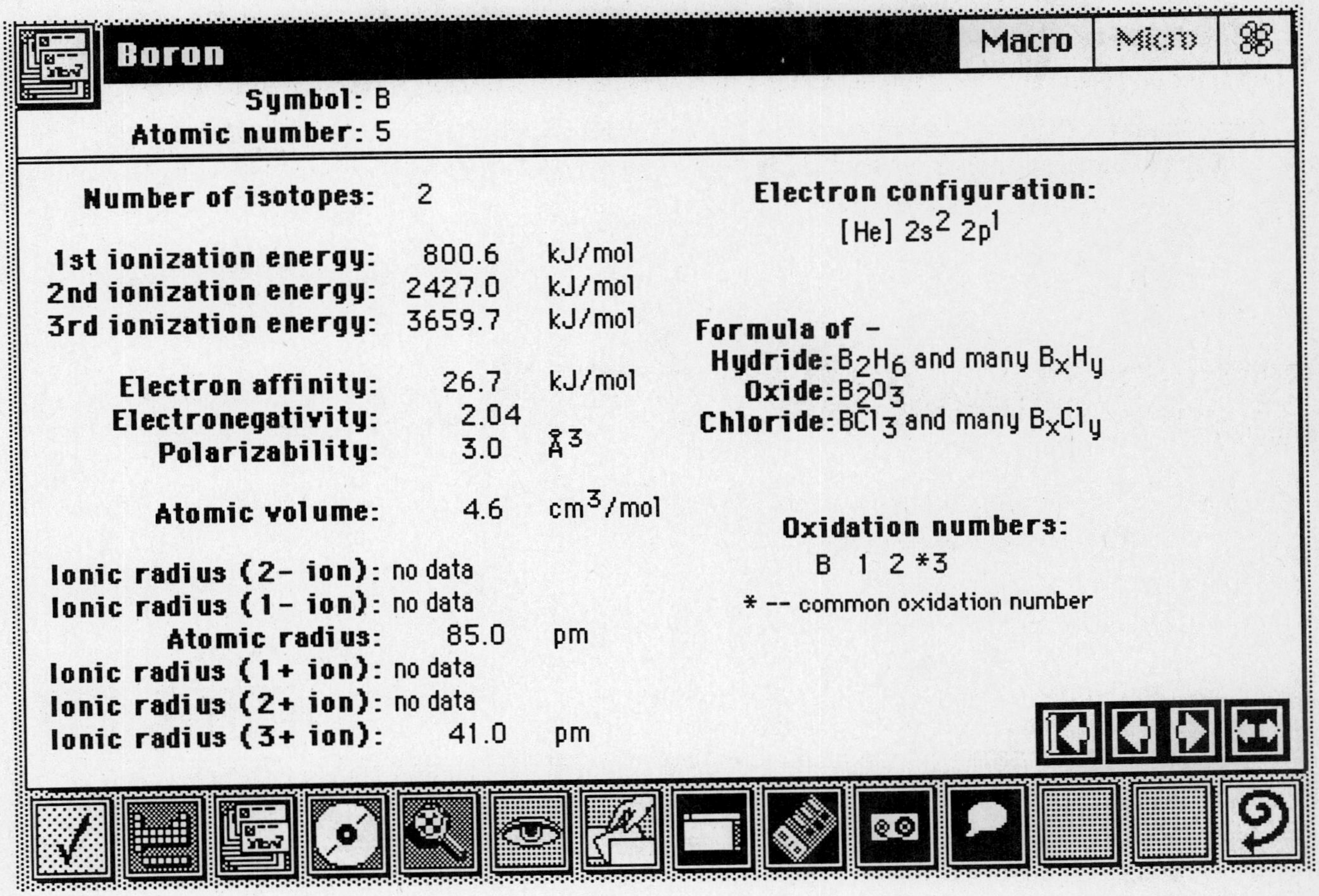

## Structure of the Element from The Periodic Table Stack

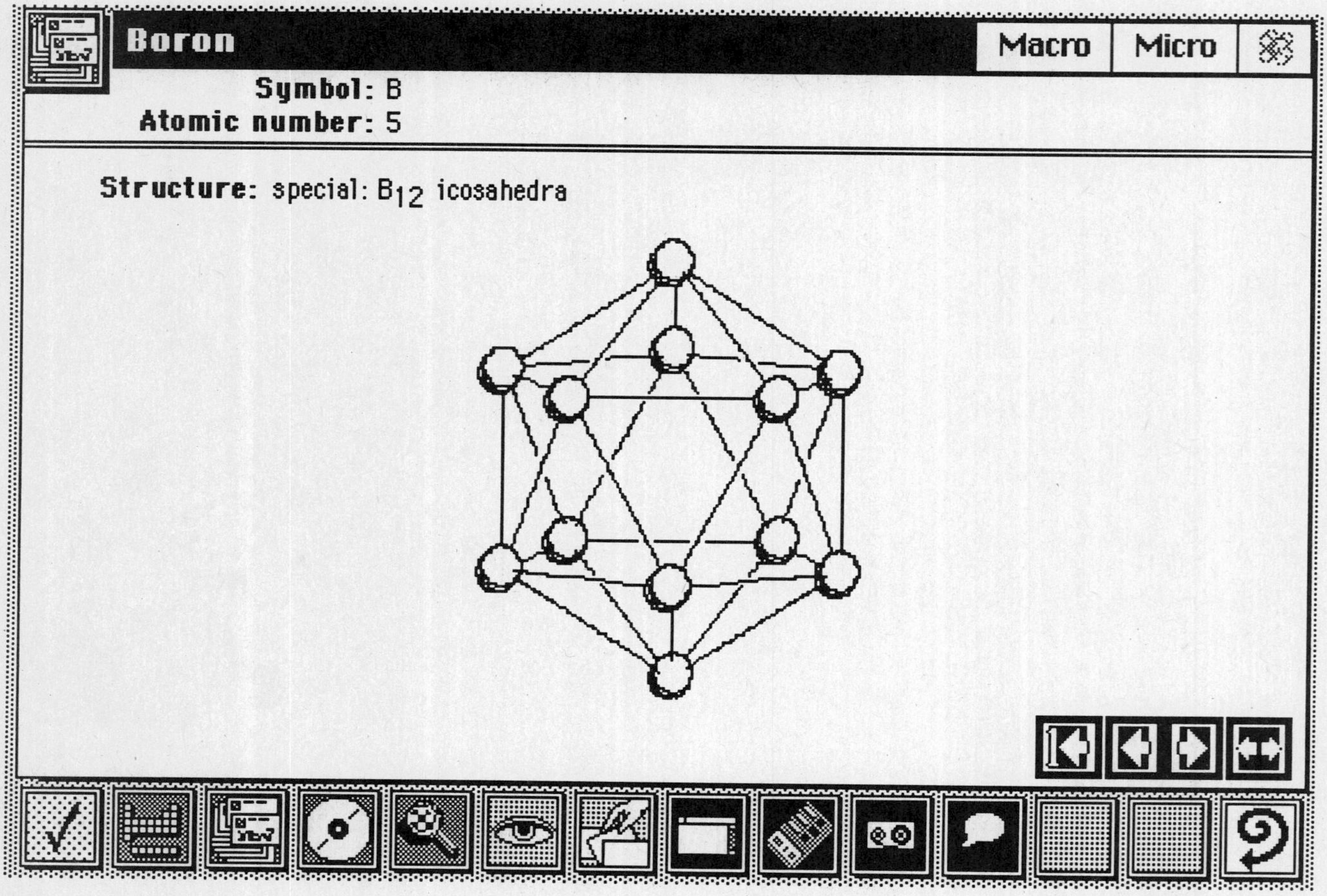

# PART I

# THE BASIC TOOLS OF CHEMISTRY

# CHAPTER 1
# THE TOOLS OF CHEMISTRY

## INSTRUCTOR'S NOTES

On beginning the course we find that Sections 1.1-1.3 can be assigned as reading and then discussed further in a recitation class. One to two lecture periods are devoted to a review of units, handling numbers, and significant figures.

## LIST OF TRANSPARENCIES FOR THIS CHAPTER

No transparencies specified for this chapter.

## SUGGESTED CLASSROOM DEMONSTRATIONS

1. To illustrate "physical change," liquid nitrogen is always a favorite of students. We freeze a banana, a hot dog, a flower, or similar object.
2. Illustration of "chemical change"
   - The reaction in Figure 1.8 was done by placing a ball of aluminum foil in aqueous copper(II) chloride. The reduction of $Cu^{2+}$(aq) to Cu(s) is relatively rapid, and the temperature of the solution increases significantly. An interesting twist to this demonstration is to use copper(II) sulfate instead of copper(II) chloride. The reduction of copper will not occur. However, if table salt is added — using a salt shaker, of course — the reaction works beautifully. The reason is that the $Cl^-$ ion apparently can diffuse through the oxide coating on the aluminum, opening a pathway for the reaction of copper(II) ion and aluminum. This is a good illustration of some principles of corrosion, and adds some practical interest to the first lecture.
   - The first lecture in the course is often begun with a "bang" by setting off several hydrogen-filled balloons in a darkened lecture room. ( See Figure 18.1.)
   - Other reactions could be done as well, depending on the facilities available. Possibilities include placing small pieces of potassium in water, the thermite reaction (Shakhashiri, volume 1, page 85), or the reaction of zinc and ammonium nitrate (Shakhashiri, volume 1,

page 51). (The latter reaction gives off a large amount of ZnO dust and other irritating fumes. It is probably not suitable for a lecture room without very good ventilation.)

• For fun, and to give some color, as well as talking about our future study of acid-base reactions, we add aqueous $NH_3$ to separate flasks containing (i) very dilute acid with phenolphthalein, (ii) $Al(NO_3)_3$, (iii) dilute $CuSO_4$.

• We often have a student contribute a penny to put into concentrated $HNO_3$. The reaction brings up a brief discussion of oxidation-reduction processes. The $NO_2$ gas generated prompts a discussion of air pollution problems, as well as the fuels used in the Lunar Lander and in the Space Shuttle. (CAUTION: $NO_2$ is a very corrosive gas. Use only in a well ventilated room. We do the reaction by putting the penny in a few milliliters of acid in a 2 liter Erlenmeyer flask that is lightly stoppered. This effectively contains the gas.)

• The **Periodic Table Videodisc** was also used to explore some reactions such as the reaction of potassium with water or the reaction of cobalt with HCl and $HNO_3$.

3. We have often noted that chemistry deals with energy and so some demonstrations in the first lecture are a thermite reaction or the use of light sticks.

4. Units of measurement:

• Add a few drop of bromcresol green to a 2-L flask before the lecture. On filling with water during lecture, the water becomes blue. When the flask is almost full, it is topped off with dilute HCl, and the solution turns yellow. Next, the water is poured into an ordinary 1-L flask or beaker that already contains some dilute base, and the solution turns blue again. This is then poured into a graduated cylinder containing some phenolphthalein. We use this sequence to comment on the relative accuracies of the different types of glassware.

• We take some containers such as soda cans so students can connect metric units with familiar objects.

• A weighed piece of fruit or some other solid gives some meaning to mass expressed in grams.

5. **KC? Discoverer** was used after solving a problem on the density of copper. We compared our value with other well known metals such as silver and gold, and with the metals of the first transition series, most of which the students have heard about.

## SOLUTIONS TO SELECTED STUDY QUESTIONS

1-2. (a) solid; (b) gas; (c) solid; (d) liquid

1-4. (a) theory

(b) law (the "law of the conservation of matter")

1-6. (a) intensive; (b) extensive; (c) intensive; (d) intensive

1-8. 18 ft, 11.5 in = 228 inches

228 in • 2.54 cm/in • 1 m/100 cm = 5.78 m

1-10. 26.2 miles • 1 km/0.62137 mile = 42.2 km

1-12. 55 miles/hr • 1 km/0.62137 mile = 89 km/hr

1-14. 8.5 in • 2.54 cm/1 in = 22 cm

11 in • 2.54 cm/1 in = 28 cm

1-16. Area = 80 $in^2$

80 $in^2$ (2.54 cm/1 in)$^2$ = 500 $cm^2$

500 $cm^2$ (1 m/100 cm)$^2$ = $5 \times 10^{-2}$ $m^2$

1-18. Volume upper compartment = $2.3 \times 10^3$ $in^3$

Volume lower compartment = $9.8 \times 10^2$ $in^3$

Total volume = $3.3 \times 10^3$ $in^3$

$3.3 \times 10^3$ $in^3$ (2.54 cm/1 in)$^3$ = $5.4 \times 10^4$ $cm^3$

$5.4 \times 10^4$ $cm^3$ (1 L/1000 $cm^3$) = 54 L

$5.4 \times 10^4$ $cm^3$ (1 m/100 cm)$^3$ = $5.4 \times 10^{-2}$ $m^3$

1-20. 2.83 g • 1 kg/1000 g = $2.83 \times 10^{-3}$ kg

2.83 g • 1 lb/453.6 g = 6.24 x $10^{-3}$ lb

1-22.

| Milligrams | Grams | Kilograms |
|---|---|---|
| 10.2 | *0.0102* | *1.02 x $10^{-5}$* |
| *1.656 x $10^{4}$* | 16.56 | *0.01656* |
| *5.45 x $10^{5}$* | *545* | 0.545 |

1-24. Total mass of tent = 3 lb (16 oz/lb) + 11 oz = 59 oz

59 oz • 453.6 g/16 oz = 1.7 x $10^3$ g

1.7 x $10^3$ g • 1 kg/1000 g = 1.7 kg

1-26. 160. lb • 453.6 g/1 lb = 7.26 x $10^4$ g (or 72.6 kg)

1-28. 120. $in^3$ (2.54 cm/1 in)$^3$ = 1.97 x $10^3$ $cm^3$ (or 1.97 L)

1-30. 676 mL = 676 $cm^3$ = 0.676 L

1-32. 0.353 g • 1 $cm^3$/3.56 g = 0.0992 $cm^3$

1-34. 2.00 g • 1 $cm^3$/0.718 g = 2.79 $cm^3$

2.79 $cm^3$ • \$2.41/1 $cm^3$ = \$6.71

1-36. 15 L • 1000 mL/1 L • 0.93 g/mL = 1.4 x $10^4$ g

1.4 x $10^4$ g • 1 lb/453.6 g = 31 lb

1-38. 6.0 g • 1 $cm^3$/10.5 g = 0.57 $cm^3$

1-40. 1.285 g solution/1 $cm^3$ • 38.08 g sulfuric acid/100 g solution

= 0.4893 g sulfuric acid/1 $cm^3$

125 g sulfuric acid req'd • 1 $cm^3$/0.4893 g sulfuric acid = 256 $cm^3$ = 256 mL

1-42. 9.93 x $10^3$ °F or 5.77 x $10^3$ K

1-44.

| | °F | °C | K |
|---|---|---|---|
| (a) | *-321* | *-196* | 77 |
| (b) | *140* | 60. | *333* |
| (c) | 1000. | *537.8* | *811.0* |

1-46. 3020. °F = 1660. °C = 1933 K

1-48. -248.6 °C = -415.5 °F = 24.6 K
-246.1 °C = -411.0 °F = 27.1 K

1-50. volume = 6.6 x $10^4$ (2 significant figures allowed)

1-52. 0.237 (3 significant figures allowed)

1-54. 2.503 x $10^3$ (4 significant figures allowed)

1-56. 8.90 x $10^{-3}$ (3 significant figures allowed)

1-58. 0.30 L • 1.05671 quarts/1 L • 2 pints/1 quart = 0.63 pints
The storage bottle contains enough fuel to refill the stove.

1-60. 0.154 nm •1 m/1 x 109 nm = 1.54 x $10^{-10}$ m
1.54 x $10^{-10}$ m • 1 Å/1 x $10^{-10}$ m = 1.54 Å

1-62. 12.5 x $10^6$ carats • 0.200 g/1 carat = 2.50 x $10^6$ g

1-64. 13 cm • 1 m/100 cm = 0.13 m

13 cm • 1 in/2.54 cm = 5.1 in

5.1 in • 1 ft/12 in = 0.43 ft

1-66. 100. m/9.95 sec • 0.62137 mile/1000. m • 3600 sec/1 hr = 22.5 miles/hour

1-68. 250. mL • 0.997 g/1 $cm^3$ = 249 g water (=249 g ice when cooled)

249 g ice • 1 $cm^3$/0.917 g = 272 $cm^3$

The water expands when frozen to ice at -10 °C and so will not fit in the 250. mL container.

1-70. 40. g platinum/100. g white gold • 6.89 g = 2.76 g platinum

Grams gold = 6.89 g - 2.76 g platinum = 4.13 g gold

1-72. [(0.125 - 0.106 g)/0.125 g]100 = 15.2%

1-74. 85.56 x $10^9$ lb • 1 T/2000 lb = 4.278 x $10^7$ T

85.56 x $10^9$ lb • 453.6 g/lb = 3.881 x $10^{13}$ g (or 3.881 x $10^{10}$ kg)

1-76. Volume of the paper = Area of paper x thickness

Volume = 6.0 $cm^3$

Density = 4.58 g/ 6.0 $cm^3$ = 0.76 g/$cm^3$

1-78. Radius = $\frac{1}{2}$ (1.90 cm) = 0.950 cm

Volume = 4/3 [3.142(0.950 cm)$^3$] = 3.59 $cm^3$

Density = 16.19 g/3.59 $cm^3$ = 4.51 g/$cm^3$

1-80. 40.0 lb • 453.6 g/lb • 6.00 g phosphorus compd/100 g

= 1.09 x $10^3$ g phosphorus compd in 40.0 pound bag

1.09 x $10^3$ g phosphorus compd • 43.64 g P/100 g phosphorus compd = 4.75 x $10^2$ g P

1-82. Volume = Area of oil spill • thickness

To find area in square meters, first convert volume and thickness to cubic meters and meters, respectively.

Volume = $1.00 \times 10^4$ L • 1000 $cm^3$/1 L • (1 m/100 cm)$^3$ = 10.0 $m^3$

Thickness = 3.0 x 102 nm • 1 m/1 x 109 nm = $3.0 \times 10^{-7}$ m

Area = Volume/Thickness = 10.0 $m^3$/$3.0 \times 10^{-7}$ m = $3.3 \times 10^7$ $m^2$

$3.3 \times 10^7$ $m^2$ (0.62137 mile/1000 m)$^2$ = 13 square miles

1-84. 150,000 people • 175 gallons water/person-day = $2.6 \times 10^7$ gallons water/day

$2.6 \times 10^7$ gallons water/day • 8.34 lb water/gallon = $2.2 \times 10^8$ lb water/day

$2.2 \times 10^8$ lb water/day • 365 day/year = $8.0 \times 10^{10}$ lb water/year

$8.0 \times 10^{10}$ lb water/year • 1 T/2000 lb = $4.0 \times 10^7$ T water/year

We now know the number of tons of water used by all the people in the city each year.

$4.0 \times 10^7$ T water/year • 1 ton fluoride/$1 \times 10^6$ T water = 40. T fluoride

40. T fluoride • 100 T sodium fluoride/45.3 T fluoride = 88 T sodium fluoride

# CHAPTER 2
# ATOMS AND ELEMENTS

## INSTRUCTOR'S NOTES

This chapter requires approximately 2 lecture periods.

## TRANSPARENCIES AVAILABLE FOR THIS CHAPTER

The periodic table from inside the front cover of the book is available as an overhead transparency.

## SUGGESTED CLASSROOM DEMONSTRATIONS

(a) Take as many samples of elements as possible to the lecture on the elements and the periodic table.

(b) To illustrate the mole, take 1 mole quantities of elements such as Mg, Al, C, Sn, Pb, Fe, and Cu.

(c) The **Periodic Table Videodisc** is ideal for showing samples of elements and illustrating their uses. See the series by Alton Banks in the *Journal of Chemical Education* titled "What's the Use?" This series describes a different element each month and gives references to the videodisc.

(d) Example 2.9 and Exercise 2.8 use mercury and sodium, respectively, in questions concerning moles. When doing examples in lecture, it is helpful to have a sample of the element available. For example, hold up a pre-weighed sample of magnesium wire and ask how many moles of metal it contains (and then ignite the wire to give a preview of the reaction chemistry described in Chapter 4.). Or, drop a pre-weighed piece of sodium metal into a dish of water on the overhead projector, and ask how many moles of sodium reacted.

(e) Use **KC? Discoverer** to plot the number of isotopes of the elements versus atomic number. Talk briefly about the fact that atoms of even atomic number have more isotopes than those of odd atomic number and note that this is related to the way the elements are formed in stars. (See Section 7.3 and Figure 7.5 and pages 351-353.)

(f) Use **KC? Discoverer** to plot the cost of the elements versus atomic number. Discuss why some elements are so costly.

(g) There are four **World of Chemistry** video tapes that would be suitable, at least in part, to use with Chapter 2. These are

i. Module 4 — The Mole

ii. Module 5 — The Atom

iii. Module 6 — The Periodic Table

iv. Module 18 — Metals

## SOLUTIONS TO SELECTED STUDY QUESTIONS

2-17. (a) manganese, (b) fluorine, (c) potassium, (d) mercury, (e) arsenic, (f) xenon, (g) tungsten, and (h) radium.

2-19. (a) Ag, (b) Al, (c) Pu, (d) Cd, (e) Sn, (f) Ba, (g) Kr, and (h) Ru.

2-21. Rn, radon, 222

2-23. (a) Fe with 30 neutrons: A = 56

(b) Am with 148 neutrons: A = 243

(c) W atom with 110 neutrons: A = 184

2-25. (a) $^{15}_{7}N$; (b) $^{64}_{30}Zn$; (c) $^{129}_{54}Xe$

2-27.

| *Atom* | *Electrons* | *Protons* | *Neutrons* |
|---|---|---|---|
| C-13 | 6 | 6 | 7 |
| Cr-50 | 24 | 24 | 26 |
| Bi-205 | 83 | 83 | 122 |

2-29.

| *Symbol* | $^{65}Cu$ | $^{37}Cl$ | $^{80}Se$ | $^{82}Kr$ |
|---|---|---|---|---|
| *Protons* | 29 | 17 | 34 | 36 |
| *Neutrons* | 36 | 20 | 46 | 46 |
| *Electrons* | 29 | 17 | 34 | 36 |
| *Name* | copper | chlorine | selenium | krypton |

2-31. Isotope 1 has A = 20, $^{20}_{10}Ne$

Isotope 2 has A = 22, $^{22}_{10}Ne$

2-33. 0.5069(78.9183) + 0.4931(80.9163) = 79.90

2-35. 0.0435 (49.9461) + 0.8379 (51.9405) + 0.0950 (52.9407) + 0.0236 (53.9389) = 52.0

2-37. Let $x$ = fractional abundance of $^{25}Mg$; then (1 - 0.7880) - $x$ = fractional abundance of $^{26}Mg$. Therefore, 0.7880(23.985) + $x$(24.986) + (0.2120 - $x$)(25.983) = 24.312, and so $x$ = 0.09687. This means that $^{25}Mg$ abundance = 9.5% and $^{26}Mg$ abundance = 0.2120 - 0.095 = 0.117 or 11.7%

2-39. titanium, Ti, element 22

2-41. selenium, Se, 34

2-43. Names: fluorine, chlorine, bromine, iodine, astatine. Common name: halogens. All are nonmetals.

2-45. See the periodic table in the front of the book for the horizontal row (period) beginning with potassium (K), calcium (Ca), scandium (Sc), and titanium (Ti), and ending with krypton (Kr).

2-47. (a) F, fluorine, nonmetal, Group 7A, 2nd period; (b) As, arsenic, metalloid, Group 5A, 4th period; (c) Cs, cesium, metal, Group 1A, 6th period; and (d) W, tungsten, metal, Group 6B, 6th period.

2-49. (a) 16.0 g Na • 1 mol/22.99 g = 0.696 mol

(b) 0.0034 g Pt • 1 mol/195.08 g = $1.7 \times 10^{-5}$ mol

(c) 1.54 g P • 1 mol/30.97 g = $4.97 \times 10^{-2}$

(d) 0.876 g As • 1 mol/74.92 g = $1.17 \times 10^{-2}$ mol

(e) 0.983 g Xe • 1 mol/131.3 g = $7.49 \times 10^{-3}$ mol

2-51. 0.00789 g • 1 mol/83.80 g = $9.42 \times 10^{-5}$ mol Kr

2-53. (a) 6.03 mol Au • 197.0 g/mol = $1.19 \times 10^{3}$ g Au

(b) 0.045 mol U • 238 g/mol = 11 g U

(c) 15.6 mol Ne • 20.18 g/mol = $3.15 \times 10^{2}$ g Ne

(d) 3.63 x 10-4 mol Pu • 244 g/mol = 0.0886 g Pu

2-55. 2.3 carat • 200. mg/carat = $4.6 \times 10^{2}$ mg

$4.6 \times 10^{2}$ mg • 1 g/1000 mg • 1 mol/12.01 g = $3.8 \times 10^{-2}$ mol

2-57. (a) 31.1 g/troy ounce • 1 mol/195.08 g = 0.159 mol/troy ounce

(b) \$5000.00 • 31.1 g/\$487 = $3.19 \times 10^{2}$ g Pt

$3.19 \times 10^{2}$ g Pt • 1 mol/195.08 g = 1.64 mol

2-59. $2.00 \times 10^5$ lb Cu • 453.6 g/lb = $9.07 \times 10^7$ g Cu

$9.07 \times 10^7$ g Cu • 1 mol/63.55 g = $1.43 \times 10^6$ mol Cu

2-61. 256 mol Li • 6.941 g/mol • 1 $cm^3$/0.534 g = $3.33 \times 10^3$ $cm^3$

Side of Li cube = $(3.33 \times 10^3\ cm^3)^{1/3}$ = 14.9 cm

2-63. $3.84 \times 10^{-3}$ mol Ar • $6.022 \times 10^{23}$ atoms/mol = $2.31 \times 10^{21}$ atoms Ar

2-65. 1.94 g Au • 1 mol/197.0 g • $6.022 \times 10^{23}$ atoms/mol = $5.93 \times 10^{21}$ atoms Au

2-67. 47.88 g/mol • 1 mol/$6.022 \times 10^{23}$ atoms = $7.951 \times 10^{-23}$ g

2-69. Length of wire = 25 feet • 12 in/ft • 2.54 cm/in = 760 cm

Volume of wire = $\pi$(radius)$^2$(length) = 3.142 • $(0.10\ cm)^2$ • 760 cm = 24 $cm^3$

24 $cm^3$ • 8.92 g/$cm^3$ = 210 g Cu

210 g Cu • 1 mol/63.6 g = 3.4 mol Cu

3.4 mol Cu • $6.022 \times 10^{23}$ atoms/mol = $2.0 \times 10^{24}$ atoms Cu

2-71. $3.2707 \times 10^{-22}$ g/atom • $6.0221367 \times 10^{23}$ atoms/mol = 196.97 g/mol

2-73. 1.0 T coal • 2000 lb/T • 454 g/lb • 0.030 g S/1 g coal • 1 mol/32.1 g = $8.5 \times 10^2$ mol S

2-75. 5.10 g coin • 0.75 g Cu/1 g coin = 3.83 g Cu

Therefore, there are 1.27 g Ni.

3.83 g Cu • 1 mol/63.55 g = 0.0602 mol Cu (= $3.62 \times 10^{22}$ atoms Cu)

1.27 g Ni • 1 mol/58.69 g = 0.0216 mol Ni (= $1.30 \times 10^{22}$ atoms Ni)

2-77. (a) \$1000.00 • 1 T/\$114.00 • $9.08 \times 10^5$ g/1 T = $7.9649 \times 10^6$ g Fe

$7.9649 \times 10^6$ g Fe • 1 mol/55.847 g = $1.4262 \times 10^5$ mol Fe

(b) $7.9649 \times 10^6$ g Fe • 1 $cm^3$/7.86 g = $1.01 \times 10^6$ $cm^3$

Side of iron cube = 100. cm

2-79. 0.018 g Hg/1 x $10^6$ g water = $1.8 \times 10^{-8}$ g Hg/g water

$1.8 \times 10^{-8}$ g Hg/g water • 237 mL • 1.00 g water/1.00 mL = $4.3 \times 10^{-6}$ g Hg

$4.3 \times 10^{-6}$ g Hg • 1 mol Hg/201 g = $2.1 \times 10^{-8}$ mol Hg

2-81. 250 x $10^6$ people • 1 atom/person-sec = $2.5 \times 10^8$ atom/sec

$2.5 \times 10^8$ atom/sec • 3600 sec/hr • 24 hr/day • 365 days/year = $7.9 \times 10^{15}$ atoms/year

$6.022 \times 10^{23}$ atoms/mol • 1 year/$7.9 \times 10^{15}$ atoms = $7.6 \times 10^7$ years or 76 million years!

2-83. $1.8 \times 10^{22}$ atoms Si • 1 mol/$6.022 \times 10^{23}$ atoms • 28.1 g/mol = 0.84 g

0.84 g Si • 1 $cm^3$/2.33 g = 0.36 $cm^3$

Radius of wafer = 2.50 in • 2.54 cm/in = 6.35 cm

Volume = $\pi r^2$(thickness)

0.36 $cm^3$ = 3.142 • $(6.35\ cm)^2$ • thickness

Thickness = $2.8 \times 10^{-3}$ cm (or 0.028 mm)

# CHAPTER 3
# COMPOUNDS AND MOLECULES

## INSTRUCTOR'S NOTES

- The material in Chapters 2-5 is treated as a unit and requires approximately 15 lectures all together. Chapter 3 itself should be completed in about 2-3 lectures.
- Some points on which students have some problems or questions are:
  (a) The rule of determining the charges on transition metal cations tells students that they can assume such ions are usually have 2+ or 3+ charges (with 2+ charges especially prominent). They are often uneasy about being given this choice. We certainly emphasize that they will see other possibilities (and that even negative charges are possible but that they will not see them in the general chemistry course).
  (b) Students have to be convinced that they have no choice but to learn the language of chemistry by memorizing the names in Table 3.1, as well as the charges on these ions.
  (c) A very common problem students have is in recognizing that $MgBr_2$, for example, is composed of $Mg^{2+}$ and two $Br^-$ ions. We have seen such combinations as $Mg^{2+}$ and $Br_2^{2-}$.

## TRANSPARENCIES AVAILABLE FOR THIS CHAPTER

| No. | Figure | Page | Subject |
|---|---|---|---|
| 1 | --- | --- | Periodic Table from inside front cover |
| 2 | 3.4 | 78 | Model of NaCl structure |

## SUGGESTED DEMONSTRATIONS

1. Burn $H_2$ in $O_2$ (and Mg in $O_2$) to show how elements are turned into compounds. Also burned Mg in $CO_2$ to show $CO_2$ is made of C and that MgO can be made another way.
2. Bring a number of common compounds that are ionic. See Figure 3.5 on page 80 and

Figure 3.6 on page 84, for example.

3. Use the **Periodic Table Videodisc** to illustrate the formation of compounds by elements reacting with $O_2$ and with acids and bases.

7. Use samples of $H_2$, $O_2$, $N_2$, and $Br_2$ to illustrate elements that are molecules.

8. Use **KC? Discoverer** to search for all the elements that are diatomic and for the sources of some of the elements.

9. Display molar quantities of NaCl, $H_2O$, sugar, common ionic compounds. Especially show some hydrated salts to emphasize the inclusion of $H_2O$ in their molar mass.

10. Display a teaspoon of water and ask how many moles, how many molecules, and how many total atoms are contained.

11. Display a piece of $CaCO_3$ and ask how many moles are contained in the piece and then how many total atoms.

12. When talking about weight percent of elements, we used $NO_2$ as an example and then made $NO_2$ from Cu + nitric acid.

13. For the discussion of analysis, we heated a sample of $CoCl_2 \cdot 6H_2O$ in a crucible to illustrate how to determine the number of waters of hydration. Also discussed the distinctive color change observed in this process.

# SOLUTIONS TO SELECTED STUDY QUESTIONS

## WRITING FORMULAS AND NAMING COMPOUNDS

3-10. (a) $C_6H_6$; (b) $C_6H_8O_6$; (c) $BaSO_4$

3-12. (a) $Co_2(CO)_8$ = 2 Co, 8 C, and 8 O

(b) $HOOCCH_2CH_2COOH$ = 4 C, 6 H, and 4 O

(c) $CH_3NH_2CHCOOH$ = 3 C, 7 H, 1 N, and 2 O

(d) $C_{10}H_9NH_2Fe$ = 10 C, 11 H, 1 N, and 1 Fe

(e) $C_6H_2CH_3(NO_2)_3$ = 7 C, 5 H, 3 N, and 6 O

3-14. (a) $SO_4^{2-}$; (b) $Ag^+$; (c) $Mn^{2+}$; (d) $V^{3+}$; (e) $MnO_4^-$; (f) $CH_3COO^-$; (g) $ClO_4^-$; (h) $S^{2-}$

3-16. (a) $(NH_4)_2SO_4$ (e) KCN

(b) $Al_2(CO_3)_3$ (f) $Cr_2O_3$

(c) $Ca_3(PO_4)_2$ (g) SnS

(d) $Ga_2O_3$ (h) $Fe(OH)_3$

3-18. (a) $PtCl_2$ (d) $Li_2O$

(b) $Ca(OCl)_2$ (e) $CaHPO_4$

(c) CsOH (f) $UF_4$

3-20.

| *Formula* | *Name* | *Number and Formulas of Ions* |
|---|---|---|
| $MgCO_3$ | magnesium carbonate | 1 $Mg^{2+}$ and 1 $CO_3^{2-}$ |
| $Fe(ClO_4)_3$ | iron(III) perchlorate | 1 $Fe^{3+}$ and 3 $ClO_4^-$ |
| $(NH_4)_2SO_3$ | ammonium sulfite | 2 $NH_4^+$ and 1 $SO_3^{2-}$ |
| $Be(CH_3COO)_2$ | beryllium acetate | 1 $Be^{2+}$ and 2 $CH_3COO^-$ |
| $Na_3PO_4$ | sodium phosphate | 3 $Na^+$ and 1 $PO_4^{3-}$ |

3-20. (continued)

| *Formula* | *Name* | *Number and Formulas of Ions* |
|---|---|---|
| $CoO$ | cobalt(II) oxide | 1 $Co^{2+}$ and 1 $O^{2-}$ |
| $Co_2O_3$ | cobalt(III) oxide | 2 $Co^{3+}$ and 3 $O^{2-}$ |
| $CrCl_3$ | chromium(III) chloride | 1 $Cr^{3+}$ and 3 $Cl^-$ |
| $SnCl_2$ | tin(II) chloride | 1 $Sn^{2+}$ and 2 $Cl^-$ |
| $SnCl_4$ | tin(IV) chloride | 1 $Sn^{4+}$ and 4 $Cl^-$ |

3-22.

| *Compound* | *Name* |
|---|---|
| $FeCl_3$ | iron(III) chloride |
| $Fe_2(SO_4)_3$ | iron(III) sulfate |
| $Fe_2S_3$ | iron(III) sulfide |
| $NaCl$ | sodium chloride |
| $Na_2SO_4$ | sodium sulfate |
| $Na_2S$ | sodium sulfide |
| $MgCl_2$ | magnesium chloride |
| $MgSO_4$ | magnesium sulfate |
| $MgS$ | magnesium sulfide |

3-24. (a) $Cl_2O$ (c) $XeF_4$

(b) $H_2S$ (d) $SO_3$

## MOLAR MASS, MOLES, AND AVOGADRO'S NUMBER

3-26. (a) $B_{10}H_{14}$, molar mass = 122.22 g/mol

(b) $C_6H_2(CH_3)(NO_2)_3$, molar mass = 227.13 g/mol

(c) $Pt(NH_3)_2Cl_2$, molar mass = 300.05 g/mol

(d) $CH_3CH_2CH_2CH_2SH$, molar mass = 90.19 g/mol

3-28. (a) 0.250 g • 1 mol/183.2 g = 0.00136 mol
(b) 0.250 g • 1 mol/236.3 g = 0.00106 mol
(c) 0.250 g • 1 mol/318.3 g = $7.85 \times 10^{-4}$ mol

3-30. $11.84 \times 10^{9}$ lb • 453.6 g/lb • 1 mol/78.115 g = $6.875 \times 10^{10}$ mol

3-32. 250 g • 1 mol/121 g • $6.022 \times 10^{23}$ molecules/mol = $1.2 \times 10^{24}$ molecules

3-34. 25.5 g • 1 mol/102.0 g = 0.250 mol
0.250 mol • $6.022 \times 10^{23}$ molecules/mol = $1.51 \times 10^{23}$ molecules
$1.51 \times 10^{23}$ molecules • 4 F atoms/molecule = $6.02 \times 10^{23}$ F atoms

3-36. 250 g • 1 mol/92.1 g = 2.7 mol
2.7 mol • $6.022 \times 10^{23}$ molecules/mol = $1.6 \times 10^{24}$ molecules
$1.6 \times 10^{24}$ molecules (3 C atoms/1 molecule) = $4.9 \times 10^{24}$ C atoms

## PERCENT COMPOSITION

3-38. (a) $MgCO_3$, molar mass = 84.31 g/mol: C = 14.25%; O = 56.93%; Mg = 28.83%
(b) $C_6H_5OH$, molar mass = 94.11 g/mol: H = 6.43%; C = 76.57%; O = 17.00%
(c) $C_2H_3O_5N$, molar mass = 121.05 g/mol: H = 2.5%; C = 19.84%; N = 11.57%; O = 66.09%
(d) $C_4H_{10}O_3NPS$, molar mass = 183.2 g/mol: H = 5.5%; C = 26.23%; N = 7.65%; O = 26.21%; P = 16.19%; S = 17.5%

3-40. Molar mass = 245.75 g/mol: H = 5.74%; N = 22.8%; O = 32.55%; S = 13.05%; Cu = 25.86%

3-42. Molar mass = 354.49 g/mol; H = 2.56%; C = 47.44%; Cl = 50.01%

## EMPIRICAL AND MOLECULAR FORMULAS

3-44. Mass of $C_2H_4NO$ unit is 58.060 g/unit. Therefore,

116.1 g/mol • 1 unit/58.060 g = 2.000 units/mol, and molecular formula = $C_4H_8N_2O_2$.

3-46. 88.5 g • 1 mol B/10.81 g = 8.19 mol B

11.5 g H • 1 mol/1.008 g = 11.4 mol H

Mole ratio:

$$\frac{11.4 \text{ mol H}}{8.19 \text{ B}} = \frac{1.39 \text{ mol H}}{1 \text{ B mol B}} \approx \frac{1.4 \text{ H}}{1 \text{ B}} = \frac{1\ 2/5 \text{ H}}{1 \text{ B}} = \frac{7/5 \text{ H}}{1 \text{ B}}$$

This means the empirical formula is $B_5H_7$.

3-48. 89.94 g C • 1 mol/12.011 g = 7.488 mol C

10.06 g H • 1 mol/1.008 g = 9.980 mol H

Mole ratio: 9.980 mol H/7.488 mol C = 1.333 mol H/1.000 mol C = 4/3 H/1 C

Empirical formula: $C_3H_4$

Molecular formula: 120.2 g/mol • 1 unit/40.07 g = 3.000 units/mol, and molecular formula = $C_9H_{12}$.

3-50. 74.03 g C • 1 mol/12.011 g = 6.164 mol C

8.70 g H • 1 mol/1.008 g = 8.63 mol H

17.27 g N • 1 mol/14.0067 g = 1.233 mol N

Mole ratios: 8.63 mol H/1.233 mol N = 7.00 mol H/1 mol N

6.164 mol C/1.233 mol N = 4.999 mol C/1 mol N

Empirical formula: $C_5H_7N$

Molecular formula: 162 g/mol • 1 unit/81.12 g = 2.00 units/mol, which gives a molecular formula of $C_{10}H_{14}N_2$

3-52. 58.77 g C • 1 mol/12.011 g = 4.893 mol C

13.81 g H • 1 mol/1.008 g = 13.70 mol H

27.42 g N • 1 mol/14.0067 g = 1.958 mol N

Mole ratios: 13.70 mol H/1.958 mol N = 6.997 mol H/1 mol N → 7 H/1 N

4.893 mol C/1.958 mol N

= 2.499 mol C/1 mol N → 2.5 C/1 H or 5 C/2 H

Empirical formula: $C_5H_{14}N_2$

Molecular formula: 102.2 g/mol (1 unit/102.2 g) = 1.00 units/mol, which gives a molecular formula of $C_5H_{14}N_2$

3-54. 14.6 g C • 1 mol/12.011 g = 1.22 mol C

39.0 g O • 1 mol/16.00 g = 2.44 mol O

46.3 g F • 1 mol/19.00 g = 2.44 mol F

Mole ratios: 2.44 mol O/1.22 mol C = 2.00 mol O/1.00 mol C

2.44 mol F/1.22 mol C = 2.00 mol F/1.00 mol C

Empirical formula: $CO_2F_2$

Molecular formula: 82 g/mol • 1 unit/82.0 g = 1.00 units/mol, which gives a molecular formula of $CO_2F_2$

3-56. 49.6 g C • 1 mol/12.011 g = 4.13 mol C

3.20 g H • 1 mol/1.008 g = 3.18 mol H

22.0 g O • 1 mol/16.00 g = 1.38 mol O

25.2 g Mn • 1 mol/54.94 g = 0.459 mol Mn

Mole ratios: 4.13 mol C/0.459 mol Mn = 9.00 mol C/1 mol Mn

3.18 mol H/0.459 mol Mn = 6.93 mol H/1 mol Mn

1.38 mol O/0.459 mol Mn = 3.01 mol O/1 mol Mn

Empirical formula: $C_9H_7O_3Mn$

MMT, which stands for methylmanganese tricarbonyl, is the organometallic compound

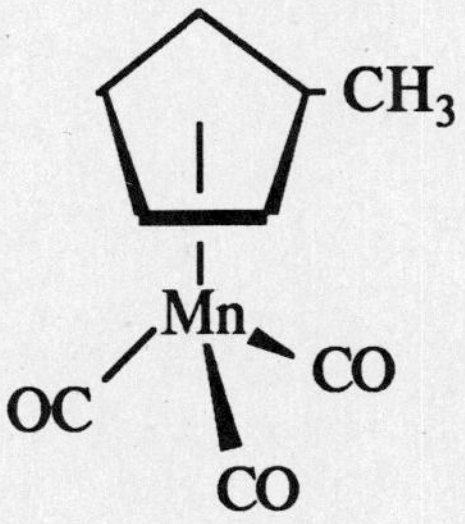

3-58. 4.74 g $KAl(SO_4)_2 \cdot x H_2O$ - 2.16 g water = 2.58 g $KAl(SO_4)_2$

2.58 g $KAl(SO_4)_2$ • 1 mol/258.2 g = 0.0100 mol

2.16 g H2O • 1 mol/18.02 g = 0.120 mol

0.120 mol H2O/0.0100 mol $KAl(SO_4)_2$ = 12 moles $H_2O$ per mole $KAl(SO_4)_2$.

Therefore, *x* is 12, and the formula of the hydrate is $KAl(SO_4)_2 \cdot 12\ H_2O$

3-60. 10.5 g • 1 mol/249.7 g = 0.0421 mol $CuSO_4 \cdot 5\ H_2O$

0.0421 mol $CuSO_4 \cdot 5\ H_2O$ • 5 mol $H_2O$/1 mol $CuSO_4 \cdot 5\ H_2O$ = 0.210 mol $H_2O$

0.210 mol $H_2O$ • 18.02 g/mol = 3.79 g $H_2O$

Mass of $CuSO_4$ = 10.5 g - 3.79 g = 6.7 g

3-62. (a) Molar mass = 169.1 g/mol

(b) 2.00 g • 1 mol/169.1 g = 0.0118 mol MSG

(c) 0.0118 mol MSG • 4 mol O/1 mol MSG = 0.0473 mol O

0.0473 mol O • $6.022 \times 10^{23}$ atom/mol = $2.85 \times 10^{22}$ atoms O

3-64. (a) Formula mass = 537.5 g/mol

(b) 0.25 g (1 mol/538 g) = $4.6 \times 10^{-4}$ mol

(c)

$$10.\ \text{g Be} \cdot \frac{1\ \text{mol Be}}{9.01\ \text{g}} \cdot \frac{1\ \text{mol } Be_3Al_2(SiO_3)_6}{3\ \text{mol Be}} \cdot \frac{538\ \text{g}}{1\ \text{mol } Be_3Al_2(SiO_3)_6}$$

= $2.0 \times 10^2$ g $Be_3Al_2(SiO_3)_6$

3-66. 0.526 g Xe • 1 mol/131.3 g = $4.01 \times 10^{-3}$ mol Xe

Grams F = 0.678 g compound - 0.526 g Xe = 0.152 g F

0.152 g F • 1 mol/19.00 g = $8.00 \times 10^{-3}$ mol F

Empirical formula = $XeF_2$

3-68. Mass of Cl in compound = 1.246 g $I_xCl_y$ - 0.678 g $I_2$ = 0.568 g Cl

0.568 g Cl • 1 mol/35.45 g = 0.0160 mol Cl

$0.678 \text{ g } I_2 \cdot 1 \text{ mol } I_2/253.8 \text{ g} \cdot 2 \text{ mol I}/1 \text{ mol } I_2 = 0.00534 \text{ mol I}$

Mole ratio: 0.0160 mol Cl/0.00534 mol I = 3.00 mol Cl/1 mol I

Empirical formula: $ICl_3$

Molecular formula: 467 g/mol • 1 unit/233.3 g = 2.00 units/mol, which gives a molecular formula of $I_2Cl_6$

## GENERAL QUESTIONS

3-70. 227 g • 5.0 g rotenone/100 g = 11 g rotenone

11 g • 1 mol/394 g = 0.029 mol rotenone

0.029 mol • $6.022 \times 10^{23}$ molecules/mol = $1.7 \times 10^{22}$ molecules

3-72. 0.050 g • 1 mol/157 g = $3.2 \times 10^{-4}$ mol $SnF_2$

$3.2 \times 10^{-4}$ mol $SnF_2$ • 1 mol $Sn^{2+}$/1 mol $SnF_2$ • $6.022 \times 10^{23}$ ions/mol

= $1.9 \times 10^{20}$ $Sn^{2+}$ ions

$3.2 \times 10^{-4}$ mol $SnF_2$ • 2 mol $F^-$/1 mol $SnF_2$ • $6.022 \times 10^{23}$ ions/mol

= $3.8 \times 10^{20}$ $F^-$ ions

3-74. 1.400 g $Mn_x(CO)_y$ - 0.394 g Mn = 1.006 g CO

1.006 g CO • 1 mol/28.011 g = 0.03591 mol CO

0.394 g Mn • 1 mol/54.94 g = 0.00717 mol Mn

Mole ratio: 5 mol CO/1 mol Mn

Empirical formula: $Mn(CO)_5$

Molecular formula: 390 g/mol • 1 unit/195 g = 2.00 units/1 mol, which gives a molecular formula of $Mn_2(CO)_{10}$.

3-76. 0.673 g N • 1 mol N/14.01 g = 0.0480 mol N

There are three times as many moles of M as N in the compound. Therefore, 1 gram of M is equivalent to 3 x 0.0480 = 0.144 mol M.

Molar mass of M = 1.00 g/0.144 mol = 6.94 g/mol

This corresponds to the molar mass of Li, so the compound is $Li_3N$ (lithium nitride).

3-78. 4.386 g MO - 4.062 g M = 0.324 g O

0.324 g O • 1 mol O/16.00 g = 0.0203 mol O

Since the formula of the compound is MO, this means that 0.0203 mol of O must be combined with 0.0203 mol M, and this mass of M is equivalent to 4.062 g of M. Therefore, the molar mass of M is

4.062 g M/0.0203 mol M = 201 g/mol

This molar mass corresponds most closely with the molar mass of Hg, so the compound is mercury(II) oxide, HgO.

3-80. 1 molecule $N_2$ < 1 penny (3 g) < 1 mol $B_2H_6$ (27.7 g)

< $6.0 \times 10^{23}$ molecules CO (28.0 g) < $3.0 \times 10^{23}$ molecules $C_4H_{10}$ (29 g)

3-82. (a) Weight percentages: H = 4.37%; C = 55.6%; O = 9.26%; and Cl = 30.77%

(b) Assume your weight is 145 pounds.

145 lb • 0.454 kg/lb = 65.8 kg

65.8 kg • 7.5 g/kg = 494 g

3-84. (a) 0.060 g vitamin • 1 mol/176 g = $3.4 \times 10^{-4}$ mol vitamin C

(b) 1.00 g • 1 mol/176 g = $5.68 \times 10^{-3}$ mol vitamin C

(c) $5.68 \times 10^{-3}$ mol vitamin C • $6.022 \times 10^{23}$ molecules/mol • 6 atoms O/molecule =

$2.05 \times 10^{22}$ atoms O

3-86. 23.5 kg • 1000 g/kg = $2.35 \times 10^4$ g $SnO_2$

$2.35 \times 10^4$ g • 1 mol $SnO_2$/150.7 g • 1 mol Sn/1 mol $SnO_2$ • 118.7 g Sn/mol)

= $1.85 \times 10^4$ g Sn

3-88. 1000. kg ore • 1000 g ore/1 kg ore • 6.75 g Ti/100 g ore = $6.750 \times 10^4$ g Ti

$6.750 \times 10^4$ g Ti • 1 mol Ti/47.88 g • 1 mol $FeTiO_3$/1 mol Ti • 151.7 g/mol $FeTiO_3$

= $2.139 \times 10^5$ g $FeTiO_3$

3-90. 18.98 g Na • 1 mol/23.000 g = 0.8252 mol Na

16.7 g Al • 1 mol/26.98 g = 0.619 mol Al

17.39 g Si • 1 mol/28.09 g = 0.6191 mol Si

7.32 g Cl • 1 mol/35.45 g = 0.207 mol Cl

39.6 g O • 1 mol/16.00 g = 2.48 mol O

Mole ratios: 0.8252 mol Na/0.207 mol Cl = 4.00 mol Na/1 mol Cl

0.619 mol Al/0.207 mol Cl = 3.00 mol Al/1 mol Cl

0.619 mol Si/0.207 mol Cl = 3.00 mol Si/1 mol Cl

2.48 mol O/0.207 mol Cl = 12.0 mol O/1 mol Cl

Empirical formula: $Na_4Al_3Si_3O_{12}Cl$ [or $Na_4Al_3(SiO_4)_3Cl$]

## SUMMARY PROBLEMS

3-92. (a) Moles of Ni

Thickness of foil = 0.550 mm • 1 cm/10 mm = 0.0550 cm

Volume of foil = 0.0550 cm • $(1.25\text{ cm})^2$ = $8.59 \times 10^{-2}$ $cm^3$

Mass of Ni foil = $8.59 \times 10^{-2}$ $cm^3$ • 8.908 g/$cm^3$ = 0.766 g

Moles of Ni = 0.766 g • 1 mol/58.69 g = 0.0130 mol

(b) Formula of nickel compound

Mass of F used = 1.261 g compound - 0.766 g Ni = 0.495 g F

Moles of F = 0.495 g • 1 mol/19.00 g = 0.0261 mol F

Mole ratio = 0.0261 mol F/0.0130 mol Ni = 2.00 mol F/1.00 mol Ni

Formula of the nickel fluoride: $NiF_2$, which is called nickel(II) fluoride

3-93. (a) Volume of Al foil = 0.0450 cm • 1.50 cm • 1.95 cm = 0.132 $cm^3$

Mass of Al foil = 0.132 $cm^3$ • 2.699 g/$cm^3$ = 0.355 g Al

Moles of Al = 0.355 g Al • 1 mol/26.98 g = 0.0132 mol Al

0.0132 mol Al • $6.022 \times 10^{23}$ atoms Al/mol = $7.93 \times 10^{21}$ atoms Al

(b) Mass of Br = 3.51 g compound - 0.355 g Al = 3.16 g Br

Moles Br = 3.16 g Br • 1 mol/79.90 g = 0.0395 mol Br

Mole ratio: 0.0395 mol Br/0.0132 mol Al = 3.00 mol Br/1.00 mol Al

Empirical formula = $AlBr_3$

(c) 533 g/mol • 1 unit/266.7 g = 2.00 units/mol

Molecular formula: $Al_2Br_6$, the name of which is aluminum(III) bromide.

3-94. (a) *Empirical formula of uranium oxide*

Moles of U = 0.169 g U (1 mol/238.0 g) = $7.10 \times 10^{-4}$ mol U

Mass of O = 0.199 g compound - 0.169 g U = 0.030 g O

Moles of O = 0.030 g O • 1 mol/16.0 g = $1.9 \times 10^{-3}$ mol O

Mole ratio = $1.90 \times 10^{-3}$ mol O/$7.10 \times 10^{-4}$ mol U = 2.6 mol O/1.0 mol U

Empirical formula = $U_3O_8$

Moles $U_3O_8$ obtained = 0.199 g • 1 mol/842.1 g = $2.36 \times 10^{-4}$ mol

(b) *Isotopes of U*

The atomic weight of U is 238.0289 amu, implying that the isotope $^{238}U$ is the most abundant.

(c) *Hydrated uranium oxide*

Mass of water of hydration = 0.865 g before heating - 0.679 g after heating = 0.186 g $H_2O$

Moles of water = 0.186 g • 1 mol/18.02 g = 0.0103 mol $H_2O$

Moles of dehydrated uranium compound = 0.679 g • 1 mol/394.0 g = 0.00172 mol

Mole ratio = 0.0103 mol $H_2O$/0.00172 mol $UO_2(NO_3)_2$

= 5.98 mol $H_2O$/1 mol $UO_2(NO_3)_2$

Formula of hydrated uranium compound: $UO_2(NO_3)_2 \cdot 6\,H_2O$

# CHAPTER 4
# CHEMICAL REACTIONS: AN INTRODUCTION

## INSTRUCTOR'S NOTES

As indicated in Chapter 3, this chapter is treated as a unit along with Chapters 3 and 5. Together they require approximately 15, 50-minute lectures.

- Although types of non-aqueous reactions are covered in this chapter, it is not absolutely essential to do this at this time. The students can work the problems on stoichiometry without knowing the reaction type. However, we like to introduce such reactions at this point because it is an opportunity to present some chemistry early in the course and to do some very visual demonstrations.
- We have written the book to stress the notion of the *stoichiometric factor* in chemical calculations. We constantly remind our students that everything funnels through this step.
- The diagram below is one we have used in class to emphasize this point.

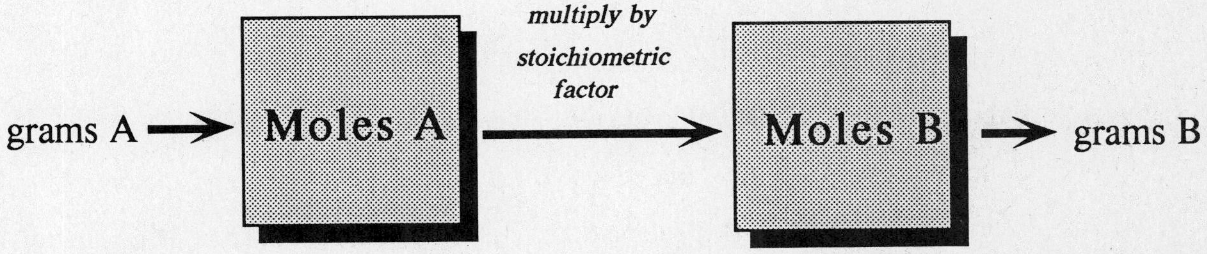


- Students have always had difficulties with limiting reagent problems. For this reason we have incorporated more solved examples in this edition.
- Students also have difficulties with determination of formulas by combustion analysis. It is well to do as many examples as possible.
- When this book was planned, we had as a major goal to introduce students early to simple classes of chemical reactions. In the many drafts of this chapter, and in teaching the material over a period of ten years or so, we have tried various approaches and tried to introduce various reaction classes. Thus, the current section 4.3 represents the culmination

of many attempts at teaching this material. In the second edition of our text, we have decided only to cover metal and nonmetal reactions with $O_2$ and simple decompositions, such as those of some oxides and carbonates. We find students easily grasp the reactions of metals and $O_2$, but we point out repeatedly that they must simply memorize the few reactions of nonmetals and $O_2$ that are introduced.

## TRANSPARENCIES AVAILABLE FOR THIS CHAPTER

| No. | Figure | Page | Subject |
|---|---|---|---|
| 3 | 4.11 | 115 | Scheme outlining stoichiometry calculations |

## SUGGESTED DEMONSTRATIONS

1. Use the **Periodic Table Videodisc** to illustrate reactions involving the combination of elements with $O_2$.
2. Take samples of metal oxides, especially those of iron, aluminum, boron, lead, and magnesium. Show sample of anodized aluminum (kitchen ware is effective) or aluminum oxide sandpaper.
3. Take a sample of silica sand and quartz crystals.
4. Take a sample of dry ice and discuss some of the chemistry of $CO_2$.
5. React hot sulfur with pure $O_2$. (See Figure 4.1, page 103.) This is especially effective in a very dark room. (Fill the bottle with pure $O_2$ before lecture and tightly stopper.)
6. Make $NO_2$ from copper (a penny) and concentrated $HNO_3$. (Place a penny in about 15 mL of concentrated $HNO_3$ in a 2 L flask and stopper lightly with a rubber stopper. Although a little $NO_2$ comes out of the flask, in a large lecture room it is not noticeable.
7. We have often begun these lectures with the *thermite reaction.*
8. It is very effective to spray metal powders into a bunsen flame using an eye dropper or other, similar device (Figure 4.6). Iron and zinc powder work especially well.
9. The classic ammonium dichromate decomposition reaction can be done safely and without a mess using the method in Figure 4.8. (See Volume 1 of the *Chemical Demonstrations* series by Shakhashiri.)
10. As an introduction to stoichiometry we have used the decomposition of ammonium nitrate

$NH_4NO_3(s) \rightarrow N_2O(g) + 2\,H_2O(g)$

This reaction is part of suggested demonstration found in Shakhashiri (Volume 1, Expt. 1.20, page 51). The overall reaction is spectacular (see the photo on page 134). However, one of the reaction products (ZnO) is irritating, so the reaction must be done in a very well ventilated room.

11. The reaction of Mg ribbon with $O_2$ (page 114)

$2\,Mg(s) + O_2(g) \rightarrow 2\,MgO(s)$

is also useful as a demonstration reaction that can be tied to an example of a simple stoichiometry reaction.

12. Another good classroom example for stoichiometry is the decomposition of $H_2O_2$ (see Figure 15.15 on page 644 as well as Figure 15.2 on page 608)).

$2\,H_2O_2(\ell) \rightarrow 2\,H_2O(g) + O_2(g)$

We use it to power a small "rocket" made as illustrated below. A small quantity of $MnO_2$ (1-2 g) is placed in the bottle as a catalyst. (We use a 100-mL bottle and find it helps if the bottle is damp.) The top is placed in the bottle, and an eye dropper containing 30% $H_2O_2$ is placed in the open hole of the 2-hole stopper. All of the $H_2O_2$ in the dropper is squirted into the bottle. The decomposition reaction is rapid and exothermic, and the steam and $O_2$ generated provide the thrust for the plastic "rocket." Under the right conditions the "rocket" will fly about 20-30 feet. (CAUTION: Wear rubber gloves when handling the $H_2O_2$ and wash the residue down the drain with a large amount of water.)

The $H_2O_2$ rocket has proved to be a very useful lecture demonstration. Not only do the students enjoy it, but it can be used to illustrate a redox reaction (Chapter 5), the concept of an exothermic reaction (Chapter 6), and catalysis (Chapter 15).

## THE HYDROGEN PEROXIDE ROCKET

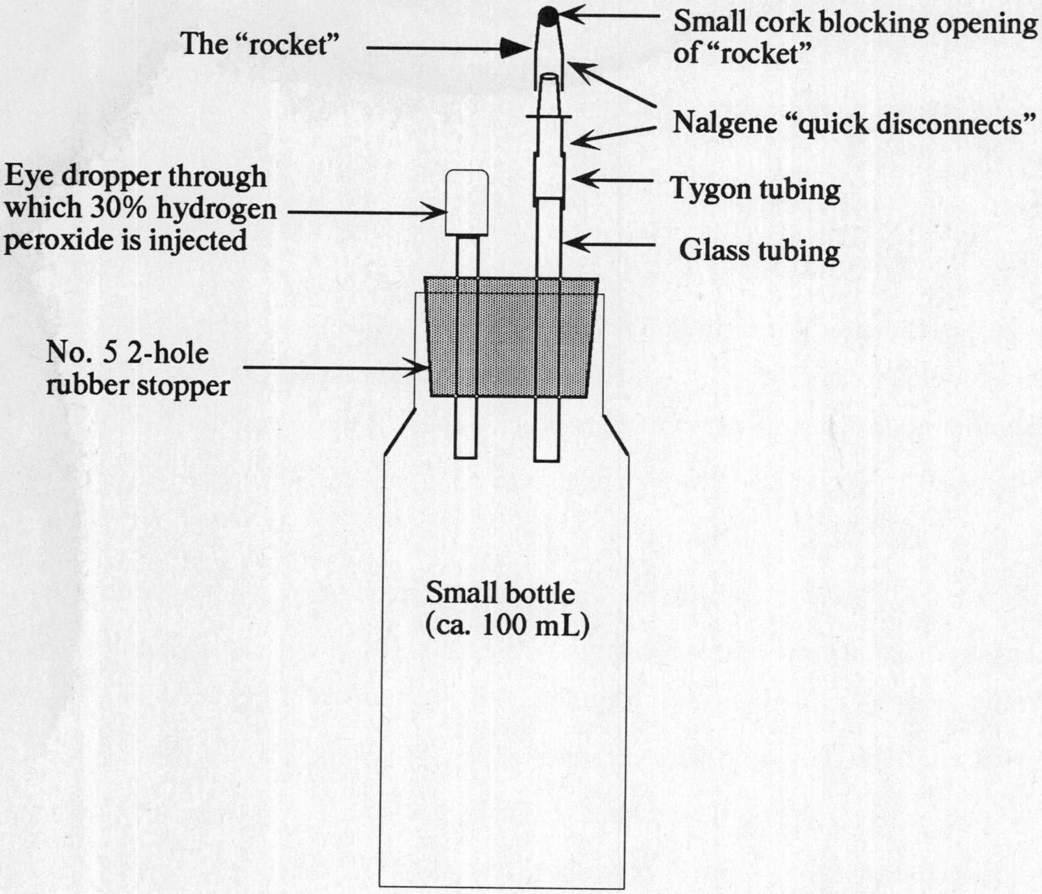


13. Limiting reagents are illustrated in the text (page ) by describing the preparation of salami sandwiches. We have found the preparation of a dry martini is very effective in class. We bring to class a cocktail shaker, some ice, a cocktail glass, a nearly full bottle of vermouth, and a gin bottle that contains only about 20 mL of "gin" (actually water, a secret never admitted to the students). While discussing the concept of limiting reagents, we proceed to mix a martini (using all the "gin" and a few milliliters of vermouth). All the while we talk about which is the limiting reagent in making a shaker full of the cocktail. Then, with appropriate flourish, we pour the mixture into the glass and consume the drink. This never fails to evoke a response from the class. In fact, students mention the demonstration even years after they have been in the class and remember that it had to do with stoichiometry.

14. Yet another useful demonstration to use when discussing a limiting reagent situation is the production of $H_2$ from Ca in water.

$$Ca(s) + 2\ H_2O(\ell) \rightarrow Ca(OH)_2(s) + H_2(g)$$

This reaction is a good demonstration because both products are visible: bubbles of $H_2$ and solid $Ca(OH)_2$. Coincidentally, you can use this reaction to illustrate two other general chemical features: the relative reactivities of Mg and Ca with water and the general insolubility of hydroxides other than those of Group 1A (a topic raised in Chapter 5). This reaction appears on the **Periodic Table Videodisc.**

15. A very simple demonstration of combustion reactions is the "plastic soda bottle rocket." Insert large nails on opposite sides of a plastic soda bottle near the bottom. Place a few milliliters of methanol in the bottle and shake to saturate the atmosphere in the bottle with alcohol. Place a cork in the top of the bottle (it will take some experimenting to see how firmly to seat the cork), and then touch one of the nails with a Tesla coil. The cork will fly out of the bottle with a loud bang.

# SOLUTIONS TO SELECTED STUDY QUESTIONS

## BALANCING EQUATIONS

4-11. (a) $2\ Al(s) + 3\ Cl_2(g) \rightarrow 2\ AlCl_3(s)$

(b) $SiO_2(s) + 2\ C(s) \rightarrow Si(s) + 2\ CO(g)$

(c) $3\ Fe(s) + 4\ H_2O \rightarrow Fe_3O_4(s) + 4\ H_2(g)$

4-13. (a) $3\ MgO(s) + 2\ Fe(s) \rightarrow Fe_2O_3 + 3\ Mg(s)$

(b) $2\ H_3BO_3(s) \rightarrow B_2O_3(g) + 3\ H_2O(\ell)$

(c) $2\ NaNO_3(s) + H_2SO_4(\ell) \rightarrow Na_2SO_4(s) + 2\ HNO_3(g)$

4-15. (a) $CaF_2(s) + H_2SO_4(\ell) \rightarrow CaSO_4(s) + 2\ HF(g)$

(b) $2\ N_2O(g) \rightarrow 2\ N_2(g) + O_2(g)$

(c) $NH_4NO_3(s) \rightarrow N_2O(g) + 2\ H_2O(g)$

4-17. (a) $CaCN_2(s) + 3\ H_2O(\ell) \rightarrow CaCO_3(s) + 2\ NH_3(g)$

(b) $2\ NaBH_4(s) + H_2SO_4(aq) \rightarrow B_2H_6(g) + 2\ H_2(g) + Na_2SO_4(aq)$

(c) $H_2S(aq) + Cl_2(aq) \rightarrow S(s) + 2\ HCl(aq)$

## CLASSES OF CHEMICAL REACTIONS

4-19. (a) $Ti(s) + O_2(g) \rightarrow TiO_2(s)$

(b) $S_8(s) + 8\ O_2(g) \rightarrow 8\ SO_2(g)$

(c) $Se(s) + O_2(g) \rightarrow SeO_2(s)$

4-21. (a) $2\ Cr(s) + 3\ Cl_2(g) \rightarrow 2\ CrCl_3(s)$

(b) $Cu(s) + Br_2(\ell) \rightarrow CuBr_2(s)$

(c) $Pt(s) + 2\ F_2(g) \rightarrow PtF_4(s)$

4-23. (a) $4\ C_{10}H_{10}Fe(s) + 53\ O_2(g) \rightarrow 2\ Fe_2O_3(s) + 40\ CO_2(g) + 20\ H_2O(g)$

(b) $2\ B_5H_9(\ell) + 12\ O_2(g) \rightarrow 5\ B_2O_3(s) + 9\ H_2O(g)$

(c) $2\ Si_2H_6(g) + 7\ O_2(g) \rightarrow 4\ SiO_2(s) + 6\ H_2O(g)$

4-25. (a) $4\ Cu(s) + O_2(g) \rightarrow 2\ Cu_2O(s)$

(b) $4\ As(s) + 3\ O_2(g) \rightarrow 2\ As_2O_3(s)$

(c) $2\ Zn(s) + O_2(g) \rightarrow 2\ ZnO(s)$

4-27. (a) $ZnCO_3(s)$ + heat $\rightarrow ZnO(s) + CO_2(g)$

(b) $MnCO_3(s)$ + heat $\rightarrow MnO(s) + CO_2(g)$

(c) $PbCO_3(s)$ + heat $\rightarrow PbO(s) + CO_2(g)$

## GENERAL STOICHIOMETRY PROBLEMS

4-29. *Step 1:* Calculate moles of NaBr

2.6 g • 1 mol/103 g = 0.025 mol

*Step 2:* Calculate moles $AgNO_3$ required

0.025 mol NaBr • 1 mol $AgNO_3$ required/1 mol NaBr = 0.025 mol $AgNO_3$

*Step 3:* Calculate mass of $AgNO_3$ required

0.025 mol $AgNO_3$ • 170. g/mol = 4.3 g $AgNO_3$

*Step 4:* Calculate mass of AgBr and $NaNO_3$ that can be obtained.

0.025 mol NaBr • 1 mol AgBr/1 mol NaBr • 188 g/mol AgBr = 4.7 g NaBr

0.025 mol NaBr • 1 mol $NaNO_3$/1 mol NaBr • 85.0 g/mol $NaNO_3$) = 2.1 g $NaNO_3$

4-31. *Step 1:* Calculate moles of Al

2.56 g Al • 1 mol/26.98 g/mol = 0.0949 mol Al

*Step 2:* Calculate moles of $Br_2$ required

0.0949 mol Al • 3 mol $Br_2$/2 mol Al = 0.142 mol $Br_2$

0.142 mol $Br_2$ • 159.8 g/mol = 22.7 g $Br_2$

*Step 3:* Grams product ($Al_2Br_6$) = grams of reactants = 2.56 g Al + 22.7 g $Br_2$ = 25.3 g

4-33. *Step 1:* Calculate grams $S_8$ required

365 g $SF_6$ • 1 mol/146.1 g = 2.50 mol $SF_6$

$$2.50 \text{ mol } SF_6 \cdot \frac{1 \text{ mol } S_8}{8 \text{ mol } SF_6} = 0.313 \text{ mol } S_8$$

0.313 mol $S_8$ • 256.5 g/mol = 80.1 g $S_8$ required

*Step 2:* Calculate grams $F_2$ required

$$2.50 \text{ mol } SF_6 \cdot \frac{24 \text{ mol } F_2}{8 \text{ mol } SF_6} = 7.49 \text{ mol } F_2$$

7.49 mol $F_2$ • 38.00 g/mol = 285 g $F_2$ required

As a check, we see that the mass of product (365 g) is equal to the total mass of the reactants (80.2 g $S_8$ + 285 g $F_2$ = 365 g).

4-35. *Step 1:* Write the balanced equation:

$4\,Al(s) + 3\,O_2(g) \rightarrow 2\,Al_2O_3(s)$

*Step 2:* Calculate moles of Al

0.569 g Al • 1 mol/26.98 g = 0.0211 mol Al

*Step 3:* Calculate mass of $O_2$ required

0.0211 mol Al • 3 mol $O_2$/4 mol Al • 32.00 g $O_2$/mol $O_2$ = 0.506 g $O_2$

*Step 4:* Calculate mass of $Al_2O_3$ produced

0.0211 mol Al • 2 mol $Al_2O_3$/4 mol Al • 102.0 g/mol = 1.08 g $Al_2O_3$

As a check, we see that the mass of product (1.08 g) is equal to the total mass of the reactants (0.569 g Al + 0.506 g $O_2$ = 1.075 g)

4-37. *Step 1:* Calculate moles of NaCl

2550 g NaCl • 1 mol/58.44 g = 43.6 mol NaCl

*Step 2:* Calculate mass of $H_2$

43.6 mol NaCl • 1 mol $H_2$/2 mol NaCl • 2.016 g/1 mol = 44.0 g $H_2$

*Step 3:* Calculate mass of $Cl_2$

43.6 mol NaCl • 1 mol $Cl_2$/2 mol NaCl • 70.91 g/1 mol = 1550 g $Cl_2$

*Step 4:* Calculate mass of NaOH

43.6 mol NaCl • 2 mol NaOH/2 mol NaCl • 40.00 g/mol = 1750 g NaOH

4-39. *Step 1:* Calculate grams of HF required.

256 g $SiO_2$ • 1 mol/60.08 g = 4.26 mol $SiO_2$

4.26 mol $SiO_2$ • 4 mol HF/1 mol $SiO_2$ = 17.0 mol HF

17.0 mol HF • 20.01 g/mol = 341 g HF

*Step 2:* Calculate grams of $SiF_4$ produced

300. g $SiO_2$ • 1 mol/60.08 g = 4.99 mol $SiO_2$

4.99 mol $SiO_2$ • 1 mol $SiF_4$/1 mol $SiO_2$ = 4.99 mol $SiF_4$

4.99 mol $SiF_4$ •104.1 g/mol = 520. g $SiF_4$

4-41. *Step 1:* Calculate mol of calcium phosphate

454 g $Ca_3(PO_4)_2$ • 1 mol/310.2 g = 1.46 mol $Ca_3(PO_4)_2$

*Step 2:* Calculate grams of silicon dioxide

1.46 mol $Ca_3(PO_4)_2$ • [6 mol $SiO_2$/2 mol $Ca_3(PO_4)_2$] • 60.08 g/1 mol $SiO_2$ = 264 g $SiO_2$

*Step 3:* Calculate grams of C

1.46 mol $Ca_3(PO_4)_2$ • [10 mol C/2 mol $Ca_3(PO_4)_2$] • 12.01 g/1 mol C = 87.9 g C

*Step 4:* Calculate grams of $P_4$

$1.46\ mol\ Ca_3(PO_4)_2 \cdot [1\ mol\ P_4/2\ mol\ Ca_3(PO_4)_2] \cdot 123.9\ g/1\ mol\ P_4 = 90.7\ g\ P_4$

4-43. *Step 1:* Calculate moles of each reactant:

$32.0\ g\ S_8 \cdot 1\ mol/256.5\ g = 0.125\ mol\ S_8$

$71.0\ g\ Cl_2 \cdot 1\ mol/70.91\ g = 1.00\ mol\ Cl_2$

*Step 2:* Compare actual mole ratio with theoretical mole ratio and decide on limiting reagent.

Actual ratio = $1.00\ mol\ Cl_2/0.125\ mol\ S_8 = 8.01\ mol\ Cl_2/1\ mol\ S_8$

Ratio from balanced equation = $4\ mol\ Cl_2/1\ mol\ S_8$

There is significantly more $Cl_2$ available per mole of $S_8$ than is required. Therefore, $S_8$ is the limiting reagent.

*Step 3:* Calculate mass of $S_2Cl_2$ expected based on $S_8$ as the limiting reagent.

$0.125\ mol\ S_8 \cdot 4\ mol\ S_2Cl_2/1\ mol\ S_8 \cdot 135.0\ g/mol = 67.4\ g\ S_2Cl_2$

*Step 4:* Calculate mass of left-over $Cl_2$.

Mass of $Cl_2$ required = $0.125\ mol\ S_8 \cdot 4\ mol\ Cl_2/1\ mol\ S_8 \cdot 71.00\ g/mol = 35.4\ g\ Cl_2$

Mass of $Cl_2$ remaining = 71.0 g - 35.4 g = 35.6 g

4-45. *Step 1:* Calculate moles of available reagents.

$1.0 \times 10^3\ g\ C_7H_6O_3 \cdot 138\ g/mol = 7.3\ mol\ C_7H_6O_3$

$1.0 \times 10^3\ g\ C_4H_6O_3 \cdot 102\ g/mol = 9.8\ mol\ C_4H_6O_3$

*Step 2:* Compare actual mole ratio with theoretical mole ratio and decide on limiting reagent.

Actual ratio = $7.3\ mol\ C_7H_6O_3/9.8\ mol\ C_4H_6O_3 = 0.74$

Ratio from balanced equation = $2\ mol\ C_7H_6O_3/1\ mol\ C_4H_6O_3$

$C_7H_6O_3$ is clearly is short supply and is the limiting reagent.

*Step 3:* Calculate the theoretical yield of aspirin

$7.3\ mol\ C_7H_6O_3 \cdot 2$ mol aspirin/2 mol $C_7H_6O_3$ = 7.3 mol aspirin

7.3 mol aspirin • 180. g/mol = 1300 g aspirin

4-47. *Step 1:* Calculate moles of available reagents.

16.3 g $SCl_2$ • 1 mol/103.0 g = 0.158 mol $SCl_2$

10.5 g NaF • 1 mol/41.99 g = 0.250 mol NaF

*Step 2:* Compare actual mole ratio with theoretical mole ratio and decide on limiting reagent.

Actual ratio = 0.250 mol NaF/0.158 mol $SCl_2$ = 1.58 mol NaF/1.00 mol $SCl_2$

Ratio from balanced equation = 4 mol NaF/3 mol $SCl_2$ = 1.33 mol NaF/1.00 mol $SCl_2$

There is more NaF available than is required. Therefore, $SCl_2$ is the limiting reagent.

*Step 3:* Calculate the theoretical yield of $S_2Cl_2$.

0.158 mol $SCl_2$ • 1 mol $S_2Cl_2$/3 mol $SCl_2$ • 135.0 g/mol = 7.11 g $S_2Cl_2$

*Step 4:* Calculate the theoretical yield of $SF_4$

0.158 mol $SCl_2$ • 1 mol $SF_4$/3 mol $SCl_2$ • 108.1 g/mol = 5.70 g $SF_4$

4-49. *Step 1:* Find moles of $SiO_2$ and $BrF_3$

1.256 g $SiO_2$ • 1 mol/60.084 g = 0.02090 mol $SiO_2$

5.893 g $BrF_3$ • 1 mol/136.90 g = 0.04307 mol $BrF_3$

*Step 2:* Determine limiting reagent

Actual mole ratio = 0.04307 mol $BrF_3$/0.02090 mol $SiO_2$ = 2.061 mol $BrF_3$/1 mol $SiO_2$

Theoretical mole ratio = 4 mol $BrF_3$/3 mol $SiO_2$ = 1.33 mol $BrF_3$/1 mol $SiO_2$

Since there is more $BrF_3$ than is required, the limiting reagent is $SiO_2$.

*Step 3:* Calculate mass of each of the products

0.02090 mol $SiO_2$ • 3 mol $SiF_4$/3 mol $SiO_2$ • 104.08 g/mol $SiF_4$ = 2.176 g $SiF_4$

0.02090 mol $SiO_2$ • 2 mol $Br_2$/3 mol $SiO_2$ • 159.81 g/mol $Br_2$ = 2.227 g $Br_2$

0.02090 mol $SiO_2$ • 3 mol $O_2$/3 mol $SiO_2$ • 31.999 g/mol $O_2$ = 0.6689 g $O_2$

## PERCENT YIELD

4-51. *Step 1:* Write balanced equation

$Ni(s) + 4\,CO(g) \rightarrow Ni(CO)_4(g)$

*Step 2:* Calculate theoretical yield of $Ni(CO)_4$

5.00 g Ni • 1 mol/58.69 g • 1 mol $Ni(CO)_4$/1 mol Ni • 170.7 g $Ni(CO)_4$/1 mol

= 14.5 g $Ni(CO)_4$

*Step 3:* Calculate grams of CO required

5.00 g Ni • 1 mol/58.69 g • 4 mol CO/1 mol Ni • 28.01/1 mol = 9.55 g CO

*Step 4:* Calculate percent yield

(9.67 g/14.5)100 = 66.5%

4-53. *Step 1:* Calculate theoretical yield of $S_2Cl_2$

5.23 g $SCl_2$ • 1 mol $SCl_2$/103.0 g • 1 mol $S_2Cl_2$/3 mol $SCl_2$ • 135.0 g $S_2Cl_2$/mol

= 2.29 g $S_2Cl_2$

*Step 2:* Calculate the percent yield

(1.19 g $S_2Cl_2$/2.29 g)100 = 52.1%

## STOICHIOMETRY AND MOLECULAR FORMULAS

4-55. *Step 1:* Calculate the moles of C and H in anthracene from the quantities of $CO_2$ and $H_2O$ evolved on combustion.

$8.64 \times 10^{-3}$ g $CO_2$ • 1 mol $CO_2$/44.01 g • 1 mol C/1 mol $CO_2$ = $1.96 \times 10^{-4}$ mol C

$1.26 \times 10^{-3}$ g $H_2O$ • 1 mol $H_2O$/18.02 g • 2 mol H/1 mol $H_2O$ = $1.40 \times 10^{-4}$ mol H

*Step 2:* Calculate the C to H ratio and derive the empirical formula

$1.96 \times 10^{-4}$ mol C/$1.40 \times 10^{-4}$ mol H = 1.40 mol C/1.00 mol H = $1\,^2/_5$ mol C/1 mol H

This leads to an empirical formula of $C_7H_5$.

*Step 3:* Derive the molecular formula

The empirical formula weight is 89.12 g/unit. This is one half of the molar mass, so the molecular formula is $C_{14}H_{10}$.

4-57. *Step 1:* Calculate the moles of C and H in isoprene from the quantities of $CO_2$ and $H_2O$ evolved on combustion.

0.507 g $CO_2$ • 1 mol $CO_2$/44.01 g • 1 mol C/1 mol $CO_2$ = 0.0115 mol C

$0.166\ g\ H_2O \cdot 1\ mol\ H_2O/18.02\ g \cdot\ 2\ mol\ H/1\ mol\ H_2O\ =\ 0.184\ mol\ H$

*Step 2:* Calculate the H to C ratio and derive the empirical formula

$0.184\ mol\ H/0.0115\ mol\ C\ =\ 1.60\ mol\ H/1.00\ mol\ C\ =\ 1\ {}^3/_5\ mol\ H/1\ mol\ C$

This leads to an empirical formula of $C_5H_8$.

*Step 3:* Derive the molecular formula

The empirical formula weight is 68.12 g/unit. This is equivalent to the molar mass, so the molecular formula is the same as the empirical formula.

4-59. *Step 1:* Calculate the moles of C and H, and then the mass of each, from the quantities of $CO_2$ and $H_2O$ evolved on combustion.

$0.300\ g\ CO_2 \cdot 1\ mol\ CO_2/44.01\ g \cdot 1\ mol\ C/1\ mol\ CO_2\ = 0.00682\ mol\ C$

$0.00682\ mol\ C \cdot 12.01\ g\ C/mol\ =\ 0.0819\ g\ C$

$0.040\ g\ H_2O \cdot 1\ mol\ H_2O/18.02\ g \cdot\ 2\ mol\ H/1\ mol\ H_2O\ =\ 0.0044\ mol\ H$

$0.0044\ mol\ H \cdot 1.01\ g\ H/mol\ =\ 0.0045\ g\ H$

*Step 2:* Calculate grams of O and then moles of O

$0.123\ g\ sample - 0.0819\ g\ C - 0.0045\ g\ H\ =\ 0.037\ g\ O$

$0.037\ g\ O \cdot 1\ mol/16.0\ g\ =\ 0.0023\ mol\ O$

*Step 3:* Find mole ratios and derive the empirical formula

$0.00682\ mol\ C/0.0023\ mol\ O\ =\ 3.0\ mol\ C/1.0\ mol\ O$

$0.0044\ mol\ H/0.0023\ mol\ O\ =\ 1.9\ mol\ H/1.0\ mol\ O\ \approx 2.0\ mol\ H/1.0\ mol\ O$

Empirical formula = $C_3H_2O$

*Step 4:* Derive the molecular formula

The empirical formula weight is 54.05 g/unit. This is one third of the molar mass of 160 g/mol. Therefore, the molecular formula is $C_9H_6O_3$.

4-61. *Step 1:* Find the moles of B and H in the compound from the masses of $B_2O_3$ and $H_2O$ formed on combustion.

$1.740\ g\ B_2O_3 \cdot 1\ mol\ B_2O_3/69.62\ g \cdot 2\ mol\ B/1\ mol\ B_2O_3 = 0.0500\ mol\ B$

$0.810\ g\ H_2O \cdot 1\ mol\ H_2O/18.02\ g \cdot\ 2\ mol\ H/1\ mol\ H_2O\ =\ 0.0900\ mol\ H$

*Step 2:* Find the ratio of H to B and derive the empirical formula

0.0900 mol H/0.0500 mol B = 9.00 mol H/5.00 mol B

Empirical formula = $B_5H_9$

4-63. *Step 1:* Find moles of Fe and moles of CO

0.799 g $Fe_2O_3$ • 1 mol $Fe_2O_3$/159.7 g • 2 mol Fe/1 mol $Fe_2O_3$ = 0.0100 mol Fe

2.200 g $CO_2$ • 1 mol/44.010 g • 1 mol CO/1 mol $CO_2$ = 0.05000 mol CO

*Step 2:* Derive the empirical formula

0.0500 mol CO/0.0100 mol Fe = 5 mol CO/1 mol Fe

Empirical formula = $Fe(CO)_5$

4-65. *Step 1:* Calculate moles of $CO_2$

0.222 g $CO_2$ • 1 mol/44.01 g = 0.00504 mol $CO_2$

*Step 2:* Calculate moles of $MCO_3$

0.000504 mol $CO_2$ • 1 mol $MCO_3$/1 mol $CO_2$ = 0.00504 mol $MCO_3$

*Step 3:* Calculate molar mass of $MCO_3$

0.598 g $MCO_3$/0.00504 mol = 119 g/mol

*Step 4:* Determine the identity of M

Mass of M + Mass of $CO_3$ = 119 g/mol

Mass of M = 119 - 60 = 59 g/mol

Therefore, M is most likely Ni, which has a molar mass of 58.69 g/mol.

## CHEMICAL ANALYSIS

4-67. *Step 1:* Calculate moles of water lost

0.873 g $H_2O$ • 1 mol $H_2O$/18.02 g = 0.0484 mol $H_2O$

*Step 2:* Calculate moles and then mass of $CuSO_4 \cdot 5\,H_2O$ in the sample

0.0484 mol $H_2O$ • 1 mol $CuSO_4 \cdot 5\,H_2O$/5 mol $H_2O$ = 0.00969 mol $CuSO_4 \cdot 5\,H_2O$

0.00969 mol $CuSO_4 \cdot 5\,H_2O$ • 249.7 g/mol = 2.42 g $CuSO_4 \cdot 5\,H_2O$

*Step 3:* Calculate the weight percent of $CuSO_4 \cdot 5\,H_2O$

$(2.42\ g\ CuSO_4 \cdot 5\ H_2O/2.628\ g\ sample)100 = 92.1\%$

4-69. *Step 1:* Calculate the moles of $C_4H_{10}$ evolved

$0.636\ g\ C_4H_{10} \bullet 1\ mol/58.12\ g = 0.0109\ mol\ C_4H_{10}$

*Step 2:* Calculate moles and then mass of $LiC_4H_9$ in the sample

$0.0109\ mol\ C_4H_{10} \bullet 1\ mol\ LiC_4H_9/1\ mol\ C_4H_{10} = 0.0109\ mol\ LiC_4H_9$

$0.0109\ mol\ LiC_4H_9 \bullet 64.06\ g/1\ mol = 0.701\ g\ LiC_4H_9$

*Step 3:* Calculate the weight percent of $LiC_4H_9$

$(0.701\ g\ LiC_4H_9/5.606\ g\ sample)100 = 12.5\%$

## GENERAL QUESTIONS

4-71. *Step 1:* Calculate moles of ammonium dichromate

$9.856\ g\ (NH_4)_2Cr_2O_7 \bullet 1\ mol/252.06\ g = 0.03910\ mol\ (NH_4)_2Cr_2O_7$

*Step 2:* Calculate moles and then mass of each product

$0.03910\ mol\ (NH_4)_2Cr_2O_7 \bullet 1\ mol\ N_2/1\ mol\ (NH_4)_2Cr_2O_7 \bullet 28.013\ g/mol = 1.095\ g\ N_2$

$0.03910\ mol\ (NH_4)_2Cr_2O_7 \bullet 1\ mol\ Cr_2O_3/1\ mol\ (NH_4)_2Cr_2O_7 \bullet 151.990\ g/mol$

$= 5.943\ g\ Cr_2O_3$

$0.03910\ mol\ (NH_4)_2Cr_2O_7 \bullet 4\ mol\ H_2O/1\ mol\ (NH_4)_2Cr_2O_7 \bullet 18.015\ g/mol$

$= 2.818\ g\ H_2O$

The sum of the product masses obtained above (9.856 g) is the same as the original mass of ammonium dichromate, as expected.

4-73. *Step 1:* Calculate the mass of fluorine combined with sulfur

$5.722\ g\ SF_x - 1.256\ g\ S = 4.466\ g\ F$

*Step 2:* Calculate moles of S and F combined in $SF_x$

$4.466\ g\ F \bullet 1\ mol/18.998\ g = 0.2351\ mol\ F$

$1.256\ g \bullet 1\ mol/32.066\ g = 0.03917\ mol\ S$

*Step 3:* Find the ratio of moles of F to moles of S and derive the empirical formula

$0.2351\ mol\ F/0.03917\ mol\ S = 6.002\ mol\ F/1.000\ mol\ S$

Therefore, the empirical formula of the compound is $SF_6$

4-75. *Step 1:* Calculate the moles of C and H, and then the mass of each, from the quantities of $CO_2$ and $H_2O$ evolved on combustion.

0.547 g $CO_2$ • 1 mol $CO_2$/44.01 g • 1 mol C/1 mol $CO_2$ = 0.0124 mol C

0.0124 mol C • 12.01 g C/mol = 0.149 g C

0.0896 g $H_2O$ • 1 mol $H_2O$/18.02 g • 2 mol H/1 mol $H_2O$ = 0.00995 mol H

0.00995 mol H • 1.01 g H/mol = 0.0100 g H

*Step 2:* Calculate grams of O and then moles of O

0.239 g sample - 0.149 g C - 0.0100 g H = 0.080 g O

0.080 g O • 1 mol/16.0 g = 0.0050 mol O

*Step 3:* Find mole ratios and derive the empirical formula

0.0124 mol C/0.0050 mol O = 2.5 mol C/1.0 mol O (or 5 C to 2 O)

0.0100 mol H/0.0050 mol O = 2.0 mol H/1.0 mol O (or 4 H to 2 O)

Empirical formula = $C_5H_4O_2$

*Step 4:* Derive the molecular formula

The empirical formula weight is 96.09 g/unit. This is equivalent to the molar mass, so the molecular formula is the same as the empirical formula.

4-77. *Step 1:* Calculate mass of water gained by sample

| | |
|---|---|
| Mass of test tube + $Ca(OH)_2$ | 11.149 g |
| - Mass of test tube + CaO | 10.860 g |
| ---------------------------------- | ----------- |
| Mass water | 0.289 g |

*Step 2:* Calculate moles of water

0.289 g • 1 mol/18.02 g = 0.0160 mol $H_2O$

*Step 3:* Calculate moles and then mass of CaO that must have been present

0.0160 mol $H_2O$ • 1 mol CaO/1 mol $H_2O$ • 56.08 g/1 mol = 0.899 g CaO

*Step 4:* Calculate the mass of the test tube

Mass of test tube and CaO - mass of CaO = 10.860 g - 0.899 g = 9.961 g

4-79. *Step 1:* Balance the equation

$Pb(NO_3)_2(s) \rightarrow N_2O_4(g) + PbO(s) + \frac{1}{2} O_2(g)$

*Step 2:* Calculate mass of $N_2O_4$ expected

$3.468 \text{ g } Pb(NO_3)_2 \cdot 1 \text{ mol } Pb(NO_3)_2/331.21 \text{ g} = 0.01047 \text{ mol } Pb(NO_3)_2$

$0.01047 \text{ mol } Pb(NO_3)_2 \cdot 1 \text{ mol } N_2O_4/1 \text{ mol } Pb(NO_3)_2 \cdot 92.011 \text{ g } N_2O_4/\text{mol}$

$= 0.9634 \text{ g } N_2O_4$

*Step 3:* Calculate the percent yield

$(0.832 \text{ g}/0.9634 \text{ g})100 = 86.4\%$

4-81. Combination reaction: $2\ Cu(s) + O_2(g) \rightarrow 2\ CuO(s)$

Decomposition reaction: $CuCO_3(s) \rightarrow CuO(s) + CO_2(g)$

4-83. *Step 1:* Calculate moles of $B_5H_9$ and $O_2$

$126 \times 10^3 \text{ g } B_5H_9 \cdot 1 \text{ mol}/63.13 \text{ g} = 2.00 \times 10^3 \text{ mol } B_5H_9$

$192 \times 10^3 \text{ g } O_2 \cdot 1 \text{ mol}/32.00 \text{ g} = 6.00 \times 10^3 \text{ mol } O_2$

*Step 2:* Find mole ratio and determine limiting reagent

Actual mole ratio $= 2 \text{ mol } O_2/1 \text{ mol } B_5H_9$

Required mole ratio $= 6 \text{ mol } O_2/1 \text{ mol } B_5H_9$

There is not sufficient $O_2$ to use up all of the $B_5H_9$, so $O_2$ is the limiting reagent. The $O_2$ tank will be emptied before the $B_5H_9$ tank.

*Step 3:* Knowing that $O_2$ is the limiting reagent, find the quantity of water formed.

$6.00 \times 10^3 \text{ mol } O_2 \cdot 9 \text{ mol } H_2O/12 \text{ mol } O_2 \cdot 32.00 \text{ g } O_2/1 \text{ mol} = 1.44 \times 10^5 \text{ g } H_2O$

4-85. *Step 1:* Calculate moles of $P_4O_{10}$ and $HNO_3$

$15.07 \text{ g } P_4O_{10} \cdot 1 \text{ mol}/283.89 \text{ g} = 0.05308 \text{ mol } P_4O_{10}$

$10.67 \text{ g } HNO_3 \cdot 1 \text{ mol}/63.013 \text{ g} = 0.1693 \text{ mol } HNO_3$

*Step 2:* Find mole ratio and determine limiting reagent

Actual mole ratio = 0.1693 mol $HNO_3$/0.05308 mol $P_4O_{10}$

= 3.190 mol $HNO_3$/1.0 mol $P_4O_{10}$

Required mole ratio = 4 mol $HNO_3$/1 mol $P_4O_{10}$

There is clearly not sufficient $HNO_3$ is use up the $P_4O_{10}$, so $HNO_3$ is the limiting reagent.

*Step 3:* Calculate theoretical yield of $N_2O_5$

0.1693 mol $HNO_3$ • 2 mol $N_2O_5$/4 mol $HNO_3$ • 108.01 g/mol = 9.143 g $N_2O_5$

*Step 4:* Calculate the percent yield of $N_2O_5$

(1.96 /9.143 g)100 = 21.4%

4-87. *Step 1:* Find moles of Fe and moles of CO

0.598 g $Fe_2O_3$ • 1 mol $Fe_2O_3$/159.7 g • 2 mol Fe/1 mol $Fe_2O_3$ = 0.00749 mol Fe

1.317 g $CO_2$ • 1 mol/44.010 g • 1 mol CO/1 mol $CO_2$ = 0.02993 mol CO

*Step 2:* Derive the empirical formula

0.02993 mol CO/0.00749 mol Fe = 3.995 mol CO/1.000 mol Fe

Empirical formula = $Fe(CO)_4$

*Step 3:* Derive molecular formula

The formula weight of the empirical formula is 167.9 g/unit. This is one third of the molar mass of 504 g/mol. Therefore, the molecular formula is $Fe_3(CO)_{12}$.

4-89. *Step 1:* Calculate the moles and masses of C and Cl from the masses of $CO_2$ and HCl.

4.400 g CO2 • 1 mol/44.010 g = 0.09998 mol $CO_2$

0.09998 mol $CO_2$ • 1 mol C/1mol $CO_2$ = 0.09998 mol C

0.09998 mol C • 12.01 g C/mol = 1.201 g C

1.825 g HCl • 1 mol/36.461 g = 0.05005 mol HCl

0.05005 mol HCl • 1 mol Cl/1 mol HCl = 0.05005 mol Cl

0.05005 mol Cl • 35.453 g Cl/1 mol = 1.775 g Cl

*Step 2:* Calculate the mass and then moles of H in the sample of vinyl chloride

3.125 g vinyl chloride - 1.201 g C - 1.775 g Cl = 0.150 g H

0.150 g H • 1 mol/1.008 g = 0.148 mol H

*Step 3:* Find the mole ratios of C, H, and Cl in the sample of vinyl chloride. Derive the empirical formula.

0.148 mol H/0.05005 mol Cl = 2.96 mol H/1.00 mol Cl ≈ 3 mol H/1 mol Cl

0.09998 mol C/0.05005 mol Cl = 2.00 mol C/1.00 mol Cl

The empirical formula is $C_2H_3Cl$

*Step 4:* Find the molecular formula

The formula weight of $C_2H_3Cl$ is 62.5 g/unit, exactly equivalent to the molar mass.

Therefore, the molecular formula of vinyl chloride is the same as the empirical formula.

4-91. Part (a)

*Step 1:* Calculate the quantity of ammonium nitrate in ton-moles

$3.00 \times 10^4$ T • 1 T-mol/80.04 T = $3.75 \times 10^2$ T-mol

*Step 2:* Calculate the quantity of each product

375 T-mol $NH_4NO_3$ • 2 T-mol $N_2$/2 T-mol $NH_4NO_3$ • 28.01 T $N_2$/T-mol

= $1.05 \times 10^4$ T $N_2$

375 T-mol $NH_4NO_3$ • 4 T-mol $H_2O$/2 T-mol $NH_4NO_3$ • 18.02 T $H_2O$/T-mol

= $1.35 \times 10^4$ T $H_2O$

375 T-mol $NH_4NO_3$ • 1 T-mol $O_2$/2 T-mol $NH_4NO_3$ • 32.00 T $O_2$/T-mol

= $6.00 \times 10^3$ T $O_2$

Total mass of product = $3.00 \times 10^4$ T. As expected it is equal to the mass of starting material.

Part (b): Calculate the mass of $NH_4NO_3$ that gives 45.0 g $H_2O$.

45.0 g $H_2O$ • 1 mol/18.02 g • 2 mol $NH_4NO_3$/4 mol $H_2O$ • 80.04 g/mol

= 99.9 g $NH_4NO_3$

Part (c): Calculate the percent yield of $O_2$

8.0 g $NH_4NO_3$ • 1 mol $NH_4NO_3$/80.04 g = 0.010 mol $NH_4NO_3$

0.010 mol $NH_4NO_3$ • 1 mol $O_2$/2 mol $NH_4NO_3$ • 32.00 g $O_2$/1 mol = 1.6 g $O_2$ expected

(1.0 g/1.6 g)100 = 63%

4-93. *Step 1:* Write a balanced equation

$6\ UO_3(s) + 8\ BrF_3(\ell) \rightarrow 6\ UF_4(s) + 4\ Br_2(\ell) + 9\ O_2(g)$

*Step 2:* Calculate moles of $O_2$ isolated

$0.0232\ g\ O_2 \cdot 1\ mol/32.00\ g = 7.25 \times 10^{-4}\ mol\ O_2$

*Step 3:* Calculate moles and then mass of U present

$7.25 \times 10^{-4}\ mol\ O_2 \cdot 6\ mol\ UO_3/9\ mol\ O_2 \cdot 1\ mol\ U/1\ mol\ UO_3 \cdot 238.0\ g/mol$

$= 0.115\ g\ U$

*Step 4:* Calculate the percentage of U in the sample

$(0.115\ g/1.485\ g)100 = 7.75\ \%$

4-95. To solve this problem, we need to write two equations. The first of these is

Mass of CuO + Mass of $Cu_2O$ = 1.000 g

or $x + y = 1.000\ g$

and the second arises from the fact that we can find the mass of Cu in either oxide from the general expression

Mass of oxide • g copper oxide/mol • mol Cu/mol oxide • 63.546 g Cu/mol

Therefore, we can write

$x(1\ mol\ CuO/79.545\ g \cdot 1\ mol\ Cu/1\ mol\ CuO \cdot 63.546\ g\ Cu/mol) +$

$y(1\ mol\ Cu_2O/143.09\ g \cdot 2\ mol\ Cu/1\ mol\ Cu_2O \cdot 63.546\ g\ Cu/mol) = 0.839\ g\ Cu$

or $x(0.799) + y(0.888) = 0.839$

Since $x + y = 1.000$ g, we now have

$0.799\ x + 0.888\ (1.000 - x) = 0.839$

$x = 0.55\ g$ = mass of CuO

Therefore, $y$ = mass of $Cu_2O$ [copper(I) oxide] = 0.45 g

## SUMMARY QUESTIONS

4-96. Part (a): Write the balanced equation

$2\ Al(s) + 3\ Br_2(\ell) \rightarrow 2\ AlBr_3(s)$

Part (b): Calculate moles of Al and $Br_2$ available

0.500 mm thick Al = 0.0500 cm thick

$5.00 \times 10^{-2}$ cm • 12.0 cm • 12 cm = 7.20 $cm^3$ Al

7.20 $cm^3$ Al • 2.70 g/$cm^3$ • 1 mol/26.98 g = 0.721 mol Al

19.50 mL $Br_2$ • 3.10 g/$cm^3$ • 1 mol/159.8 g = 0.378 mol $Br_2$

Part (c): Find the theoretical yield of $AlBr_3$

Actual mole ratio = 0.378 mol $Br_2$/0.721 mol Al = 0.525 mol $Br_2$/1.00 mol Al

Theoretical mole ratio = 3 mol $Br_2$/2 mol Al = 1.5 mol $Br_2$/1.0 mol Al

There is clearly a deficiency of $Br_2$, so $Br_2$ is the limiting reagent.

0.378 mol $Br_2$ • 2 mol $AlBr_3$/3 mol $Br_2$ • 266.7 g/mol = 67.3 g $AlBr_3$ should be produced

4-97. *Step 1:* Balance the equations

$2\ PbS(s) + 3\ O_2(g) \rightarrow 2\ PbO(s) + 2\ SO_2(g)$

$PbO(s) + C(s) \rightarrow Pb(s) + CO(g)$

| **Compound** | **Name** |
|---|---|
| PbS | lead(II) sulfide |
| PbO | lead(II) oxide |
| $SO_2$ | sulfur dioxide |
| CO | carbon monoxide |
| $O_2$ | oxygen (sometimes called dioxygen) |

*Step 2:* Calculate the mass of Pb obtainable

1.00 lb PbS • 454 g/lb • 1 mol/239.3 g = 1.90 mol PbS

1.90 mol PbS • 2 mol PbO/2 mol PbS • 1 mol Pb/1 mol PbO • 207.2 g/mol = 393 g Pb

Step 3: Calculate the mass of C required

1.90 mol PbS • 2 mol PbO/2 mol PbS • 1 mol C/1 mol PbO • 12.01 g/mol = 22.8 g C

*Step 4:* Calculate the thickness of a sheet of lead that has a mass of 393 g and dimensions of 9.60 cm by 34.5 cm

393 g Pb • 1 $cm^3$/11.34 g = 34.7 $cm^3$

34.7 $cm^3$ = 9.60 cm • 34.5 cm • thickness

Thickness = 0.105 cm

# CHAPTER 5
# REACTIONS IN AQUEOUS SOLUTION

## INSTRUCTOR'S NOTES

With the renewed emphasis on chemical reactivity by those of us teaching chemistry, we consider Chapter 5 an important early chapter in the book. We cover this material early in our course because our laboratory emphasizes solution stoichiometry. However, *we emphasize that it is not crucial that this material be covered at this time in the course.* In particular, *it would be easy to cover the subject of redox reactions at some other point in the year,* such as when covering electrochemistry later in the year. *The book is constructed so that solution stoichiometry is not needed until one begins Part III of the book.*

Chapter 5 begins with an introduction to the properties of ionic compounds in aqueous solution, including general guidelines to predict the aqueous solubility of simple ionic compounds. Although one can readily find many exceptions to the guidelines in Figure 5.5, we do find them useful for students in a beginning course. We hope that the color photos of soluble and insoluble compounds in Figure 5.6 will help students remember some of these guidelines.

The introduction to acids and bases beginning on page 144 is necessarily brief at this stage, but it does enable one to use common acids and bases in examples and in the laboratory with the knowledge that students have some familiarity with them.

The section on net ionic equations beginning on page 145 can be a difficult one for our students. We admit it does take time and effort to help the students come to grips with this concept. However, it is a useful one, since (1) many reactions are best seen in this fashion and (2) balancing redox reactions in Section 5.5 is made easier. We hope that the use of color in this section will be especially helpful. Students can generally write balanced equations for exchange reactions (metathesis), but they sometimes have a difficult time turning them into net ionic equations. Time and effort on everyone's part is needed here.

A few more comments on redox reactions (Section 5.5) are appropriate here.

- We emphasize to our students that we have introduced oxidation numbers *only* as a way of telling if a redox reaction has occurred. We do *not* use them to balance the equation. Once the decision has been made that the reaction involves electron transfer, the student can forget about oxidation numbers.
- There are of course many errors students commonly make when balancing redox reactions, but we have tried to anticipate these. Typically, they do not put charges on ions or electrons, and they use or produce O atoms or H atoms in reactions (instead of $H_2O$, $OH^-$, or $H^+$, as appropriate). The method outlined has been used in the classroom for about 12 years. It has been especially successful when teaching students to balance equations in basic solution.
- After considerable debate, we decided to use $H^+$ instead of $H_3O^+$ in balancing redox reactions. Our approach can be adapted readily to the use of the hydronium ion, but using this ion can be cumbersome and, for the beginner, it can obscure the process. Later in the book, when balanced redox reactions appear they are balanced with the hydronium ion, where appropriate.

## TRANSPARENCIES AVAILABLE FOR THIS CHAPTER

| No. | Figure | Page | Subject |
|---|---|---|---|
| 4 | 5.2 | 138 | Dissolving NaCl in water |
| 5 | 5.5 | 141 | Solubility guidelines |
| 6 | 5.18 | 161 | Scheme outlining solution stoichiometry calculations |

## SUGGESTED DEMONSTRATIONS

1. We always illustrate the conductivity, and the difference between strong and weak electrolytes, using an apparatus such as that in Figure 5.3 or 5.4. We use solutions such as $CuSO_4$, $Na_2CO_3$, vinegar, and pure water. It is interesting to compare the behavior of these solutions with the conductivity of a 7-up or Coke.
2. The reaction of Mg with HCl (Figure 5.1) is representative of reactions that we illustrate in a large classroom using an overhead projector. A petri dish containing HCl is placed on the projector and a strip of metal added. The bubbles are clearly visible. (If the amount of acid and metal are adjusted correctly, an indicator could be added to give a dramatic color.) (Alternatively, the reaction of Zn and HCl works well.) Other reactions that can be

illustrated in a dish on an overhead projector include Figure 5.11 (a piece of chalk in acid) and Figure 5.24 (redox reaction of Cu and $Ag^+$). The latter is especially striking, since the tiny crystals of silver are magnified and can be seen clearly on the screen. Further, the blue color from aqueous $Cu^{2+}$ ia clearly visible.

3. Precipitation reactions *cannot* be illustrated on an overhead projector. The precipitate is seen only as a dark blob. However, even in a large lecture room, students can see the reaction if it is done in a large flask (1-2 L).

4. The **Periodic Table Videodisc** can be used to illustrate many oxidation-reduction reactions.

5. It is very important to illustrate the solubility of salts. Most recently, our demonstrations have consisted of the following:
   - A gram or so of KI is dissolved in a few hundred milliliters of water in one flask, and $Pb(NO_3)_2$ is dissolved in water in another flask. The students can clearly see the salts dissolve. On mixing the solutions, though, a bright yellow precipitate of $PbI_2$ appears, clearly demonstrating one of the exceptions to the general rule that halide salts are water-soluble.
   - A solution of $(NH_4)_2S$ is mixed with a solution of $Cd(NO_3)_2$ to show that metal sulfides (bright yellow CdS in this case) are generally not soluble.
   - $BaCl_2$ is dissolved in water while discussing halide solubility and mixed with a solution of $CuSO_4$ (which shows the usual solubility of sulfates). Precipitation of $BaSO_4$ on mixing the solutions then demonstrates the exceptions to the guideline regarding sulfates.
   - Our attempt at dissolving a piece of blackboard chalk fails, clearly showing that many carbonates are not soluble.

   Throughout the demonstration the solubility guidelines are projected onto the screen in the lecture room.

6. Other solubility demonstrations include:
   - Dissolve NaCl in water and then precipitate AgCl with $AgNO_3$.
   - Dissolve $Na_2CO_3$ and compare this with chalk.
   - Show a sample of fool's gold (iron pyrite) to illustrate the insolubility of metal sulfides.

- Add NaOH to solutions of $Fe(NO_3)_3$ and $CuSO_4$ to get insoluble hydroxides.

7. Although is it easy for students to see precipitation reactions done in large flasks, we find that acid-base reactions are difficult to demonstrate, since common ones are not spectacular. However, a suggested demonstration would be to dissolve a precipitate of $Ca(OH)_2$. At least the solid would be seen to dissolve on adding acid.
8. We illustrate Example 5.14 in class with the following procedure: (1) Dissolve a piece of Cu metal in nitric acid that already contains some $Ag^+$. (2) Add HCl to the solution to show how AgCl precipitates. (3) Filter the solution to show how to collect the precipitate.
9. One of the biggest problems in teaching stoichiometry is to get the students to connect the words on paper with actual operations in the laboratory. To help in making this connection, we have included photos in the text of (a) the preparation of a solution starting with a solid compound (Figure 5.15), preparing a solution by dilution of a more concentrated one (Figure 5.17), and acid-base and redox titrations (Figures 5.19 and 5.25, respectively). We find it very helpful to demonstrate these processes in class as well.
10. **Oxidation-reduction chemistry:** We introduce the topic of redox reactions using the **Periodic Table Videodisc** to illustrate such reactions. After introducing the ideas of redox and the concept of oxidation numbers, we do a demonstration on the oxidation states of vanadium.

    *Directions:* Dissolve 1 g of ammonium vanadate in 200 mL of water. (Has to be warmed to dissolve completely.)

    (a) Add 3 M $H_2SO_4$ (20-50 mL), and the solution turns yellow owing to an acid-base reaction

    $VO_3^- + 2\,H^+ \rightarrow VO_2^+ + H_2O$

    (b) Add a handful of zinc chips. These reduce the V(V) to V(IV).

    $2\,VO_2^+ + Zn + 4\,H^+ \rightarrow 2\,VO^{2+}$(vanadyl, blue) $+ 2\,H_2O + Zn^{2+}$

    (c) The solution will slowly turn emerald green owing to another reduction step.

    $2\,VO^{2+}$ (blue) $+ Zn + 4\,H^+ \rightarrow 2\,V^{3+}$(green) $+ 2\,H_2O + Zn^{2+}$

    (d) The solution will finally turn violet. We allow the reaction to continue throughout the lecture with a stopper on the flask. This excludes air well enough that the violet, aqueous vanadium(II) ion can form. (The vanadium(II) ion is readily oxidized by oxygen in air to $V^{3+}$. Therefore, to see the violet color air must be excluded.)

$2\ V^{3+}(\text{green}) + Zn \rightarrow 2\ V^{2+} + Zn^{2+}$

11. To introduce the idea of half-reactions, KI is electrolyzed in a cell on the overhead projector. We use a petri dish and make electrodes out of nichrome. Bend the wire so the electrodes lie on the bottom of the dish. Place an aqueous solution of KI in the dish and add a few drops of phenolphthalein. Attach the electrodes to a battery (or better, a toy train transformer) and within a few seconds $I_2$ will appear at one electrode and $H_2$ and $OH^-$ at the other.

$2\ I^-(aq) \rightarrow I_2(s) + 2e^-$

$2\ H_2O(\ell) + 2e^- \rightarrow H_2(g) + 2\ OH^-(aq)$

---

$2\ I^-(aq) + 2\ H_2O(\ell) \rightarrow I_2(s) + H_2(g) + 2\ OH^-(aq)$

We trace the changes in oxidation numbers, talk about which is an oxidation and which is a reduction, and then show that these half reactions can be added to obtain the overall reaction.

# SOLUTIONS TO SELECTED STUDY QUESTIONS

## PROPERTIES OF IONIC COMPOUNDS

5-17. (a) $Pb(NO_3)_2$; (b) All $Na^+$ salts are soluble; (c) KBr and $AlBr_3$.

5-19. (a) There are relatively few simple chlorides that are not soluble. The insoluble chlorides that we usually think of are AgCl, $PbCl_2$, and $Hg_2Cl_2$. Virtually all other chlorides are soluble.

(b) Soluble hydroxides are obtained with the alkali metal cations. Most other hydroxides are only poorly soluble. For example, insoluble hydroxides might be $Fe(OH)_3$, $Ni(OH)_2$, $Ca(OH)_2$, and $Al(OH)_3$.

(c) Soluble carbonates are obtained with the alkali metal cations. Most other carbonates are only poorly soluble. For example, insoluble carbonates might be $Fe_2(CO_3)_3$, $NiCO_3$, and $CaCO_3$.

5-21. (a) $KOH \rightarrow K^+(aq) + OH^-(aq)$

(b) $K_2CO_3 \rightarrow 2\ K^+(aq) + CO_3^{2-}(aq)$

(c) $NaNO_3 \rightarrow Na^+(aq) + NO_3^-(aq)$

(d) $(NH_4)_2SO_4 \rightarrow 2\ NH_4^+(aq) + SO_4^{2-}(aq)$

5-23. (a) Sodium carbonate is soluble in water. $Na_2CO_3 \rightarrow 2\ Na^+(aq) + CO_3^{2-}(aq)$

(b) Copper(II) sulfate is soluble in water. $CuSO_4 \rightarrow Cu^{2+}(aq) + SO_4^{2-}(aq)$

(c) Cadmium sulfide is not soluble in water.

(d) Calcium bromide is water-soluble. $CaBr_2 \rightarrow Ca^{2+}(aq) + 2\ Br^-(aq)$

5-25. Water soluble $Ca^{2+}$ compounds: $CaCl_2$, $CaBr_2$, or $Ca(CH_3COO)_2$

Water insoluble $Ca^{2+}$ compounds: $CaCO_3$ or $Ca(OH)_2$

## ACIDS AND BASES

5-27. $HNO_3(aq) + H_2O(\ell) \rightarrow H_3O^+(aq) + NO_3^-(aq)$

## NET IONIC EQUATIONS

5-29. (a) Complete equation: $(NH_4)_2S(aq) + Cu(NO_3)_2(aq) \rightarrow CuS(s) + 2\ NH_4NO_3(aq)$

Net ionic equation: $S^{2-}(aq) + Cu^{2+}(aq) \rightarrow CuS(s)$

(b) Complete equation: $Pb(NO_3)_2(aq) + 2\ HCl(aq) \rightarrow PbCl_2(s) + 2\ HNO_3(aq)$

Net ionic equation: $Pb^{2+}(aq) + 2\ Cl^-(aq) \rightarrow PbCl_2(s)$

(c) Complete equation: $BaCO_3(s) + 2\ HCl(aq) \rightarrow BaCl_2(aq) + H_2O(\ell) + CO_2(g)$

Net ionic equation: $BaCO_3(s) + 2\ H^+(aq) \rightarrow Ba^{2+}(aq) + H_2O(\ell) + CO_2(g)$

Net ionic equation using hydronium ion:

$$BaCO_3(s) + 2\ H_3O^+(aq) \rightarrow Ba^{2+}(aq) + 3\ H_2O(\ell) + CO_2(g)$$

5-31. (a) Complete equation: $ZnCl_2(aq) + 2\ KOH(aq) \rightarrow 2\ KCl(aq) + Zn(OH)_2(s)$

Net ionic equation: $Zn^{2+}(aq) + 2\ OH^-(aq) \rightarrow Zn(OH)_2(s)$

(b) Complete equation: $AgNO_3(aq) + KI(aq) \rightarrow AgI(s) + KNO_3(aq)$

Net ionic equation: $Ag^+(aq) + I^-(aq) \rightarrow AgI(s)$

(c) Complete equation: $2\ NaOH(aq) + FeCl_2(aq) \rightarrow Fe(OH)_2(s) + 2\ NaCl(aq)$

Net ionic equation: $2\ OH^-(aq) + Fe^{2+}(aq) \rightarrow Fe(OH)_2(s)$

## TYPES OF REACTIONS

5-33. (a) Precipitation reaction: $Cu(NO_3)_2(aq) + Na_2S(aq) \rightarrow 2\ NaNO_3(aq) + CuS(s)$

(b) Precipitation reaction: $Pb(NO_3)_2(aq) + 2\ KCl(aq) \rightarrow PbCl_2(s) + 2\ KNO_3(aq)$

(c) Acid-base reaction: $LiOH(aq) + HClO_4(aq) \rightarrow LiClO_4(aq) + H_2O(\ell)$

5-35. (a) Acid-base reaction: $2\ HNO_3(aq) + Ca(OH)_2(s) \rightarrow Ca(NO_3)_2(aq) + 2\ H_2O(\ell)$

(b) Acid-base reaction: $2\ HCl(aq) + Fe(OH)_2(s) \rightarrow FeCl_2(aq) + 2\ H_2O(\ell)$

(c) Precipitation reaction: $SrCl_2(aq) + Na_2CO_3(aq) \rightarrow SrCO_3(s) + 2\ NaCl(aq)$

(d) Gas-forming reaction:

$$Na_2CO_3(aq) + 2\ HBr(aq) \rightarrow 2\ NaBr(aq) + H_2O(\ell) + CO_2(g)$$

## PREPARATION OF SALTS

5-37. (a) $NaOH(aq) + CH_3COOH(aq) \rightarrow NaCH_3COO(aq) + H_2O(\ell)$

(b) $NH_3(aq) + HNO_3(aq) \rightarrow NH_4NO_3(aq)$

(c) $Mg(OH)_2(s) + H_2SO_4(aq) \rightarrow MgSO_4(s) + 2\ H_2O(\ell)$

(d) $3\ NaOH(aq) + H_3PO_4(aq) \rightarrow Na_3PO_4(aq) + 3\ H_2O(\ell)$

5-39. (a) $FeCl_2(aq) + Na_2CO_3(aq) \rightarrow FeCO_3(s) + 2\ NaCl(aq)$

(b) $3\ AgNO_3(aq) + H_3PO_4(aq) \rightarrow Ag_3PO_4(s) + 3\ HNO_3(aq)$

(c) $ZnCl_2(aq) + Na_2S(aq) \rightarrow 2\ NaCl(aq) + ZnS(s)$

(d) $KCl(aq) + AgNO_3(aq) \rightarrow KNO_3(aq) + AgCl(s)$

5-41. (a) $Ag_2CO_3(s) + 2\ HNO_3(aq) \rightarrow 2\ AgNO_3(aq) + H_2O(\ell) + CO_2(g)$

(b) $3\ Na_2CO_3(aq) + 2\ H_3PO_4(aq) \rightarrow 2\ Na_3PO_4(aq) + 3\ H_2O(\ell) + 3\ CO_2(g)$

(c) $MnCO_3(s) + H_2SO_4(aq) \rightarrow MnSO_4(aq) + H_2O(\ell) + CO_2(g)$

5-43. (a) *Sodium sulfate*

Acid-base reaction: $2\ NaOH(aq) + H_2SO_4(aq) \rightarrow Na_2SO_4(aq) + 2\ H_2O(\ell)$

Gas-forming reaction: $Na_2CO_3(aq) + H_2SO_4(aq)$

$$\rightarrow Na_2SO_4(aq) + H_2O(\ell) + CO_2(g)$$

(b) *Lead(II) iodide*

Precipitation reaction: $Pb(NO_3)_2(aq) + 2\ NaI(aq) \rightarrow PbI_2(s) + 2\ NaNO_3(aq)$

Acid-base reaction: $Pb(OH)_2(s) + 2\ HI(aq) \rightarrow PbI_2(s) + 2\ H_2O(\ell)$

Gas-forming reaction: $PbCO_3(s) + 2\ HI(aq) \rightarrow PbI_2(s) + H_2O(\ell) + CO_2(g)$

(c) *Cadmium sulfide*

Precipitation reaction: $CdCl_2(aq) + Na_2S(aq) \rightarrow CdS(s) + 2\ NaCl(aq)$

(d) *Magnesium chloride*

Acid-base reaction: $Mg(OH)_2(s) + 2\ HCl(aq) \rightarrow MgCl_2(aq) + 2\ H_2O(\ell)$

Gas-forming reaction: $MgCO_3(s) + 2\ HCl(aq)$

$\rightarrow MgCl_2(aq) + H_2O(\ell) + CO_2(g)$

(e) *Sodium perchlorate*

Acid-base reaction: $NaOH(aq) + HClO_4(aq) \rightarrow NaClO_4(aq) + H_2O(\ell)$

Gas-forming reaction: $Na_2CO_3(s) + 2\ HClO_4(aq)$

$\rightarrow 2\ NaClO_4(aq) + H_2O(\ell) + CO_2(g)$

## CONCENTRATION OF SOLUTIONS

5-45. (a) $(1.98\ g\ KMnO_4)(1\ mol/158.0\ g) = 0.0125\ mol$

$0.0125\ mol/0.250\ L = 0.0501\ M$

(b) $(0.500\ g\ KMnO_4)(1\ mol/158.0\ g) = 3.16 \times 10^{-3}\ mol$

$(3.16 \times 10^{-3}\ mol)(1\ L/0.0145\ mol) = 0.218\ L$ or 218 mL

5-47. (a) $(0.25\ L)(0.050\ mol/L) = 0.013\ mol\ Na_2C_2O_4$

$(0.0125\ mol\ Na_2C_2O_4)(134\ g/mol) = 1.7\ g\ Na_2C_2O_4$

(b) $(0.125\ L)(0.15\ mol/L) = 0.019\ mol\ K_2Cr_2O_7$

$(0.019\ mol\ K_2Cr_2O_7)(294\ g/mol) = 5.5\ g\ K_2Cr_2O_7$

(c) $(0.50\ L)(0.10\ mol/L) = 0.050\ mol\ KHCO_3$

$(0.050\ mol\ KHCO_3)(100.\ g/mol) = 5.0\ g\ KHCO_3$

5-49. (a) (0.55 g NaOH)(1 mol/40.0 g) = 0.014 mol NaOH
0.014 mol NaOH/0.050 L = 0.28 M NaOH
(b) (153 g $Na_2CO_3$)(1 mol/106.0 g) = 1.44 mol $Na_2CO_3$
1.44 mol $Na_2CO_3$/1.55 L = 0.931 M $Na_2CO_3$
(c) (1.256 g $KMnO_4$)(1 mol/158.03 g) = $7.948 \times 10^{-3}$ mol $KMnO_4$
$7.948 \times 10^{-3}$ mol $KMnO_4$/0.250 L = 0.0318 M $KMnO_4$

5-51. (a) 0.10 M $KClO_4$ → 0.10 M $K^+$(aq) + 0.10 M $ClO_4^-$(aq)
(b) 0.00298 M $KMnO_4$ → 0.00298 M $K^+$(aq) + 0.00298 M $MnO_4^-$(aq)
(c) 0.385 M $Na_2CO_3$ → 0.770 M $Na^+$(aq) + 0.385 M $CO_3^{2-}$(aq)

5-53. (0.500 L)(0.33 M NaOH) = (? L)(1.25 M NaOH)
? L = 0.13 L or 130 mL

5-55. (0.250 L)(0.0366 M $KMnO_4$) = (2.00 L)(? M)
? M = 0.00458 mol/L

5-57. (0.1000 L)(0.0584 M NaOH) = (0.250 L)(? M)
? M = 0.0234 mol/L

5-59. (2.5 L)(0.15 mol/L) = (? L)(17.4 mol/L)
? L = 0.022 L
Take 22 mL of 17.4 M acetic acid and dilute to 2.5 L to have a 0.15 M solution.

## STOICHIOMETRY OF REACTIONS IN SOLUTION

5-61. (0.050 L)(0.15 M $H_2S$) = 0.0075 mol $H_2S$
0.0075 mol $H_2S$ • 1 mol PbS/1 mol $H_2S$ • 239 g PbS/mol = 1.8 g PbS
0.0075 mol $H_2S$ • 1 mol $Pb(NO_3)_2$/1 mol $H_2S$ • 331 g $Pb(NO_3)_2$/mol
= 2.5 g $Pb(NO_3)_2$

5-63. *Step 1:* Calculate moles of $KMnO_4$

$0.125\ L \cdot 0.0525\ mol/L = 0.00656\ mol\ KMnO_4$

*Step 2:* Calculate volume of $H_2O_2$ solution required.

$0.00656\ mol\ KMnO_4 \cdot 3\ mol\ H_2O_2/2\ mol\ KMnO_4 = 0.00984\ mol\ H_2O_2$

$0.00984\ mol\ H_2O_2 \cdot 1\ L/0.88\ mol$

$= 0.011\ L\ H_2O_2$ (or 11 mL of the $H_2O_2$ solution)

*Step 3:* Calculate grams of $MnO_2$ produced.

$0.00656\ mol\ KMnO_4 \cdot 2\ mol\ MnO_2/2\ mol\ KMnO_4 \cdot 86.94\ g/mol = 0.571\ g\ MnO_2$

5-65. $31.1\ g\ Au \cdot 1\ mol/197.0\ g = 0.158\ mol\ Au$

$0.158\ mol\ Au \cdot 8\ mol\ NaCN/4\ mol\ Au = 0.316\ mol\ NaCN$

$0.316\ mol\ NaCN \cdot 1\ L/0.55\ mol = 0.57\ L\ NaCN$ (or 570 mL)

$0.158\ mol\ Au \cdot 4\ mol\ NaAu(CN)_2/4\ mol\ Au \cdot 272.0\ g\ NaAu(CN)_2/mol$

$= 42.9\ g\ NaAu(CN)_2$

5-67. $46.3\ g\ Zn \cdot 1\ mol/65.39\ g = 0.708\ mol\ Zn$

$0.065\ L \cdot 2.45\ mol\ K_2Cr_2O_7/L = 0.159\ mol\ K_2Cr_2O_7$

Ratio of available moles: $0.708\ mol\ Zn/0.159\ mol\ K_2Cr_2O_7$

$= 4.45\ mol\ Zn/1\ mol\ K_2Cr_2O_7$

Ratio of required moles: $4\ mol\ Zn/1\ mol\ K_2Cr_2O_7$

Zn is present in excess, so $K_2Cr_2O_7$ is the limiting reagent and will be consumed completely.

## ACID-BASE TITRATIONS

5-69. Balanced equation: $HCl(aq) + NaOH(aq) \rightarrow NaCl(aq) + H_2O(\ell)$

$0.0365\ L \cdot 0.100\ mol\ NaOH/L = 0.00365\ mol\ NaOH$

$0.00365\ mol\ NaOH \cdot 1\ mol\ HCl/1\ mol\ NaOH \cdot 1\ L/0.250\ mol\ HCl = 0.0146\ L\ HCl$

$0.0146\ L\ HCl \rightarrow 14.6$ mL of 0.250 M HCl

5-71. *Step 1:* Write balanced equation

$Ba(OH)_2(s) + 2\ HCl(aq) \rightarrow BaCl_2(aq) + 2\ H_2O(\ell)$

*Step 2:* Calculate moles of $Ba(OH)_2$

2.50 g $Ba(OH)_2$ • 1 mol/171.3 g = 0.0146 mol $Ba(OH)_2$

*Step 3:* Calculate moles of HCl required

0.0146 mol $Ba(OH)_2$ • 2 mol HCl/1 mol $Ba(OH)_2$ = 0.0292 mol HCl

*Step 4:* Calculate volume of HCl solution

0.0292 mol HCl • 1 L/0.110 mol = 0.266 L (or 266 mL HCl)

5-73. *Step 1:* Calculate moles of potassium acid phthalate

0.902 g • 1 mol/204.2 g = 0.00442 mol

*Step 2:* Calculate moles of NaOH required

0.00442 mol acid • 1 mol NaOH/1 mol acid = 0.00442 mol NaOH

*Step 3:* Calculate molarity of NaOH

0.00442 mol NaOH/0.3945 L = 0.0112 mol/L

5-75. *Step 1:* Calculate moles of NaOH

0.03651 L • 0.0102 mol/L = $3.72 \times 10^{-4}$ mol NaOH

*Step 2:* Calculate moles of citric acid

$3.72 \times 10^{-4}$ mol NaOH • 1 mol citric acid/3 mol NaOH

= $1.24 \times 10^{-4}$ mol citric acid

Step 3: Calculate mass of citric acid

$1.24 \times 10^{-4}$ mol citric acid • 192.1 g/mol = 0.0239 g citric acid in 100.0 mL

*Step 4:* Calculate concentration of citric acid

$1.24 \times 10^{-4}$ mol citric acid/0.100 mL = 0.00124 M

NOTE: In the first printing of the book the formula of sodium citrate was given incorrectly as $Na_2C_6H_5O_7$. The formula should be $Na_3C_6H_5O_7$.

5-77. *Step 1:* Calculate moles NaOH used

0.02595 L • 0.467 mol/L = 0.0121 mol NaOH

*Step 2:* Calculate grams of tartaric acid

0.0121 mol NaOH • 1 mol acid/2 mol NaOH • 150.1 g acid/mol = 0.910 g acid

*Step 3:* Calculate weight percent acid

(0.910 g acid/1.036 g sample)100 = 87.8% tartaric acid

## ANALYSIS BY PRECIPITATION REACTIONS

5-79. 0.288 g $CaC_2O_4$ • 1 mol/128.1 g = $2.25 \times 10^{-3}$ mol $CaC_2O_4$

$2.25 \times 10^{-3}$ mol $CaC_2O_4$ • 1 mol Ca/1 mol $CaC_2O_4$ • 40.0 g Ca/mol = 0.899 g Ca

Weight percent Ca = (0.899 g Ca/75.0 g)100 = 0.120 %

5-81. 0.4659 g $Ni(C_4H_7N_2O_2)_2$ • 1 mol/288.93 g = $1.613 \times 10^{-3}$ mol $Ni(C_4H_7N_2O_2)_2$

$1.613 \times 10^{-3}$ mol $Ni(C_4H_7N_2O_2)_2$ • 1 mol Ni/1 mol $Ni(C_4H_7N_2O_2)_2$ • 58.69 g/mol

= 0.0946 g Ni

(0.0946 g Ni/0.4734 g)100 = 19.9 % Ni

## OXIDATION NUMBERS

5-83. (a) C = 0, H = +1, and H = -2

(b) H = +1, O = -2, and P = +5

(c) H = +1 and C = -8/3

(d) H = +1, O = -2, and As = +5

(e) H = +1, O = -2, and C = 0

(f) O = -2 and Xe = +6

## OXIDATION-REDUCTION REACTIONS

5-85. (a) Ca is oxidized to $Ca^{2+}$ and H is reduced from $H^+$ to $H_2$.

(b) Sn is oxidized from $Sn^{2+}$ to $Sn^{4+}$, and Cr in $Cr_2O_7^{2-}$ is reduced from an oxidation number of +6 to +3 in $Cr^{3+}$.

(c) Fe is oxidized from $Fe^{2+}$ (in FeS), S is oxidized from -2 (in FeS) to +6 in $SO_4^{2-}$, and N is reduced from +5 in $NO_3^{2-}$ to +2 in NO.

## HALF REACTIONS

5-87. (a) $Br_2(aq) + 2e^- \rightarrow 2\,Br^-(aq)$

$Br_2$ is an oxidizing agent; reaction is a reduction.

(b) $H_2S(aq) \rightarrow S(s) + 2\,H^+(aq) + 2e^-$

$H_2S$ is a reducing agent; reaction is an oxidation of S from -2 to 0.

(c) $VO^{2+}(aq) + 2\,H^+(aq) + e^- \rightarrow V^{3+}(aq)$

$VO^{2+}$ is an oxidizing agent; reaction is a reduction of V from +5 to +3.

(d) $U^{4+}(aq) + 2\,H_2O(\ell) \rightarrow UO_2^+(aq) + 4\,H^+(aq) + e^-$

$U^{4+}$ is a reducing agent; it is oxidized from +4 to +5.

5-89. (a) $HOI(aq) + H^+(aq) + 2e^- \rightarrow I^-(aq) + H_2O(\ell)$

HOI is an oxidizing agent; the I atom is reduced from +1 in HOI to -1.

(b) $NO(g) + H_2O(\ell) \rightarrow HNO_2(aq) + H^+(aq) + e^-$

NO is a reducing agent; N is oxidized from +2 in NO to +3 in $HNO_2$.

(c) $C_6H_5CH_3(aq) + 2\,H_2O(\ell) \rightarrow C_6H_5COOH(aq) + 6\,H^+(aq) + 6e^-$

$C_6H_5CH_3$, toluene, is a reducing agent; it is oxidized to benzoic acid.

(d) $SO_4^{2-}(aq) + 4\,H^+(aq) + 2e^- \rightarrow H_2SO_3(aq) + H_2O(\ell)$

The sulfate ion is an oxidizing agent; S is reduced from +6 in sulfate to +4 in sulfurous acid.

5-91. (a) $CrO_2^-(aq) + 4\,OH^-(aq) \rightarrow CrO_4^{2-}(aq) + 2\,H_2O(\ell) + 3e^-$

The $CrO_2^-(aq)$ ion is a reducing agent; Cr is oxidized from +3 to +6.

(b) $Br_2(aq) + 12\,OH^-(aq) \rightarrow 2\,BrO_3^-(aq) + 6\,H_2O(\ell) + 10\,e^-$

$Br_2$ is a reducing agent in basic solution; the Br atoms are oxidized from 0 to +5.

(c) $Ni(OH)_2(s) + 2\,OH^-(aq) \rightarrow NiO_2(s) + 2\,H_2O(\ell) + 2e^-$

Nickel(II) hydroxide is a reducing agent in basic solution; the Ni is oxidized from +2 to +4.

## BALANCING OXIDATION-REDUCTION EQUATIONS

5-93. (a) $Hg^{2+}(aq) + 2e- \rightarrow Hg(\ell)$ oxidizing agent

$Cu(s) \rightarrow Cu^{2+}(aq) + 2e^-$ reducing agent

------------------------------------

$Hg^{2+}(aq) + Cu(s) \rightarrow Cu^{2+}(aq) + Hg(\ell)$

(b) $MnO_2(s) + 4\,H^+(aq) + 2e^- \rightarrow Mn^{2+}(aq) + 2\,H_2O(\ell)$ oxidizing agent

$2\,Cl^-(aq) \rightarrow Cl_2(g) + 2\,e^-$ reducing agent

------------------------------------------------------------

$MnO_2(s) + 2\,Cl^-(aq) + 4\,H^+(aq) \rightarrow Mn^{2+}(aq) + Cl_2(g) + 2\,H_2O(\ell)$

(c) $3[Br_2(aq) + 2e^- \rightarrow 2\,Br^-(aq)]$ oxidizing agent

$I^-(aq) + 3\,H_2O(\ell) \rightarrow IO_3^-(aq) + 6\,H^+(aq) + 6\,e^-$ reducing agent

------------------------------------------------------------

$I^-(aq) + 3\,Br_2(aq) + 3\,H_2O(\ell) \rightarrow IO_3^-(aq) + 6\,Br^-(aq) + 6\,H^+(aq)$

(d) $4[Zn(s) \rightarrow Zn^{2+}(aq) + 2e^-]$ reducing agent

$2\,NO_3^-(aq) + 10\,H^+(aq) + 8e^- \rightarrow N_2O(g) + 5\,H_2O(\ell)$ oxidizing agent

------------------------------------------------------------

$4\,Zn(s) + 2\,NO_3^-(aq) + 10\,H^+(aq) \rightarrow 4\,Zn^{2+}(aq) + N_2O(g) + 5\,H_2O(\ell)$

5-95. (a) $2\,[MnO_4^-(aq) + 8\,H^+(aq) + 5e^- \rightarrow Mn^{2+}(aq) + 4\,H_2O(\ell)]$

$5\,[HSO_3^-(aq) + H_2O(\ell) \rightarrow SO_4^{2-}(aq) + 3\,H^+(aq) + 2e^-]$

------------------------------------------------------------

$2\,MnO_4^-(aq) + 5\,HSO_3^-(aq) + H^+(aq) \rightarrow 2\,Mn^{2+}(aq) + 5\,SO_4^{2-}(aq) + 3\,H_2O(\ell)$

(b) $Cr_2O_7^{2-}(aq) + 14\ H^+(aq) + 6e^- \rightarrow 2\ Cr^{3+}(aq) + 7\ H_2O(\ell)$

$6\ [Fe^{2+}(aq) \rightarrow Fe^{3+}(aq) + e^-]$

---

$Cr_2O_7^{2-}(aq) + 6\ Fe^{2+}(aq) + 14\ H^+(aq) \rightarrow 2\ Cr^{3+}(aq) + 6\ Fe^{3+}(aq) + 7\ H_2O(\ell)$

(c) $Ag(s) \rightarrow Ag^+(aq) + e^-$

$NO_3^-(aq) + 2\ H^+(aq) + e^- \rightarrow NO_2(g) + H_2O(\ell)$

---

$Ag(s) + NO_3^-(aq) + 2\ H^+(aq) \rightarrow NO_2(g) + Ag^+(aq) + H_2O(\ell)$

(d) $Br_2(aq) + 2e^- \rightarrow 2\ Br^-(aq)$

$C_6H_8O_6(aq) \rightarrow C_6H_6O_6(aq) + 2\ H^+(aq) + 2e^-$

---

$Br_2(aq) + C_6H_8O_6(aq) \rightarrow 2\ Br^-(aq) + C_6H_6O_6(aq) + 2\ H^+(aq)$

(e) $OsO_4(s) + 8\ H^+(aq) + 4e^- \rightarrow Os^{4+}(aq) + 4\ H_2O(\ell)$

$2\ [C_4H_8(OH)_2(aq) \rightarrow C_4H_8O_2(aq) + 2\ H^+(aq) + 2e^-]$

---

$OsO_4(s) + 2\ C_4H_8(OH)_2(aq) + 4\ H^+(aq) \rightarrow Os^{4+}(aq) + 2\ C_4H_8O_2(aq) + 4\ H_2O(\ell)$

5-97. (a) $3\ [2\ Fe(OH)_2(s) + 2\ OH^-(aq) \rightarrow Fe_2O_3(s) + 3\ H_2O(\ell) + 2e^-]$

$2[\ CrO_4^{2-}(aq) + 4\ H_2O(\ell) + 3e^- \rightarrow Cr(OH)_4^-(aq) + 4\ OH^-(aq)]$

---

$6\ Fe(OH)_2(s) + 2\ CrO_4^{2-}(aq)$

$\rightarrow 3\ Fe_2O_3(s) + 2\ Cr(OH)_4^-(aq) + 2\ OH^-(aq) + H_2O(\ell)$

(b) $PbO_2(s) + 2\ H_2O(\ell) + 2e^- \rightarrow Pb(OH)_3^-(aq) + OH^-(aq)$

$Cl^-(aq) + 2\ OH^-(aq) \rightarrow ClO^-(aq) + H_2O(\ell) + 2e^-$

---

$PbO_2(s) + Cl^-(aq) + OH^-(aq) + H_2O(\ell) \rightarrow ClO^-(aq) + Pb(OH)_3^-(aq)$

(c) $2\ [Al(s) + 4\ OH^-(aq) \rightarrow Al(OH)_4^-(aq) + 3e\text{-}]$

$3\ [2\ H_2O(\ell) + 2e^- \rightarrow H_2(g) + 2\ OH^-(aq)]$

---

$2\ Al(s) + 2\ OH^-(aq) + 6\ H_2O(\ell) \rightarrow 2\ Al(OH)_4^-(aq) + 3\ H_2(g)$

(d) $3\ [CN^-(aq) + 2\ OH^-(aq) \rightarrow CNO^-(aq) + H_2O(\ell) + 2e^-]$

$2\ [CrO_4^{2-}(aq) + 4\ H_2O(\ell) + 3e^- \rightarrow Cr(OH)_4^-(aq) + 4\ OH^-(aq)]$

---

$3\ CN^-(aq) + 2\ CrO_4^{2-}(aq) + 5\ H_2O(\ell)$

$\rightarrow 3\ CNO^-(aq) + 2\ Cr(OH)_4^-(aq) + 2\ OH^-(aq)$

## STOICHIOMETRY: OXIDATION-REDUCTION TITRATIONS

5-99. 6.50 g $I_2$ • 1 mol/253.8 g = 0.0256 mol $I_2$

0.0256 mol $I_2$ • 2 mol $S_2O_3^{2-}$/1 mol $I_2$ = 0.0512 mol $S_2O_3^{2-}$

• 1 L/0.0525 mol $S_2O_3^{2-}$ = 0.976 L (→ 976 mL)

5-101. *Step 1:* Write the balanced equation [see Study Question 5-95 (b)].

$Cr_2O_7^{2-}(aq) + 6\ Fe^{2+}(aq) + 14\ H^+(aq) \rightarrow 2\ Cr^{3+}(aq) + 6\ Fe^{3+}(aq) + 7\ H_2O(\ell)$

*Step 2:* 0.742 g iron compd. • 1 mol/392.13 g = $1.89 \times 10^{-3}$ mol iron compound

Since each mole of $Fe(NH_4)_2(SO_4)_2 \cdot 6\ H_2O$ produces one mole of $Fe^{2+}$, this means the solution contains $1.89 \times 10^{-3}$ mol of iron(III) ion.

*Step 3:* Calculate moles of $K_2Cr_2O_7$ required.

$1.89 \times 10^{-3}$ mol $Fe^{2+}$ • 1 mol $Cr_2O_7^{2-}$/6 mol $Fe^{2+}$ = $3.15 \times 10^{-4}$ mol $Cr_2O_7^{2-}$

*Step 4:* $3.15 \times 10^{-4}$ mol $Cr_2O_7^{2-}$ • 1 L/0.0124 mol $K_2Cr_2O_7$ = 0.0254 L

Volume required = 25.4 mL

5-103. *Step 1:* Calculate moles of $KMnO_4$ used

0.02345 L • 0.0250 mol/L = $5.86 \times 10^{-4}$ mol $MnO_4^-$

*Step 2:* Calculate moles of oxalic acid required

$5.86 \times 10^{-4}$ mol $MnO_4^-$ • 5 mol $H_2C_2O_4$/2 mol $MnO_4^-$ = $1.47 \times 10^{-3}$ mol $H_2C_2O_4$

*Step 3:* Calculate mass of oxalic acid

$1.47 \times 10^{-3}$ mol $H_2C_2O_4$ • 90.04 g/mol = 0.132 g

*Step 4:* Calculate the weight percent of oxalic acid

(0.132 g/15.67 g)100 = 0.842%

5-105. *Step 1:* Calculate the mass and moles of Au in the sample.

1.050 g • 0.0300 g Au/100 g sample = $3.15 \times 10^{-4}$ g Au

$3.15 \times 10^{-4}$ g Au • 1 mol/197.0 g = $1.60 \times 10^{-6}$ mol Au

*Step 2:* Calculate moles of $Na_2S_2O_3$ required

$1.60 \times 10^{-6}$ mol Au • 1 mol $I_2$/1 mol Au • 2 mol $Na_2S_2O_3$/1 mol $I_2$

= $3.20 \times 10^{-6}$ mol $Na_2S_2O_3$

*Step 3:* Calculate volume of $Na_2S_2O_3$ solution required

$3.20 \times 10^{-6}$ mol $Na_2S_2O_3$ • 1 L/$1.00 \times 10^{-4}$ mol = 0.0320 L or 32.0 mL

## GENERAL QUESTIONS

5-107. (a) Precipitation reaction: $Na_2SO_4(aq) + BaCl_2(aq) \rightarrow 2\ NaCl(aq) + BaSO_4(s)$

The NaCl can be isolated by filtering off the precipitate of barium sulfate.

Acid-base reaction: $NaOH(aq) + HCl(aq) \rightarrow NaCl(aq) + H_2O(\ell)$

(b) Precipitation reaction: $Pb(NO_3)_2(aq) + 2\ KCl(aq) \rightarrow PbCl_2(s) + 2\ KNO_3(aq)$

(c) Barium sulfate can be prepared by the reaction given in part (a).

(d) Precipitation reaction: $AgNO_3(aq) + KI(aq) \rightarrow AgI(s) + KNO_3(aq)$

5-109. (a) $Na_2CO_3(aq) + 2\ HCl(aq) \rightarrow 2\ NaCl(aq) + CO_2(g) + H_2O(\ell)$

(b) $BaCO_3(s) + 2\ HNO_3(aq) \rightarrow Ba(NO_3)_2(aq) + CO_2(g) + H_2O(\ell)$

(c) $\text{heat} + BaCO_3(s) \rightarrow BaO(s) + CO_2(g)$

5-111. 0.250 mol/L • 0.100 L = 0.0250 mol $MgCl_2$

0.0250 mol $MgCl_2$ • 95.21 g/mol = 2.38 g $MgCl_2$

5-113. (a) 454 g $N_2H_4$ • 1 mol/32.05 g = 14.2 mol $N_2H_4$

$[N_2H_4]$ = 14.2 mol/400. L = 0.0354 M

(b) 14.2 mol $N_2H_4$ • 1 mol $H_2SO_4$/2 mol $N_2H_4$ • 1 L/18.0 mol $H_2SO_4$

= 0.394 L $H_2SO_4$

0.394 L $H_2SO_4$ • 1 bottle/2.5 L = 0.16 bottle of $H_2SO_4$

5-115. *Step 1:* Calculate moles of $I_2$ used

0.04021 L • 0.246 mol/L = $9.89 \times 10^{-3}$ mol $I_2$

*Step 2:* Calculate moles of $Na_2S_2O_3$ required

$9.89 \times 10^{-3}$ mol $I_2$ • 2 mol $Na_2S_2O_3$/1 mol $I_2$ = 0.0198 mol $Na_2S_2O_3$

*Step 3:* Calculate weight percent of $Na_2S_2O_3$

0.0198 mol $Na_2S_2O_3$ • 158.1 g/mol = 3.13 g $Na_2S_2O_3$

$(3.13 \text{ g } Na_2S_2O_3/3.232 \text{ g})100 = 96.8\%$

5-117. *Step 1:* Calculate the moles of KMnO4 used

$0.00100 \text{ mol/L} \cdot 0.01163 \text{ L} = 1.16 \times 10^{-5} \text{ mol } KMnO_4$

*Step 2:* Calculate moles of $Ca^{2+}$ in sample

$1.16 \times 10^{-5} \text{ mol } KMnO_4 \cdot (5 \text{ mol } CaC_2O_4/2 \text{ mol } KMnO_4) \cdot 1 \text{ mol Ca}/1 \text{ mol } CaC_2O_4$

$= 2.91 \times 10^{-5} \text{ mol Ca}$

*Step 3:* Calculate mass of Ca

$2.91 \times 10^{-5} \text{ mol Ca} \cdot 40.08 \text{ g}/1 \text{ mol} = 1.17 \times 10^{-3} \text{ g Ca}$

*Step 4:* Calculate mass of Ca in 100 mL of blood

$(1.17 \times 10^{-3} \text{ g Ca}/5 \text{ mL}) \cdot 100 \text{ mL} = 2.33 \times 10^{-2} \text{ g Ca}$ or 23.3 mg Ca in 100 mL

5-119. *Step 1:* Calculate moles of CaO and mass of Ca

$0.959 \text{ g CaO} \cdot 1 \text{ mol}/56.08 \text{ g} = 0.0171 \text{ mol CaO}$

$0.0171 \text{ mol CaO} \cdot 1 \text{ mol Ca}/1 \text{ mol CaO} = 0.0171 \text{ mol Ca}$

$0.0171 \text{ mol Ca} \cdot 40.08 \text{ g/mol} = 0.685 \text{ g Ca}$

*Step 2:* Calculate moles and mass of Cl that are in the sample in the form of $CaCl_2$

$0.0171 \text{ mol Ca} \cdot 2 \text{ mol Cl}/1 \text{ mol Ca} = 0.0342 \text{ mol Cl}$

$0.0342 \text{ mol Cl} \cdot 35.45 \text{ g/mol} = 1.21 \text{ g Cl}$

*Step 3:* Calculate mass of $CaCl_2$ present in the sample

Mass of $CaCl_2 = 0.685 \text{ g Ca} + 1.21 \text{ g Cl} = 1.90 \text{ g } CaCl_2$

*Step 4:* Calculate weight percent of $CaCl_2$

(1.90 g CaC12/4.22 g sample)100 = 44.9%

5-121. *Step 1:* Calculate moles of Mg in the sample

$0.725 \text{ g } MgNH_4PO_4 \cdot 1 \text{ mol}/137.3 \text{ g} \cdot 1 \text{ mol Mg}/1 \text{ mol } MgNH_4PO_4$

$= 5.28 \times 10^{-3} \text{ mol Mg}$

*Step 2:* Calculate moles of Cl in the sample

$1.514 \text{ g AgCl} \cdot 1 \text{ mol}/143.32 \text{ g} \cdot 1 \text{ mol Ag}/1 \text{ mol AgCl} = 0.01056 \text{ mol Cl}$

*Step 3:* Find the mole ratio of Cl to Mg

0.01056 mol Cl/0.00528 mol Mg = 2.00 mol Cl/1.00 mol Cl

Therefore, the formula of the anhydrous salt is $MgCl_2$.

*Step 4:* Find the number of moles of $MgCl_2$ and of water.

0.503 g $MgCl_2$ • 1 mol/95.21 g = $5.28 \times 10^{-3}$ mol $MgCl_2$

1.072 g hydrated salt - 0.503 g $MgCl_2$ = 0.569 g $H_2O$

0.569 g $H_2O$ • 1 mol/18.02 g = 0.0316 mol $H_2O$

*Step 5:* Find the mole ratio of water to $MgCl_2$

0.0316 mol $H_2O$/0.00528 mol $MgCl_2$ = 5.98 mol $H_2O$/1.00 mol $MgCl_2$

Therefore, the formula of the hydrated salt is $MgCl_2 \cdot 6\ H_2O$.

5-123. To solve for two unknowns here, we need two equations. If M is the mass of a given compound, we have equation 1.

*Equation 1:* $M_{KBr} + M_{NaBr} = 0.560$ g

The mass of Br contained in AgBr and thus in the mixture of KBr and NaBr is given by the equation

0.970 g AgBr • 1 mol AgBr/187.8 g • 1 mol Br/1 mol AgBr • 79.90 g Br/1 mol = 0.413 g Br

Therefore, we can write the second equation we need.

*Equation 2:* $M_{\text{Br as KBr}} + M_{\text{Br as NaBr}} = 0.413$ g

or

$x + y = 0.413$ g

We know the composition of KBr and NaBr. Therefore, we know the following:

(a) 119 g KBr/79.9 g Br = 1.49 g KBr/g Br

(b) 103 g NaBr/79.9 g Br = 1.29 g NaBr/g Br

This means, we can write the equation

$x$(1.49 g KBr/g Br) + $y$(1.29 g NaBr/g Br) = 0.560 g of mixture

and this equation can now be solved.

$1.49\,x + 1.29\,y = 0.560$

$1.49\,x + 1.29\,(0.413 - x) = 0.560$

$1.49\,x + 0.533 - 1.29\,x = 0.560$

$0.20\ x = 0.027$

$x$ = mass of Br as KBr = 0.14 g

This now allows us to solve for the mass of KBr.

0.14 g Br • 1 mol Br/79.9 g • 1 mol KBr/1 mol Br • 119 g KBr/1 mol = 0.20 g KBr

This means KBr represents (0.20/0.560) or 0.36 of the sample weight, so NaBr is 0.64 of the sample weight.

5-125. *Step 1:* Calculate moles of KOH used and of $H_3O^+$ supplied by the acid.

0.02725 L • 0.532 mol/L = 0.0145 mol KOH

Therefore, the sample of acid must have supplied 0.0145 mol of $H_3O^+$.

*Step 2:* Calculate moles of acid

0.855 g acid • 1 mol/118.1 g = 0.00724 mol acid

*Step 3:* Calculate the ratio of moles of $H_3O^+$ supplied per mole of acid.

0.0145 mol $H_3O^+$/0.00724 mol acid = 2 moles $H_3O^+$ supplied per mole of acid.

*Step 4:* Write the balanced equation

$C_4H_6O_4(aq) + 2\ KOH(aq) \rightarrow K_2C_4H_4O_4(aq) + 2\ H_2O(\ell)$

## SUMMARY PROBLEM

5-127. (a) $1.000 \times 10^6$ g rock • 0.019 g Au/100. g rock = 190 g Au

190 g Au • 1 mol Au/197 g = 0.96 mol Au

0.96 mol Au • 8 mol NaCN/4 mol Au = 1.9 mol NaCN

1.9 mol NaCN • 1 L/0.075 mol = 26 L

(b) Balanced equation:

$2\ [Au(CN)_2]^-(aq) + Zn(s) \rightarrow Zn^{2+}(aq) + Au(s) + 4\ CN^-(aq)$

0.96 mol Au • 1 mol Zn/2 mol Au • 65.4 g Zn/1 mol = 32 g Zn (or 0.032 kg)

(c) 190 g Au • 1 $cm^3$/19.3 g = 9.8 $cm^3$

9.8 $cm^3$/15.0 cm = cross section area of rod = 0.66 $cm^2$ = $\pi r^2$

radius = 0.46 cm, so the diameter is 0.92 cm.

# CHAPTER 6
# THERMOCHEMISTRY

## INSTRUCTOR'S NOTES

This topic is normally covered in 3-4 lectures. Our experience is that students will have some difficulty with thermochemistry, primarily because they have problems with the concept of energy. Some demonstrations, described below, can help them.

This chapter need not be covered at this point in the course. However, we have used the notion of energy changes in our discussions of atomic properties and in the discussion of bond energies in the first chapter on bonding (Chapter 10).

## TRANSPARENCIES AVAILABLE FOR THIS CHAPTER

| No. | Figure | Page | Subject |
|---|---|---|---|
| 7 | 6.9 | 211 | Enthalpy diagram for the interconversion of water vapor, liquid water, and the elements of water. |
| 8 | 6.11 | 221 | Combustion calorimeter |

## SUGGESTED DEMONSTRATIONS

In spite of their experience, some students do not have a good "feeling" for heat and energy. Therefore, demonstrations are quite useful here. In particular, we open the lecture on heat transfer, thermometers, and heat capacity with the demonstration in Figure 6.3.

A number of demonstrations involving exo- and endothermic reactions and processes are presented in Chapter 1 of Volume 1 of the Shakhashiri series on chemical demonstrations. We have used the following experiments:

- the endothermic reaction of barium hydroxide and ammonium salts
- the nonburning towel
- the reaction of calcium oxide with water
- the burning of magnesium (page 114)
- the reaction of zinc and ammonium nitrate (page 134)

- the thermite reaction (page 837)

In addition, the demonstration with the balloon and dry ice in Figure 6.6 is very effective and very easy to do. Furthermore, we often do the peroxide rocket again (see Chapter 4 in this *Manual*) and blow up a balloon containing hydrogen.

Finally, when discussing calorimetry, we always take a laboratory calorimeter to the lecture so the students can see the various parts and get a better notion of how they work in practice.

# SOLUTIONS TO SELECTED STUDY QUESTIONS

## ENERGY UNITS

6-10. (a) 1440 cal/mol • 4.184 J/cal = $6.03 \times 10^3$ J/mol

(b) $6.025 \times 10^3$ J/mol • 1 mol/18.02 g = 334 J/g

(c) 1440 cal or $6.03 \times 10^3$ J are evolved when 1 mol of liquid water freezes to ice.

6-12. (a) 297 kJ • 1000 J/kJ = $2.97 \times 10^5$ J

(b) $2.97 \times 10^5$ J • 1 cal/4.184 J = $7.10 \times 10^4$ cal

(c) $7.10 \times 10^4$ cal • 1 kcal/1000 cal = 71.0 kcal

## SPECIFIC HEAT AND HEAT CAPACITY

6-14. $q$ = heat absorbed by aluminum = (20.0 g)(0.902 J/K · mol)(55.0 - 4.4)K

$q = 9.13 \times 10^2$ J

6-16. (a) $q$ for water = (1.00 g)(4.184 J/K · mol)(100. - 25)K = 314 J

$q$ for glycol = (1.00 g)(2.42 J/K · mol)(100. - 25)K = 182 J

As predicted from its higher heat capacity, water requires more heat to warm it from 25 °C to 100. °C.

(b) 5.00 quart • 0.946 L/1 quart • 1000 $cm^3$/1 L • 1 g/$cm^3$ = $4.73 \times 10^3$ g

q for water = ($4.73 \times 10^3$ g)(4.184 J/K · mol)(100. - 25)K = $1.48 \times 10^6$ J

$q$ for glycol = ( $4.73 \times 10^3$ g)(2.42 J/K · mol)(100. - 25)K = $8.59 \times 10^5$ J

6-18. *Step 1:* Calculate quantity of benzene in moles

$1.00 \times 10^3$ g • 1 mol/78.11 g = 12.8 mol

*Step 2:* Calculate the quantity of heat the benzene absorbs on warming from 20.0 °C to 80.1 °C

$q$ = (12.8 mol)(136 J/mol . K)(80.1 - 20.0)K = $1.05 \times 10^5$ J

*Step 3:* Calculate the quantity of heat to vaporize the benzene.

$q$ = $(1.00 \times 10^3$ g)(395 J/g) = $3.95 \times 10^5$ J

*Step 4:* Add the two quantities of heat to find the total heat absorbed.

Total heat = $1.05 \times 10^5$ J + $3.95 \times 10^5$ J = $5.00 \times 10^5$ J

6-20. Heat lost by copper = - heat gained by water

(192 g)(0.385 J/g . K)($T_{final}$ - 100.0) = - (750. g)(4.184 J/K • g)($T_{final}$ - 4.0)

$T_{final}$ = 6.2 °C

6-22. *Step 1:* Calculate the quantity of mercury

1.00 mL • 13.6 g/cm$^3$ = 13.6 g Hg

*Step 2:* Calculate the quantity of heat the mercury transfers on cooling from 23.0 °C to -39 °C

$q$ = (13.6 g)(0.138 J/K • g)(23.0 - (-39))K = 116 J

*Step 3:* Calculate the quantity of heat transferred on freezing the mercury.

$q$ = (13.6 g)(11 J/g) = 150. J

*Step 4:* Add the two quantities of heat to find the total heat transferred.

Total heat = 116 J + 150. J = 266 J

6-24. (a) (43.93 J/K · mol)(1 mol/26.038 g) = 1.687 J/g· K

(b) (120.5 J/K · mol)(1 mol/ 187.38 g) = 0.6431 J/g· K

**ENTHALPY**

6-26. Heat is a product of the reaction, so it is exothermic.

6-28. 3.56 g $P_4$ • 1 mol/123.9 g = 0.0287 mol $P_4$

86.5 kJ/0.0287 mol = 3,010 kJ/mol

6-30. (a) Calculate the heat evolved by 10.0 g Al

10.0 g Al • 1 mol/26.98 g = 0.371 mol Al

0.371 mol Al • 852 kJ/2 mol Al = 158 kJ

(b) Use 158 kJ to warm 500. g of water from 20.0 °C to a higher temperature.

$158 \times 10^3$ J = (500. g)(4.184 J/K • g)($T_{final}$ - 20.0)K

$T_{final}$ = 95.5 °C

The water is not brought to a boil.

6-32. $PbCl_2(s) \rightarrow Pb(s) + Cl_2(g)$ ΔH = +359.4 kJ

$Pb(s) + 2\ Cl_2(g) \rightarrow PbCl_4(\ell)$ ΔH = -329.3 kJ

---

$PbCl_2(s) + Cl_2(g) \rightarrow PbCl_4(\ell)$ ΔH = +30.1 kJ

6-34. $SiO_2(s) + 2\ C(s) \rightarrow Si(s) + 2\ CO(g)$ ΔH = +689.9 kJ

$Si(s) + 2\ Cl_2(g) \rightarrow SiCl_4(g)$ ΔH = - 657.0 kJ

$SiCl_4(g) + 2\ Mg(s) \rightarrow 2\ MgCl_2(s) + Si(s)$ ΔH = -625.6 kJ

---

$SiO_2(s) + 2\ C(s) + 2\ Cl_2(g) + 2\ Mg \rightarrow$
$2\ MgCl_2(s) + 2\ CO + Si(s)$ ΔH = -592.7 kJ

6-36. $C\text{-}N(g) \rightarrow C(graphite) + \frac{1}{2}\ N_2(g)$ ΔH = -416 kJ

$C(graphite) \rightarrow C(g)$ ΔH = +717 kJ

$\frac{1}{2}[N_2(g) \rightarrow 2\ N(g)]$ $\frac{1}{2}$[ΔH = +946 kJ]

---

$C\text{—}N(g) \rightarrow C(g) + N(g)$ ΔH = +774 kJ

## STANDARD ENTHALPIES OF FORMATION

6-38. (a) Since $\Delta H_f^\circ$ is negative, the reaction is exothermic.

(b) $6\ C(graphite) + 6\ H_2(g) + 3\ O_2(g) \rightarrow C_6H_{12}O_6(s)$

6-40. $\Delta H^\circ_{rxn} = \Delta H^\circ_f[FeBr_3(s)] - \Delta H^\circ_f[FeBr_2(s)] = -268.2 \text{ kJ} - (-249.8 \text{ kJ}) = -18.4 \text{ kJ}$

6-42. $\Delta H^\circ_{rxn} = 4\,\Delta H^\circ_f[NO(g)] + 6\,\Delta H^\circ_f[H_2O(g)] - \{4\,\Delta H^\circ_f[NH_3(g)] + 5\,\Delta H^\circ_f[O_2(g)]$

$= 4 \text{ mol } (90.25 \text{ kJ/mol}) + 6 \text{ mol } (-241.818 \text{ kJ/mol})$

$- [4 \text{ mol } (-46.11 \text{ kJ/mol}) + 5 \text{ mol } (0 \text{ kJ/mol})]$

$= -905.47 \text{ kJ}$

6-44. $\Delta H^\circ_{rxn} = \Delta H^\circ_f[UF_4(s)] + 2\,\Delta H^\circ_f[H_2O(g)] - \{\Delta H^\circ_f[UO_2(s)] + 4\,\Delta H^\circ_f[HF(g)]\}$

$= 1 \text{ mol } (-1914 \text{ kJ/mol}) + 2 \text{ mol } (-241.818 \text{ kJ/mol})$

$- [1 \text{ mol } (-1085 \text{ kJ/mol}) + 4 \text{ mol } (-271.1 \text{ kJ/mol})]$

$= -228 \text{ kJ}$

6-46. $\Delta H^\circ_{rxn} = +139.5 \text{ kJ} = \Delta H^\circ_f[SiCl_4(g)] + 2\,\Delta H^\circ_f[H_2O(g)]$

$- \{[\Delta H^\circ_f[SiO_2(s)] + 4\,\Delta H^\circ_f[HCl(g)]\}$

$+139.5 \text{ kJ} = \Delta H^\circ_f[SiCl_4(g)] + 2 \text{ mol } (-241.818 \text{ kJ/mol})$

$- [1 \text{ mol } (-910.94 \text{ kJ}) + 4\,(-92.307 \text{ kJ})]$

$+139.5 \text{ kJ} = \Delta H^\circ_f[SiCl_4(g)] + 769.53 \text{ kJ}$

$\Delta H^\circ_f[SiCl_4(g)] = -657.0 \text{ kJ}$

6-48. The enthalpy of formation of iron(III) oxide is -824.2 kJ/mol. Therefore, the reaction of iron and oxygen is exothermic.

$2\,Fe(s) + 3/2\,O_2(g) \rightarrow Fe_2O_3(s) + 824.2 \text{ kJ}$

$5.58 \text{ g Fe} \cdot 1 \text{ mol}/55.847 \text{ g} = 0.0999 \text{ mol Fe}$

$0.0999 \text{ mol Fe} \cdot 824.2 \text{ kJ}/2 \text{ mol Fe} = 41.2 \text{ kJ}$ for 5.58 g of Fe

6-50. *Step 1:* Find the enthalpy change for the decomposition of ammonium nitrate

$\Delta H^\circ_{rxn} = \Delta H^\circ_f[N_2O(g)] + 2\,\Delta H^\circ_f[H_2O(g)] - \Delta H^\circ_f[NH_4NO_3(s)]$

$= 1 \text{ mol } (82.05 \text{ kJ}) + 2 \text{ mol } (-241.818 \text{ kJ/mol}) - 1 \text{ mol } (-365.56 \text{ kJ/mol})$

$= -36.03 \text{ kJ}$

*Step 2:* Calculate moles of ammonium nitrate

$1.00 \times 10^3$ g • 1 mol/80.04 g = 12.5 mol

*Step 3:* Find the heat transferred by 1.00 kg of ammonium nitrate (= 12.5 mol)

12.5 mol • 36.03 kJ/mol = 450. kJ

6-52. *Step 1:* Find the enthalpy of reaction and then the enthalpy of reaction per mole of nitroglycerin.

$\Delta H^\circ_{rxn} = 6\ \Delta H^\circ_f[CO_2(g)] + 5\ \Delta H^\circ_f[H_2O(g)] - 2\ \Delta H^\circ_f[C_3H_5(NO_3)_3(\ell)]$

$= 6$ mol (-393.509 kJ/mol) + 5 mol (-241.818 kJ/mol) - 2 mol (-364 kJ/mol)

= -2842 kJ for 2 moles of nitroglycerin

Therefore, $\Delta H^\circ_{rxn}$ for 1 mole of nitroglycerin is -1421 kJ.

*Step 2:* Calculate moles of nitroglycerin equivalent to 10.0 pounds

10.0 pounds • 454 g/pound • 1 mol/227.1 g = 20.0 mol

*Step 3:* Calculate the heat transferred by 10.0 pounds of nitroglycerin

20.0 mol • 1421 kJ/mol = 2841 kJ

6-54. Part (a): Calculate the molar enthalpy of formation of the fat.

37.8 kJ/1.00 g • 885.4 g/mol = $3.35 \times 10^4$ kJ/mol

$\Delta H^\circ_{rxn} = -3.35 \times 10^4$ kJ

$= 57\ \Delta H^\circ_f[CO_2(g)] + 52\ \Delta H^\circ_f[H_2O(\ell)] - \Delta H^\circ_f[C_{57}H_{104}O_6(s)]$

$-3.35 \times 10^4$ kJ = 57 mol (-393.509 kJ/mol) + 52 mol (-285.830 kJ/mol) $- \Delta H^\circ_f[C_{57}H_{104}O_6(s)]$

$-3.35 \times 10^4$ kJ = -37,293.173 kJ - $\Delta H^\circ_f[C_{57}H_{104}O_6(s)]$

$\Delta H^\circ_f[C_{57}H_{104}O_6(s)] = -3.83 \times 10^3$ kJ/mol

Part (b): Calculate moles of fat. First find the number of moles of fat in 1.00 pound and then multiply by the enthalpy of combustion of 1.00 of fat.

454 g fat • 1 mol/885.5 g = 0.513 mol fat

0.513 mol fat • $3.35 \times 10^4$ kJ/mol = $1.72 \times 10^4$ kJ

17200 kJ • 1 kcal/4.184 kJ • 1 Calorie/1 kcal = 4100 Calories

6-56. $2\,MnO(s) + O_2(g) \rightarrow 2\,MnO_2(s)$ $\Delta H^\circ = -264$ kJ

$MnO_2(s) + Mn(s) \rightarrow 2\,MnO(s)$ $\Delta H^\circ = -240$ kJ

------------------------------------------ --------------------

$Mn(s) + O_2(g) \rightarrow MnO_2(s)$ $\Delta H^\circ = -504$ kJ

**CALORIMETRY**

6-58. Step 1: Calculate the quantity of heat transferred to the bomb

$q$ = 923 J/deg (26.72 - 21.25)deg = 5.05 x $10^3$ J

Step 2: Calculate the quantity of heat transferred to the water

$q$ = (815 g)(4.184 J/K • g)(26.72 - 21.25)deg = 1.87 x $10^4$ J

*Step 3:* Determine the total heat transferred to the surroundings

5.05 x $10^3$ J + 1.87 x $10^4$ J = 2.37 x $10^4$ J or 23.7 kJ

*Step 4:* Find the number of moles of S and then the number of moles of $SO_2$ formed

2.56 g S • 1 mol/32.066 g = 0.0798 mol

0.0798 mol S • 1 mol $SO_2$/1 mol S = 0.0798 mol $SO_2$

*Step 5:* Calculate the quantity of heat transferred per mole of $SO_2$ formed

23.7 kJ/0.0798 mol = 297 kJ/mol of $SO_2$

6-60. *Step 1:* Calculate the quantity of heat transferred to the bomb

$q$ = 893 J/K (31.69 - 22.50)K = 8210 J (or 8.21 kJ)

Step 2: Calculate the quantity of heat transferred to the water

$q$ = (775 g)(4.184 J/K • g)(31.69 - 22.50)K = 29,800 J (or 29.8 kJ)

*Step 3:* Calculate the total heat transferred to the surroundings

8.21 kJ + 29.8 kJ = 38.0 kJ

*Step 4:* Calculate moles of benzoic acid

1.500 g acid • 1 mol/122.1 g = 0.0123 mol

*Step 5:* Calculate the quantity of heat transferred per mole of acid

38.0 kJ/0.0123 mol = 3090 kJ/mol

6-62. *Step 1:* Calculate moles of CsOH and HF and the total mass of final solution.

Moles of CsOH = 0.125 L • 0.250 mol/L = 0.0313 mol

Moles of HF = 0.050 L • 0.625 mol/L = 0.0313 mol

Therefore, we begin with equal quantities of CsOH and HF.

The total mass of the solutions is 125 g for CsOH + 50.0 g for HF = 175 g

*Step 2:* Calculate the heat transferred by the reaction

$q$ = (175 g)(4.20 J/K • g)(24.40 - 21.50)K = 2130 J

*Step 3:* Calculate the heat transferred per mole of CsOH

2130 J/0.0313 mol = 68,100 J (or 68.1 kJ)

## GENERAL PROBLEMS

6-64. *Step 1:* Calculate the quantity of heat transferred from the water

q = (100.0 g)(4.184 J/K • g)(0.00 - 60.0)K = $2.51 \times 10^4$ J

*Step 2:* Calculate the quantity of ice that must have melted to absorb $2.51 \times 10^4$ J

$2.51 \times 10^4$ J • 1 g/333 J = 75.4 g ice melted

6-66. $\Delta H^\circ_{rxn} = \Delta H^\circ_f[B_2O_3(s)] + 3\,\Delta H^\circ_f[H_2O(g)] - \Delta H^\circ_f[B_2H_6(g)]$

$-1941$ kJ = 1 mol($-1273$ kJ/mol) + 3 mol($-241.818$ kJ/mol) - $\Delta H^\circ_f[B_2H_6(g)]$

$\Delta H^\circ_f[B_2H_6(g)] = -57.5$ kJ/mol

6-68. *Step 1:* Calculate the moles of $Fe_2O_3$ available

2.0 g • 1 mol/160. g = 0.013 mol

*Step 2:* Calculate the mass of products

0.013 mol $Fe_2O_3$ • 2 mol Fe/1 mol $Fe_2O_3$ • 55.847 g Fe/mol = 1.4 g Fe

0.013 mol $Fe_2O_3$ • 1 mol $Al_2O_3$/1 mol $Fe_2O_3$ • 102 g $Al_2O_3$/mol = 1.3 g $Al_2O_3$

*Step 3:* Calculate the quantity of heat transferred by the reaction if only 0.013 mol of $Al_2O_3$ is produced.

0.013 mol $Al_2O_3$ • 852 kJ/mol $Al_2O_3$ = 10.6 kJ

*Step 4:* Calculate the quantity of heat required to raise the temperature of a mixture of 1.4 g of Fe and 1.3 g of $Al_2O_3$ to the melting point of iron (1530 °C). Assume we begin at room temperature (25 °C).

$$q = (1.3\text{ g }Al_2O_3)(0.775\text{ J/K}\cdot\text{g})(1530 - 25)\text{K}$$
$$+ (1.4\text{ g Fe})(0.451\text{ J/K}\cdot\text{g})(1530 - 25)\text{K}$$
$$q = 2.5 \times 10^3 \text{ (or 2.5 kJ)}$$

The quantity of heat required to raise the temperature of the mixture to 1530 °C is much less than that evolved by the reaction producing 0.013 mol of $Al_2O_3$.

6-70. *Step 1:* Find the heat transferred by the reaction of MgO and HCl.

$$q = (101\text{ g})(4.20\text{ J/K}\cdot\text{g})(5.85\text{ K})$$
$$= 2.48 \times 10^3\text{ J (or 2.48 kJ) for 0.700 g of MgO}$$

0.700 g MgO • 1 mol/40.30 g = 0.0174 mol MgO

2.48 kJ/0.0174 mol = 143 kJ/mol

*Step 2:* Add chemical equations and values of ΔH to find ΔH for reaction (iv).

| | |
|---|---|
| $Mg(s) + 2\,HCl(aq) \rightarrow MgCl_2(aq) + H_2(g)$ | ΔH = -466 kJ |
| $MgCl_2(aq) + H_2O(\ell) \rightarrow MgO(s) + 2\,HCl(aq)$ | ΔH = +143 kJ |
| $H_2(g) + \frac{1}{2}\,O_2(g) \rightarrow H2O(\ell)$ | ΔH = -285.83 kJ |
| ------------------------------------------------------- | --------------------- |
| $Mg(s) + \frac{1}{2}\,O_2(g) \rightarrow MgO(s)$ | ΔH = -609 kJ |

6-72. $\Delta H^\circ_{rxn} = -270\text{ kJ} = \Delta H^\circ_f[HCOOH(g)] - \Delta H^\circ_f[H_2CO(g)]$

$-270\text{ kJ} = 1\text{ mol}(-379\text{ kJ/mol}) - \Delta H^\circ_f[H_2CO(g)]$

$\Delta H^\circ_f[H_2CO(g)] = -109\text{ kJ/mol}$

6-74. Part (a): Heat transferred at constant pressure by decomposition of $NH_4NO_3$

$\Delta H^\circ_{rxn} = \Delta H^\circ_f[H_2O(g)] + \Delta H^\circ_f[N_2O(g)] - \Delta H^\circ_f[NH_4NO_3(s)]$

$\Delta H^\circ_{rxn} = 1\text{ mol }(-241.818\text{ kJ/mol}) + 1\text{ mol }(82.0\text{ kJ/mol}) - 1\text{ mol }(-365.56\text{ kJ/mol})$

$= -36.1\ kJ$

Part (b): The first step is to find the enthalpy change for the second reaction.

$\Delta H^\circ_{rxn} = 4\ \Delta H^\circ_f[H_2O(g)]\ -\ 2\ \Delta H^\circ_f[NH_4NO_3(s)]$

$= 4\ mol\ (-241.818\ kJ/mol)\ -\ 2\ mol\ (-365.56\ kJ/mol)$

$= -967.272\ kJ\ +\ 731.12\ kJ$

$= -236.15\ kJ$

The quantity of heat transferred in the second reaction is 236.15 kJ/2 mol $NH_4NO_3$.

Therefore, for 8.00 kg: $8.00 \times 10^3\ g \cdot 1\ mol/80.04\ g \cdot 236.15\ kJ/2\ mol = 1.18 \times 10^4\ kJ$

6-76. $\Delta H^\circ_{rxn} = -8703\ kJ = 5\ \Delta H^\circ_f[B_2O_3(s)]\ +\ 9\ \Delta H^\circ_f[H_2O(g)]\ -\ 2\ \Delta H^\circ_f[B_5H_9(\ell)]$

$-8703\ kJ = 5\ mol(-1273\ kJ/mol)\ +\ 9\ mol(-241.818\ kJ/mol)\ -\ 2\ \Delta H^\circ_f[B_5H_9(\ell)]$

$\Delta H^\circ_f[B_5H_9(g)] = 80.8\ kJ/mol$

6-78. *Step 1:* $SO_2(g) + 2\ H_2O(g) + Br_2(g) \rightarrow H_2SO_4(\ell) + 2\ HBr(g)$

$\Delta H^\circ = -106.3\ kJ$

*Step 2:* $H_2SO_4(\ell) \rightarrow H_2O(g) + SO_2(g) + \frac{1}{2}\ O_2(g)$

$\Delta H^\circ = 275.3\ kJ$

*Step 3:* $2\ HBr(g) \rightarrow H_2(g) + Br_2(g)$ $\Delta H^\circ = 72.8\ kJ$

---

*Sum:* $H_2O(g) \rightarrow H_2(g) + \frac{1}{2}\ O_2(g)$ $\Delta H^\circ = 241.8\ kJ$

The sum of the reactions is the chemical equation for the decomposition of water to its elements. Therefore, we expect ΔH° for the reaction to be the opposite of the enthalpy of formation of water, which has a value of -241.8 kJ/mol.

6-80. *Step 1:* Calculate the quantity of heat that must be transferred to melt 56.0 g of ice and then warm it to 75.0 °C.

$q$ for melting ice = 56.0 g (333 J/g) = $1.86 \times 10^4$ J

$q$ to warm water from 0 °C to 75.0 °C = (56.0 g)(4.184 J/K • g)(75.0 °C)

$= 1.76 \times 10^4$ J

Total heat energy required = $3.62 \times 10^4$ J

*Step 2:* Calculate the quantity of heat derived from 1.00 g of propane

$C_3H_8(g) + 5\,O_2(g) \rightarrow 3\,CO_2(g) + 4\,H_2O(g)$

$\Delta H^\circ_{rxn} = 4\,\Delta H^\circ_f[H_2O(g)] + 3\,\Delta H^\circ_f[CO_2(g)] - \Delta H^\circ_f[C_3H_8(g)]$

$= 4$ mol (-241.818 kJ/mol) + 3 mol (-393.509 kJ/mol) - 1 mol (-103.8 kJ/mol)

= -2044.0 kJ/mol

2044.0 kJ/mol • 1 mol/44.097 g = 46.352 kJ/g (or $46.352 \times 10^3$ J/g)

*Step 3:* Calculate the quantity of propane required to transfer $4.22 \times 10^4$ J

$3.62 \times 10^4$ J • 1 g/$46.352 \times 10^3$ J = 0.781 g propane

6-82. Let us begin with reaction 3:

$UF_4(s) + F_2(g) \rightarrow UF_6(s)$ $\Delta H^\circ = -283.1$ kJ

If we can find the enthalpy of formation of $UF_4(s)$, then we can derive that for $UF_6(s)$, since we know that

$\Delta H^\circ_{rxn} = -283.1\text{ kJ} = \Delta H^\circ_f[UF_6(s)] - \Delta H^\circ_f[UF_4(s)]$

However, the value of $\Delta H^\circ_f[UF_4(s)]$ can be derived from reaction 2.

$UO_2(s) + 4\,HF(g) \rightarrow UF_4(s) + 2\,H_2O(g)$ $\Delta H^\circ = -232.2$ kJ

since we know the enthalpy of reaction and we know all of the enthalpies of formation involved, except that of $UF_4$.

$\Delta H^\circ_{rxn} = \Delta H^\circ_f[UF_4(s)] + 2\,\Delta H^\circ_f[H_2O(g) - \{\Delta H^\circ_f[UO_2(s)] + 4\,\Delta H^\circ_f[HF(g)]$

-232.2 kJ = $\Delta H^\circ_f[UF_4(s)]$ + 2 mol(-241.818 kJ/mol)

- [1 mol(-1081.2 kJ/mol) + 4 mol (-271.1 kJ/mol)

$\Delta H^\circ_f[UF_4(s)]$ = -1914.2 kJ/mol

Now, this is substituted into the equation

$\Delta H^\circ_{rxn} = -283.1\text{ kJ} = \Delta H^\circ_f[UF_6(s)] - \Delta H^\circ_f[UF_4(s)]$

and this gives the enthalpy of formation of $UF_6(s)$ as

$\Delta H_f^\circ[UF_6(s)]$ = -2197.3 kJ/mol

An alternative solution to this problem is to add the equation given in such a way that you obtain the net equation

$U(s) + 3\ F_2(g) \rightarrow UF_6(s)$

| | | |
|---|---|---|
| *Reaction 1:* | $U(s) + O_2(g) \rightarrow UO_2(s)$ | $\Delta H^\circ$ = -1081.2 kJ |
| *Reaction 2:* | $UO_2(s) + 4\ HF(g) \rightarrow UF_4(s) + 2\ H_2O(g)$ | $\Delta H^\circ$ = -232.2 kJ |
| *Reaction 3:* | $UF_4(s) + F_2(g) \rightarrow UF_6(s)$ | $\Delta H^\circ$ = -283.1 kJ |
| 4 x *Reaction 4:* | $4[\frac{1}{2}\ H_2(g) + \frac{1}{2}\ F_2(g) \rightarrow HF(g)]$ | 4[$\Delta H^\circ$ = -271.1 kJ] |
| -2 x *Reaction 5:* | $2\ [H_2O(g) \rightarrow H_2(g) + \frac{1}{2}\ O_2(g)]$ | -2 [$\Delta H^\circ$ = -241.8 kJ] |
| *Sum:* | $U(s) + 3\ F_2(g) \rightarrow UF_6(s)$ | $\Delta H^\circ$ = -2197.3 kJ |

# PART II

# ATOMIC AND MOLECULAR STRUCTURE

# CHAPTER 7

# NUCLEAR CHEMISTRY

## INSTRUCTOR'S NOTES

As noted in the preface to the text, the discussion of nuclear chemistry is typically left to one of the final chapters in chemistry textbooks. However, we believe the recent reactor accidents in the United States and the Soviet Union, and the ongoing debate in the United States concerning nuclear waste disposal, make it desireable to give this topic more prominence. In addition, its importance is indicated by the fact that about 20% of the papers abstracted in *Chemical Abstracts* annually in the general areas of physical, inorganic, and analytical chemistry concern nuclear chemistry. It is for this reason that the Nuclear Chemistry chapter is included as an introduction to the section on the structure of atoms and molecules where it logically fits. In spite of this, *this chapter is not at all crucial to the development of Part II; it is certainly possible to leave the chapter for a later point in the course.* Nonetheless, it is our hope that the material will be discussed in class to some extent, or that students will read the chapter out of interest in the subject.

You should be aware that there is a discussion of first order kinetics in this chapter. The discussion in Section 7.4 ("Rates of Disintegration Reactions") is written to stand on its own; we believe there is enough information here so that you do not need to have covered the general subject of kinetics before talking about nuclear reactions. Furthermore, you do not need to refer to Chapter 7 when covering first order reactions in the chapter on kinetics (Chapter 15); first order reactions are discussed again in Chapter 15 using standard chemical reactions. However, when covering Chapter 15 you should be able easily to refer back to Section 7.4 on "Rates of Disintegration Reactions."

## LIST OF TRANSPARENCIES FOR THIS CHAPTER

| No. | Figure | Page | Subject |
|---|---|---|---|
| 9 | 7.2 | 244 | Behavior of subatomic particles in electric field |
| 10 | 7.4 | 250 | Uranium radioactive series |
| 11 | 7.5 | 252 | Naturally occurring isotopes |
| 12 | 7.10 | 262 | The carbon cycle in the biosphere |
| 13 | 7.14 | 266 | A nuclear chain reaction |
| 14 | 7.15 | 267 | Essential parts of a nuclear power plant |

## SUGGESTED DEMONSTRATIONS AND READINGS

The database of demonstrations available from Project SERAPHIM lists at least two suitable demonstrations for nuclear chemistry.

- Para, A.F.; Lazzarini, E. "Some Simple Classroom Experiments on the Monte Carlo Method," *J. Chem. Educ.* **1974**, *51*, 336.
- Smith, W.T.; Wood, J.H. "A Half-life Experiment for General Chemistry Students," (measurement of the half-life of bismuth-210), *J. Chem. Educ.* **1959**, *36*, 492.

Since radon in homes is very much in the news, it might be interesting to have the class test some area of the chemistry building or dormitory for radon using a home testing kit available at most hardware stores (see page 272).

In addition, there are some very good sources of reading for students. The marginal notes on pages 243, 263, and 270 give good references, and the *Something More About ...* boxes on pages 269 and 272 give others. In addition, a symposium on "Topics in Nuclear and Radiochemistry for College Curricula and High School Science Programs" was held at the 11th Biennial Conference on Chemical Education in Atlanta, GA in August 1990. The excellent booklet of readings may still be available from the conference organizer, Dr. Patricia A. Baisden, Nuclear Chemistry Division, Lawrence Livermore National Laboratory, P.O. Box 808, L-234, Livermore, CA 94551.

## SOLUTIONS TO SELECTED STUDY QUESTIONS

### NUCLEAR REACTIONS

7-13. (a) $^{9}_{4}Be + ^{1}_{1}H \rightarrow ^{6}_{3}Li + ^{4}_{2}He$

(b) $^{27}_{13}Al + ^{1}_{0}n \rightarrow ^{24}_{11}Na + ^{4}_{2}He$

(c) $^{40}_{20}Ca + ^{1}_{0}n \rightarrow ^{40}_{19}K + ^{1}_{1}H$

(d) $^{241}_{95}Am + ^{4}_{2}He \rightarrow ^{243}_{97}Bk + 2\ ^{1}_{0}n$

(e) $^{246}_{96}Cm + ^{12}_{6}C \rightarrow 4\ ^{1}_{0}n + ^{254}_{102}No$

(f) $^{238}_{92}U + ^{16}_{8}O \rightarrow ^{249}_{100}Fm + 5\ ^{1}_{0}n$

7-15. (a) $^{19}_{10}Ne \rightarrow ^{0}_{+1}\beta + ^{19}_{9}F$

(b) $^{56}_{26}Fe \rightarrow ^{0}_{-1}\beta + ^{56}_{27}Co$

(c) $^{40}_{19}K \rightarrow ^{0}_{-1}\beta + ^{40}_{20}Ca$

(d) $^{37}_{18}Ar + ^{0}_{-1}e$ (K-capture) $\rightarrow ^{37}_{17}Cl$

(e) $^{56}_{26}Fe + ^{0}_{-1}e$ (K-capture) $\rightarrow ^{56}_{25}Mn$

(f) $^{26}_{13}Al \rightarrow ^{25}_{12}Mg + ^{1}_{1}H$

7-17. See answer given in Appendix L for question 7-16.

7-19. *Step 1:* Calculate the binding energy for $^{30}_{15}P$

$15\ ^{1}_{1}H = 15 \cdot 1.00783 =$ 15.11745 g/mol

$15\ ^{1}_{0}n = 15 \cdot 1.00867 =$ 15.13005 g/mol

------------------

30.24750 g/mol

$\Delta mass = 29.9880 - 30.24750 = -0.2595$ g/mol

$\Delta E = -(2.595 \times 10^{-4}\ kg/mol)(3.00 \times 10^{8}\ m/s)^2(1\ J/1\ kg \cdot m^2/s^2)$

$\Delta E = -2.34 \times 10^{13}$ J/mol or $-2.34 \times 10^{10}$ kJ/mol

*Step 2:* Calculate the binding energy for $^{31}_{15}P$

$15\ ^{1}_{1}H = 15 \cdot 1.00783 = \quad 15.11745$ g/mol

$16\ ^{1}_{0}n = 16 \cdot 1.00867 = \quad 16.13872$ g/mol

------------------

31.25617 g/mol

$\Delta mass = 30.97376 - 31.25617 = -0.28241$ g/mol

$\Delta E = -(2.8241 \times 10^{-4}\ kg/mol)(3.00 \times 10^{8}\ m/s)^2(1\ J/1\ kg \cdot m^2/s^2)$

$\Delta E = -2.54 \times 10^{13}$ J/mol or $-2.54 \times 10^{10}$ kJ/mol

*Step 3:* Compare binding energies.

A comparison of binding energies shows that $^{31}_{15}P$ is the more stable isotope.

## RATES OF DISINTEGRATION REACTIONS

7-21. 10.8 days • 1 half-life/2.7 days = 4.0 half-lives

After 4.0 half-lives, the quantity of gold isotope has gone from 5.6 mg to 0.35 mg. That is, after four half-lives only $(\frac{1}{2})^4$ or 1/16 of the sample remains.

5.6 mg —*1st half-life*→ 2.8 mg —*2nd half-life*→ 1.4 mg —*3rd half-life*→ 0.70 mg —*4th half-life*→ 0.35 mg

7-23. (a) $^{32}_{15}P \rightarrow ^{0}_{-1}e + ^{32}_{16}S$

(b) 28.6 days • 1 half-life/14.3 days = 2 half-lives

After 2 half-lives, the sample has decreased to 1/4 of its original quantity. Therefore, after 28.6 days only 2.4 mg remain.

7-25. 13 days • 24 hours/day • 1 half-life/77.9 hours = 4.0 half-lives

After 4 half-lives, 1/16 of the sample remains (see Question 7-21). Therefore, 1/16 of 0.15 mg or $9.4 \times 10^{-3}$ mg remain.

7-27. (a) $^{222}_{86}Rn \rightarrow ^{4}_{2}He + ^{118}_{84}Po$

(b) $t_{1/2}$ = 3.82 days = 0.693/$k$

$k$ = 0.181/day

Using ln [A] / [A]$_0$ = -$k \cdot t$, we have [A] = 0.10 [A]$_0$ and $k$ = 0.181/day.

ln (0.10) = -(0.181/day)$t$

$t$ = 12.7 days

7-29. $t_{1/2}$ = 27.7 years = 0.693/$k$

$k$ = 0.0250/year

Since activity is proportional to the concentration of radioactive isotopes, we have

ln (2.5/80.) = -(0.0250/year)$t$

$t$ = 139 years = 5.00 half-lives

Year when activity reaches 2.5 dps: 1990 + 139 = 2129

7-31. (a) $^{239}_{94}Pu \rightarrow ^{4}_{2}He + ^{235}_{92}U$

(b) $k = 0.693/t_{1/2}$ = 0.693/24,390 years = 2.841 x $10^{-5}$/year

2.841 x $10^{-5}$/year • 1 year/365 days • 1 day/24 hours • 1 hour/60 min

= 5.406 x $10^{-11}$/min

1.0 x $10^{-3}$ g Pu • 1 mol/244 g • 6.022 x $10^{23}$ atoms/mol = 2.5 x $10^{18}$ atoms Pu

Activity = $kN$ = (5.406 x $10^{-11}$/min · atom)(2.5 x $10^{18}$ atoms Pu)

Activity = 1.4 x $10^{8}$ disintegrations per minute

7-33. $k = 0.693/t_{1/2}$ = 0.693/5.73 x $10^3$ years = 1.21 x $10^{-4}$/year

ln (0.72/1) = - (1.21 x $10^{-4}$/year)$t$

$t$ = 2.72 x $10^3$ years

The bone fragment is about 2700 years old.

## NUCLEAR TRANSFORMATIONS

7-35. $^{239}_{94}Pu + ^{4}_{2}He \rightarrow ^{240}_{95}Am + ^{1}_{1}H + 2\,^{1}_{0}n$

7-37. $^{40}_{20}Ca + ^{248}_{96}Cm \rightarrow ^{288}_{116}X$

## NUCLEAR FISSION AND POWER

7-39. (a) $^{238}_{92}U + ^{1}_{0}n \rightarrow ^{239}_{92}U$

(b) $^{239}_{92}U \rightarrow ^{0}_{-1}e + ^{239}_{93}Np$

(c) $^{239}_{93}Np \rightarrow ^{0}_{-1}e + ^{239}_{94}Pu$

(d) $^{239}_{94}Pu + ^{1}_{0}n \rightarrow 2\ ^{1}_{0}n$ + fission products

## USES OF RADIOISOTOPES

7-41. We shall assume that the O atom of the alcohol is "tagged" with radioactive oxygen (= ◎). If the O in the water comes from the —OH of the acid, the water is free of the radioactive ◎ isotope.

$H_3C{-}C({=}O){-}O{-}H \;+\; H{-}◎{-}CH_3 \longrightarrow H_3C{-}C({=}O){-}◎{-}H \;+\; H{-}O{-}H$

If the O in the water comes from the alcohol, however, then the water will contain radioactive oxygen.

$H_3C{-}C({=}O){-}O{-}H \;+\; H{-}◎{-}CH_3 \longrightarrow H_3C{-}C({=}O){-}O{-}H \;+\; H{-}◎{-}H$

## GENERAL QUESTIONS

7-43. $_{82}Pb + _{24}Cr \rightarrow _{106}Q$ (= Unh)

$_{98}Cf + _{8}O \rightarrow _{106}Q$ (= Unh)

$_{106}Q$ (= Unh) $\rightarrow ^{4}_{2}\alpha + _{104}\Delta$ (= Unq)

$_{104}\Delta \rightarrow ^{4}_{2}\alpha + _{102}\Sigma$ (= No)

$_{102}\Sigma \rightarrow ^{4}_{2}\alpha + _{100}\Pi$ (= Fm)

7-45. $k = 0.693/t_{1/2} = 0.693/7.04 \times 10^8 \text{ years} = 9.84 \times 10^{-10}/\text{year}$

$\ln(0.72/3.0) = -(9.84 \times 10^{-10}/\text{year})t$

$t = 1.45 \times 10^9$ years

Approximately 1.45 billion years ago natural uranium was 3.0% U-235.

# CHAPTER 8
# ATOMIC STRUCTURE

## INSTRUCTOR'S NOTES

We consider this chapter and the next one (Chapter 9) to be of some importance in giving the students a firm foundation for understanding of the theories of chemistry. Therefore, we devote a total of about 9 lecture hours to these chapters, most of that time devoted to electron configurations — particularly of ions — and to periodic properties.

The historical background to our current ideas of atomic structure is interesting to most instructors and, often, to the students as well. Not only is it replete with interesting people, but it provides a good discussion of the way science advances. For this reason, some aspects of the history of physics early in this century are included. However, we recognize that not everyone is as interested in this subject, so it is set off in a *Something More About ...* box at the end of the chapter. Indeed, in our own lectures the historical aspects occupy only about 10% of the total time.

## LIST OF TRANSPARENCIES FOR THIS CHAPTER

| No. | Figure | Page | Subject |
|---|---|---|---|
| 15 | 8.3 | 282 | Electromagnetic spectrum |
| 16 | 8.6 | 286 | Photoelectric effect |
| 17 | 8.8 | 288 | H atom line spectrum |
| 18 | 8.11 | 293 | Electronic transitions in the excited H atom |
| 19 | 8.13 | 302 | Different views of the 1*s* orbital |
| 20 | --- | 303 | Millikan oil drop experiment |
| 21 | --- | 304 | Rutherford experiment |
| 22 | 8.15 | 306 | Probability density plots for 1*s* and 2*p* |
| 23 | 8.16 | 307 | The *d* atomic orbitals |

## SUGGESTED DEMONSTRATIONS

One place where our students have some difficulty is in understanding the ideas of "line spectra." This subject is not so important when discussed in terms of the Bohr model as it is in understanding the general idea of the absorption and emission of energy by atoms or molecules. To help in this regard, we do the following classroom demonstration. When the students arrive for the lecture on this topic, each is given a small diffraction grating. (Large sheets of gratings were obtained from a scientific supply house. The sheets were cut into small squares and then each piece was mounted in a piece of cardboard and held in place with a staple.) Using these gratings, the students view the spectra of light from the discharge tubes in Figure 8.7 and can clearly see the various lines of color emitted by these sources.

To illustrate the shapes of atomic orbitals, we use a set of wooden orbital models. These are currently obtainable from Klinger Educational Products Corp., 112-19 14th Road, College Point, New York, 11356-1453.

Project SERAPHIM and *JCE: Software* has available at least one set of programs suitable for this topic: "The One-Computer Classroom" has simulations of the Rutherford experiment and a nice demonstration of wave properties. The programs run on an Apple IIE, and the output can be projected on a screen using an LCD projection device.

# SOLUTIONS TO SELECTED STUDY QUESTIONS

## ELECTROMAGNETIC RADIATION

8-23. (a) Radar

(b) Microwaves

(c) Visible Light

8-25. $\lambda = c/\nu = (3.00 \times 10^8 \text{ m/s}) / (91.7 \times 10^6 \text{ sec}^{-1}) = 3.27 \text{ m}$

8-27. (a) $\lambda = 6.8 \times 10^2 \text{ nm} \cdot 1 \text{ m}/1.00 \times 10^9 \text{ nm} = 6.8 \times 10^{-7} \text{ m}$

$\nu = c/\lambda = (3.00 \times 10^8 \text{ m/s}) / (6.8 \times 10^{-7} \text{ m}) = 4.4 \times 10^{14} \text{ /s}$

(b) $E = h\nu = (6.6261 \times 10^{-34} \text{ J} \cdot \text{s})(4.4 \times 10^{14} \text{ /s}) = 2.9 \times 10^{-19} \text{ J/photon}$

Energy per mole $= 2.9 \times 10^{-19}$ J/photon $\cdot$ $6.022137 \times 10^{23}$ photons/mol

$= 1.8 \times 10^5$ J/mol (or 180 kJ/mol)

8-29. (a) $\lambda = c/\nu = (3.00 \times 10^8 \text{ m/s}) / (1.00 \times 10^{11} \text{ sec}^{-1}) = 9.00 \times 10^{-3} \text{ m}$

(b) $E = h\nu = (6.6261 \times 10^{-34} \text{ J . s})(1.00 \times 10^{11} \text{ /s}) = 6.6 \times 10^{-24} \text{ J/photon}$

8-31. $\nu = c/\lambda = (3.00 \times 10^8 \text{ m/s}) / (0.50 \times 10^{-2} \text{ m}) = 6.0 \times 10^{10} \text{ /s}$

$E = h\nu = (6.6261 \times 10^{-34} \text{ J . s})(6.0 \times 10^{10} \text{ /s}) = 4.0 \times 10^{-23} \text{ J/photon}$

Energy per mole $= 4.0 \times 10^{-23}$ J/photon $\cdot$ $6.022137 \times 10^{23}$ photons/mol

$= 24$ J/mol

8-33. FM music < microwaves < green light < x-rays

*—increasing energy→*

8-35. To answer this question we shall calculate the wavelength of light corresponding to the required energy.

$\nu = E/h = (6.7 \times 10^{-19} \text{ J/photon}) / (6.6261 \times 10^{-34} \text{ J} \cdot \text{s}) = 1.0 \times 10^{15} \text{ /s}$

$\lambda = c/\nu = (3.00 \times 10^8 \text{ m/s}) / (1.00 \times 10^{15} \text{ sec}^{-1}) = 3.0 \times 10^{-7} \text{ m}$

$3.0 \times 10^{-7} \text{ m} \cdot 1 \text{ nm}/1.0 \times 10^{-9} \text{ m} = 300 \text{ nm}$

The device requires light with a wavelength of no more than 300 nm. However, all light of wavelengths of 540 nm or less is filtered out. Therefore, the device will not work.

## ATOMIC SPECTRA AND THE BOHR ATOM

8-37. (a) There are 10 possible transitions from higher energy levels to lower levels.

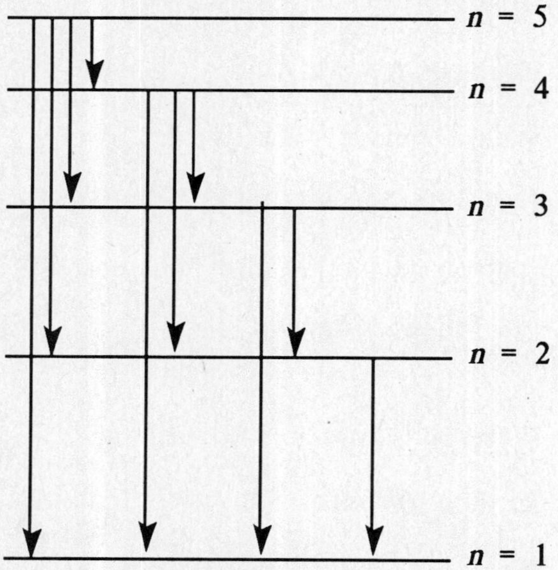


(b) Photons of the *highest* frequency are emitted by a transition between energy levels with the *greatest* possible energy difference. In the case, this means that $n_{initial} = 5$ and $n_{final} = 1$.

(c) Light with the *longest* wavelength will be emitted by a transition between energy levels with the *smallest* possible energy difference. In this case, this means that $n_{initial} = 5$ and $n_{final} = 4$.

(d) Light with the *shortest* wavelength will be emitted by a transition between energy levels with the *largest* possible energy difference. In this case, this means that $n_{initial} = 5$ and $n_{final} = 1$.

8-39. (a) The photon with the smaller energy will be emitted in a transition involving the smaller energy difference between energy levels. This would correspond to a transition from $n = 4$ to $n = 3$ (where $\Delta E = 0.14Rhc$), rather than a transition from $n = 3$ to $n = 1$ (where $\Delta E = 0.89Rhc$).

(b) The photon with the greater energy will be emitted in a transition involving the greater energy difference between energy levels. $\Delta E = 0.94Rhc$ for the transition from $n = 4$ to $n = 1$, while it is only $0.21Rhc$ for the transition from $n = 5$ to $n = 2$.

8-41. To find the value of $n_{initial}$ for a transition in the Balmer series, the simplest method is to use the Rydberg equation (page 293).

$$\frac{1}{\lambda} = R\left(\frac{1}{n^2_{final}} - \frac{1}{n^2_{initial}}\right)$$

where $n_{final} = 2$. Taking R as $1.09737 \times 10^7\ m^{-1}$ (page 288), we find that $n_i = 6$. This same result is found in Figure 8.11 on page 293.

8-43. Here we shall use a modification of the Rydberg:

$$\frac{1}{\lambda} = R\left(\frac{1}{n^2_{final}} - \frac{1}{n^2_{initial}}\right)$$

Taking $R = 1.09737 \times 10^7\ m^{-1}$, $n_{final} = 5$, and $n_{initial} = 6$, we find $\lambda = 7.46 \times 10^{-6}$ m or 7460 nm.

$\nu = c/\lambda = (3.00 \times 10^8\ m/s) / (7.46 \times 10^{-6}\ m) = 4.02 \times 10^{13}\ /s$

## DE BROGLIE AND MATTER WAVES

8-45. $E = \frac{1}{2}mv^2$

$v^2 = 2E/m = 2\,(1.60 \times 10^{-19}\ J/eV)(100.\ eV) / (9.1094 \times 10^{-31}\ kg) = 3.51 \times 10^{13}\ m^2/s^2$

$v = 5.93 \times 10^6\ m/s$

$\lambda = h/mv = (6.6261 \times 10^{-34}\ kg \cdot m^2/s^2)(s) / (9.1094 \times 10^{-31}\ kg)(5.93 \times 10^6\ m/s)$

$\lambda = 1.23 \times 10^{-10}$ m

$\lambda = (1.23 \times 10^{-10}$ m$)(10^9$ nm/m$) = 0.123$ nm

8-47. *Step 1:* Calculate the velocity of the bullet in meters per second

700. miles / hr • 1 km / 0.62137 mile • 1000 *m*/ km • 1 hr / 3600 s = $3.13 \times 10^2$ m/s

*Step 2:* Calculate the wavelength of the bullet

$\lambda = h/mv = (6.6261 \times 10^{-34}\ \text{kg}\cdot\text{m}^2/\text{s}^2)(\text{s}) / (1.50 \times 10^{-3}\ \text{kg})(313\ \text{m/s})$

$\lambda = 1.41 \times 10^{-33}$ m

## QUANTUM MECHANICS

8-49.

| **Description** | **Quantum Number** |
|---|---|
| Energy of electron in H atom | $n$ |
| Orbital shape | $\ell$ |
| Orbital size | $n$ |
| Orbital orientation | $m_\ell$ |

8-51. (a) One of the 4*d* orbitals.

(b) The $n = 6$ shell contains $n^2 = 36$ orbitals arranged in 6 subshells ($\ell$ = 0, 1, 2, 3, 4, and 5). The subshell labels are

| **Value of $\ell$** | 0 | 1 | 2 | 3 | 4 | 5 |
|---|---|---|---|---|---|---|
| **Subshell Label** | *s* | *p* | *d* | *f* | *g* | *h* |
| **Number of Orbitals** | 1 | 3 | 5 | 7 | 9 | 11 |

(c) Any electron shell contains *n* subshells, and any subshell contains $2\ell + 1$ orbitals. Therefore, a *g* subshell contains 2(4) + 1 = 9 orbitals (see the table just above).

8-53. For a 4*d* orbital, *n* must be 4 and $\ell$ must be 2. The value of $m_\ell$ can be any value in the range -2, -1, 0, 1, 2.

8-55. When $n = 5$, the shell contains $n^2 = 25$ orbitals arranged in 5 subshells (since $\ell = 0, 1, 2, 3$, and 4).

8-57. (a) Not a valid set, since $\ell$ must be no larger than $n - 1$.

(b) Valid set

(c) Valid set

(d) Not a valid set, since $m_\ell$ must be in the range $-\ell \ldots 0 \ldots +\ell$.

8-59. (a) No orbitals. Not a valid set of quantum numbers.

(b) 3 orbitals (a *p* subshell is indicated)

(c) 11 orbitals (since the number of orbitals in a subshell is $2\ell + 1$)

(d) 1 orbital

8-61.

| Orbital | Number of Planar Nodes |
|---|---|
| 4*f* | 3 |
| 2*p* | 1 |
| 6*s* | 0 |

8-63.

| Orbital Label | Comment |
|---|---|
| 3*p* | Valid set of quantum numbers ($n = 3$ and $\ell = 1$) |
| 4*s* | Valid set of quantum numbers ($n = 4$ and $\ell = 0$) |
| 2*f* | Not valid. Here $n$ would be 2 and $\ell$ would be 3. The "rules" tell us that $\ell$ must be less than $n$. |
| 5*g* | Valid set of quantum numbers ($n = 5$ and $\ell = 4$) |
| 1*p* | Not valid. Here $n$ would be 1 and $\ell$ would be 1. The "rules" tell us that $\ell$ must be less than $n$. |

8-65.

| Subshell | $n$ | $\ell$ | $m_\ell$ |
|---|---|---|---|
| $5f$ | 5 | 3 | Any value in the range -3 ...0...+3 |
| $4d$ | 4 | 2 | Any value in the range -2...0...+2 |
| $2s$ | 2 | 0 | 0 |

8-67. See Figures 8.14 and 8.16.

8-69. A $4d$ orbital has $n = 4$, $\ell = 2$, and $m_\ell = -2$.

## GENERAL QUESTIONS

8-71. $2.7 \times 10^9$ miles • 1 km/0.62137 mile • 1000 m/km = $4.3 \times 10^{12}$ m

$4.3 \times 10^{12}$ m • 1 s/$3.00 \times 10^8$ m = $1.4 \times 10^4$ s (or 4.0 hours)

8-73. The theory tells us that the energy (or momentum) and position of the electron can be known simultaneously.

8-75. We begin with the equation for the difference in energy between two H atom energy levels.

$$\Delta E = -Rhc\left(\frac{1}{n^2_{\text{final}}} - \frac{1}{n^2_{\text{initial}}}\right)$$

$$\Delta E = -Rhc\left(\frac{1}{\text{infinity}^2} - \frac{1}{1^2}\right)$$

$\Delta E = Rhc$

From Example 8.3, we know that $Rhc$ is 1312.8 kJ/mol. Therefore, this is the ionization energy for the H atom.

8-77. The maximum probability density is reached at a distance of $2a_0$ from the nucleus. This is equivalent to 2 • 0.0529 nm or 0.1058 nm.

8-79. (a) The quantum number $n$ describes the **size** of an atomic orbital.

(b) The shape of an atomic orbital is given by the quantum number $\ell$

(c) A photon of orange light has **less** energy than a photon of yellow light.

(d) The maximum number of electrons that may be associated with the following set of quantum numbers $n = 4$ and $\ell = 3$ is **14**.

(e) If a completed electron subshell has 14 electrons, the $\ell$ value for the subshell is **3**.

(f) The maximum number of electrons that may be associated with the quantum number set $n = 3$, $\ell = 2$, and $m_\ell = -2$ is **2**.

(g) Label each of the orbital pictures below with the appropriate letter:

*p* **orbital**

*d* **orbital**

(h) When $n = 5$, the possible values of $\ell$ are **0, 1, 2, 3, and 4**.

(i) The maximum number of electrons that can be assigned to the $n = 4$ shell is **32**.

# CHAPTER 9

# ATOMIC ELECTRON CONFIGURATIONS AND PERIODICITY

## INSTRUCTOR'S NOTES

As noted in the previous chapter, about 9 lectures are devoted to the material in Chapter 8 and 9.

One problem we have had in writing this chapter is the definition of *electron affinity.* We believe the definition we have used is the one most consistent with current practice and with common thermodynamic conventions. That is, we define EA as the energy involved with an atom in the gas phase acquires an electron. Therefore, the electron affinity for F is -328 kJ/mol, for example. This leaves us a problem when discussing periodic trends. That is, as one proceeds across a period the *affinity for an electron* increases (note the wording). The manifestation of this is that the values of EA become more negative or decrease. To get around this problem we used the phrase *affinity for an electron* in the text.

Our definition of electron affinity is not universal. (We would have preferred to refer to the ionization energy of a negative ion as a measure of electron affinity, since the sign of the energy change would follow thermodynamic convention and because an increases in the affinity of the atom for an electron would be signaled by an increasingly positive value.) In **KC? Discoverer** (see below), electron affinity is defined as in our text, but *the sign of the value is positive when the process is exothermic.*

## LIST OF TRANSPARENCIES FOR THIS CHAPTER

| No. | Figure | Page | Subject |
|---|---|---|---|
| 24 | 9.4 | 319 | Magnetic balance |
| 25 | 9.7 | 323 | Orbital energies |
| 26 | Table 9.2 | 326 | Atomic electron configurations |
| 27 | 9.8 | 327 | Order of orbital filling |
| 28 | Table 9.3 | 328 | Electron configurations for H to Ne |
| 29 | 9.9 | 328 | Electron configurations and the periodic table |
| 30 | Table 9.4 | 333 | Orbital box diagrams for the elements Ca through Zn |
| 31 | 9.11 | 339 | Atomic radii |
| 32 | 9.13 | 342 | Plot of first ionization energies |
| 33 | Table 9.5 | 342 | Table of first ionization energies |
| 34 | 9.14 | 344 | Plot of electron affinities |
| 35 | Table 9.6 | 344 | Table of electron affinities |
| 36 | 9.15 | 346 | Radii of some common ions |
| 37 | --- | 352 | Abundances of the lighter elements as a function of atomic number |

## SUGGESTED DEMONSTRATIONS

The program **KC? Discoverer** and the **Periodic Table Videodisc** from Project SERAPHIM and *JCE: Software* are most valuable with this chapter. See the information in the Preface to this *Manual* and the following article by J. Kotz: *J. Chem. Educ.,* **1989**, *66*, 750.

We use **KC? Discoverer**, which runs on either an MS-DOS machine or Macintosh, in class to illustrate periodic properties by projecting the output of the computer onto a screen in the lecture room. Several universities are also trying the program as a laboratory exercise.

The **Periodic Table Videodisc** can be used to illustrate periodic trends in reactions. The videodisc is an image data base that contains images of the elements reacting with air, water, acids, and a base. In addition, it has information on the uses of the elements. We use it, for example, to illustrate the trend in reactivity of the alkali and alkaline earth metals with air and water. In addition, we carry out the reactions of some of the alkali and alkaline earth metals with water in petri dishes on the overhead projector.

# SOLUTIONS TO SELECTED STUDY QUESTIONS

## WRITING ELECTRON CONFIGURATIONS

9-21. *Aluminum*

Spectroscopic: $1s^2\ 2s^2\ 2p^6\ 3s^2\ 3p^1$

Orbital box:

| 1s | 2s | 2p | | | 3s | 3p | | |
|---|---|---|---|---|---|---|---|---|
| ↑↓ | ↑↓ | ↑↓ | ↑↓ | ↑↓ | ↑↓ | ↑ | | |

*Sulfur*

Spectroscopic: $1s^2\ 2s^2\ 2p^6\ 3s^2\ 3p^4$

| 1s | 2s | 2p | | | 3s | 3p | | |
|---|---|---|---|---|---|---|---|---|
| ↑↓ | ↑↓ | ↑↓ | ↑↓ | ↑↓ | ↑↓ | ↑↓ | ↑ | ↑ |

9-23. *Chromium:* $1s^2\ 2s^2\ 2p^6\ 3s^2\ 3p^6\ 3d^5\ 4s^1$

*Iron:* $1s^2\ 2s^2\ 2p^6\ 3s^2\ 3p^6\ 3d^6\ 4s^2$

9-25. *Krypton:* $1s^2\ 2s^2\ 2p^6\ 3s^2\ 3p^6\ 3d^6\ 4s^2 4p^6$

9-27. *Tungsten:* [Xe] $4f^{14}\ 5d^4\ 6s^2$

*Platinum:* [Xe] $4f^{14}\ 5d^8\ 6s^2$

9-29. *Americium:* [Rn] $5f^7\ 7s^2$

9-31. *Element 104, Unq:* [Rn] $5f^{14}\ 6d^2\ 7s^2$

9-33. (a) $Mg^{2+}$

| 1s | 2s | 2p | | | 3s | 3p | | |
|---|---|---|---|---|---|---|---|---|
| ↑↓ | ↑↓ | ↑↓ | ↑↓ | ↑↓ | | | | |

(b) $Si^{4+}$

| 1s | 2s | 2p | | | 3s | 3p | | |
|---|---|---|---|---|---|---|---|---|
| ↑↓ | ↑↓ | ↑↓ | ↑↓ | ↑↓ | | | | |

(c) $O^{2-}$

| 1s | 2s | 2p | | |
|---|---|---|---|---|
| ↑↓ | ↑↓ | ↑↓ | ↑↓ | ↑↓ |

Note that all ions above are isoelectronic. All have the same configuration as neon.

9-35. Vanadium

| | five 3d orbitals | | | | | 4s |
|---|---|---|---|---|---|---|
| [Ar] | ↑ | ↑ | ↑ | | | ↑↓ |

Vanadium(II) ion, $V^{2+}$

| | five 3d orbitals | | | | | 4s |
|---|---|---|---|---|---|---|
| [Ar] | ↑ | ↑ | ↑ | | | |

Vanadium(III), $V^{3+}$

| | five 3d orbitals | | | | | 4s |
|---|---|---|---|---|---|---|
| [Ar] | ↑ | ↑ | | | | |

Both of the ions are paramagnetic.

9-37. Cobalt(II), $Co^{2+}$ $[Ar]3d^7$

| | five 3d orbitals | | | | | 4s |
|---|---|---|---|---|---|---|
| [Ar] | ↑↓ | ↑↓ | ↑ | ↑ | ↑ | |

Cobalt(III), $Co^{3+}$ $[Ar]3d^6$

| | five 3d orbitals | | | | | 4s |
|---|---|---|---|---|---|---|
| [Ar] | ↑↓ | ↑ | ↑ | ↑ | ↑ | |

Both ions are paramagnetic.

9-39. Platinum configuration — derived by following the "rules": $[Xe]\ 4f^{14}\ 5d^8\ 6s^2$

| | seven 4f orbitals | | | | | | | five 5d orbitals | | | | | 6s |
|---|---|---|---|---|---|---|---|---|---|---|---|---|---|
| [Xe] | ↓↑ | ↓↑ | ↓↑ | ↓↑ | ↓↑ | ↓↑ | ↓↑ | ↓↑ | ↓↑ | ↓↑ | ↑ | ↑ | ↓↑ |

Does not agree with the configuration in Table 9.2. The actual configuration has 9 electrons assigned to 5d orbitals and only 1 electron assigned to 6d.

Platinum(II), $Pt^{2+}$: $[Xe]\ 4f^{14}\ 5d^8$

| | seven 4f orbitals | | | | | | | five 5d orbitals | | | | | 6s |
|---|---|---|---|---|---|---|---|---|---|---|---|---|---|
| [Xe] | ↓↑ | ↓↑ | ↓↑ | ↓↑ | ↓↑ | ↓↑ | ↓↑ | ↓↑ | ↓↑ | ↓↑ | ↑ | ↑ | |

This configuration would be derived no matter whether we began with the actual configuration of Pt given in Table 9.2 or with the configuration we derived by following the rules.

9-41. Element 113: $[Rn]\ 5f^{14}\ 6d^{10}\ 7s^2\ 7p^1$

Element 113 is in Group 3A, the group that includes boron and aluminum.

9-43. $Al^{3+}$: Diamagnetic; configuration is [Ne].

$Ba^{2+}$: Diamagnetic; configuration is [Xe].

$V^{3+}$: Paramagnetic; configuration is $[Ar]3d^2$.

9-45. Four unpaired electrons in their 2+ ions:

(a) $Cr^{2+}$: $[Ar]\ 3d^4$

(b) $Fe^{2+}$: $[Ar]\ 3d^6$

## ELECTRON CONFIGURATIONS AND QUANTUM NUMBERS

9-47. Silicon configuration is [Ne] $3s^2\ 3p^2$

| | 1 2 | 3 4 | | |
|---|---|---|---|---|
| [Ne] | ↑↓ | ↑ | ↑ | |
| | 3s | 3p | | |

| Electron | $n$ | $\ell$ | $m_\ell$ | $m_s$ |
|---|---|---|---|---|
| 1 | 3 | 0 | 0 | $+\frac{1}{2}$ |
| 2 | 3 | 0 | 0 | $-\frac{1}{2}$ |
| 3 | 3 | 1 | +1 | $+\frac{1}{2}$ |
| 4 | 3 | 1 | 0 | $+\frac{1}{2}$ |

9-49. Gallium

| [Ar] | ↑↓ | ↑↓ | ↑↓ | ↑↓ | ↑↓ | ↑↓ | ↑ | | |
|---|---|---|---|---|---|---|---|---|---|
| | five 3d orbitals | | | | | 4s | 4p | | |

A possible set of quantum numbers for the 4p electron is $n = 4$, $\ell = 1$, $m_\ell = +1$, and $m_s = +\frac{1}{2}$.

9-51. (a) When $\ell = 3$ (for $n = 4$) this indicates a set of 4*f* orbitals. Therefore, 14 electrons can be associated with these quantum numbers.

(b) This set of quantum numbers indicates just one of the three possible 6*p* orbitals. Therefore, 2 electrons are indicated.

(c) This set of quantum numbers is not allowed, since $\ell$ cannot equal $n$.

(d) This set of four quantum numbers indicates only a single electron.

9-53. (a) For the O atom, the valence electrons are the 6 electrons of n = 2 ($2s^2\ 2p^4$). The first set is not appropriate since $n = 4$. In addition, an $\ell$ value of 2 indicates d orbitals,

orbitals that are not involved here.

(b) Here the value of $m_\ell$ is not within the appropriate range of -1, 0, and +1.

9-55. The electron configuration for scandium is

| | 1 | | | | | | 2,3 |
|---|---|---|---|---|---|---|---|
| [Ar] | ↑ | | | | | | ↑↓ |
| | five 3d orbitals | | | | | | 4s |

| Electron | $n$ | $\ell$ | $m_\ell$ | $m_s$ |
|---|---|---|---|---|
| 1 | 3 | 2 | -2 | $+\frac{1}{2}$ |
| 2 | 4 | 0 | 0 | $-\frac{1}{2}$ |
| 3 | 4 | 0 | 0 | $+\frac{1}{2}$ |

9-57. The estimated bond length of the Xe—F bond is 72 pm + 131 pm = 203 pm

9-59. P (110 pm) < Ge (122 pm) < Ca (197 pm) < Sr (215 pm) < Rb (248 pm)

9-61. (a) Cs > Rb; (b) $O^{2-}$ > O; (c) As > Br

9-63. Mg 738 kJ/mol) < Al (578 kJ/mol) < P (1012 kJ/mol) < F (1681 kJ/mol)

9-65. K (419 kJ/mol) < Ca (599 kJ/mol) < Si (786 kJ/mol) < As (947 kJ/mol)

9-67. Na has the largest difference between the 1st and 2nd ionization energies because the 2nd electron removed is from an electron shell of lower *n*. This is in contrast to an element such as Mg where both electrons removed have the same *n* value. Some values (in kJ/mol) from **KC? Discoverer** are:

| **Element** | **1st IE** | **2nd IE** | **ΔIE** |
|---|---|---|---|
| Na | 495.8 | 4562.4 | 4066.6 |
| Mg | 737.7 | 1450.6 | 712.9 |

9-69. (a) Most metallic character: Al

(b) Largest atomic radius: Al

(c) Most negative EA: C

(d) Increasing IE: Al (578 kJ/mol) < B (801 kJ/mol) < C (1086 kJ/mol)

9-71. (a) Increasing atomic radius: F < O < S

(b) Based on a knowledge of first-order periodic trends, we would predict that S should have the largest IE of the group P, Si, S, and Se. However, we recall from the discussion on pages 341-342 that the O atom IE is smaller than that of N, so it is not surprising that the same effect carries over into the third period. That is, The order of ionization energies is Si < Se < S < P. (See Table 9.5.)

(c) Increasing radius: Ne < $F^-$ < $O^{2-}$ < $N^{3-}$

(d) Increasing IE: Cs < Ba < Sr

9-73. Group 3A and, to some extent, Group 5A.

## GENERAL QUESTIONS

9-75. (a) Atomic number = 19

(b) Total number of *s* electrons = 7

(c) Total number of *p* electrons = 12

(d) Total number of *d* electrons = 0

9-77. 8*s*

9-79. The $n$ = 3 shell is finally filled at copper, which has the configuration [Ar] $3d^{10}\,4s^1$.

9-81. Sets (a) and (b) are not allowed sets of quantum numbers to describe the ground state of an atom. In set (a) $n$ and $\ell$ are equal, and this is not allowable. In set (b) two electrons (the third and fourth) of the same spin have been assigned to the same orbital. Set (c), however, corresponds to a pair of electrons assigned to the 2*s* orbital.

9-83. (a) Sulfur, S
(b) Radium, Ra
(c) Nitrogen, N
(d) Ruthenium, Ru
(e) Astatine, At
(f) Copper, Cu

9-85. (a) A is a metal.
(b) B should have the larger ionization energy since it is a *p*-block element whereas A is an *s*-block element.
(c) B, the nonmetal, should have the more negative electron affinity.
(d) A, the *s*-block element, should have the larger atoms.

9-87. • $In^{4+}$ is an unlikely ion. For indium to lose 4 electrons would mean that one would have to begin removing 4*d* electrons, after having removed two 5*s* and one 5*p* electrons.

• We also consider a "bare" $Fe^{6+}$ to be unlikely. In this case we would have to remove two 4*s* and four 3*d* electrons (to leave two 3*d* electrons). Creating an ion of such high charge requires an enormous energy expenditure.

• $Sn^{5+}$ is unlikely. Here we would have to remove two 5*s* electron, two 5*p* electrons, and one 4*d* electron. To begin removing *d* electrons after having removed four other electrons would again require a large energy expenditure.

9-89. $Cl^-$ < Ca (599 kJ/mol) < Zn (906 kJ/mol) < $Ca^{2+}$

The ionization energy of a negative ion is almost always less than the energy required to produce a positive ion. Here the electron affinity of Cl is -349 kJ/mol. Therefore, the energy required to remove the electron is +349 kJ/mol. The $Ca^{2+}$ ion has the largest ionization energy because further ionization of the ion would require removing an electron of an inner shell.

9-91. (a) Sulfur, S
(b) Chloride ion, $Cl^-$

(c) Sodium, Na

(d) Oxygen, O

(e) Nitride ion, $N^{3-}$

9-93. Size: $K^+ < Cl^- < S^{2-}$

Ionization energy: $S^{2-} < Cl^- < K^+$

Electron affinity: $K^+ < Cl^- < S^{2-}$

9-95. Element 113

(a) Electron configuration: [Rn] $5f^{14}\ 6d^{10}\ 7s^2\ 7p^1$

(b) Element 113 is in Group 3A.

(c) $Et_2O_3$ and $EtCl_3$. (Since thallium commonly forms compounds where Tl is +1, other possible compounds are $Et_2O$ and EtCl.)

# CHAPTER 10

# BASIC CONCEPTS OF CHEMICAL BONDING

## INSTRUCTOR'S NOTES

About 5 or 6 hours are devoted to this chapter. We pay particular attention to the following topics:

- drawing electron dot structures for main group elements. (Relatively little attention is given to molecules with expanded valence.)
- resonance structures for simple molecules.
- bond properties (order, length, energy, polarity).
- VSEPR theory.
- molecular polarity.

With regard to VSEPR theory, we have used the new term "structural-pair geometry" to designate the positions of the stereochemically active pairs. We find our students then differentiate between the geometry of the electron pairs (structural-pair geometry) and the positions of the atoms in the molecule or ion (molecular geometry). However, there is still some problem getting students to see that, while the structural-pair geometry determines the positions of the atoms, the word description given the molecular geometry reflects only the atom positions. We also pay attention to the values of bond angles as predicted by VSEPR, but we generally do not discuss the finer details (as outlined in the box on page 397).

Finally, we give very little coverage to oxidation numbers and formal charges, which is unfortunate because the "formal charge concept" is very useful.

## LIST OF TRANSPARENCIES FOR THIS CHAPTER

| No. | Figure | Page | Subject |
|---|---|---|---|
| 38 | Table 10.5 | 384 | Electronegativity values |
| 39 | Table 10.6 | 391 | Perspective molecular sketches |
| 40 | Table 10.7 | 394 | Structural pair geometries for 2-4 structural pairs |

| | | | |
|---|---|---|---|
| 41 | Table 10.8 | 398 | Structural pair geometries for 5 and 6 structural pairs |

## SUGGESTED DEMONSTRATIONS

The most useful demonstrations here are models of structural-pair and molecular geometries. We find it useful to illustrate the former with balloons as indicated in Figure 10.12. Purchase a bag of round "party balloons" at a stationary or drug store. After blowing them up, tie them together in pairs; additionally, make at least one set of three tied together. All of the models in Figure 10.12 can be assembled easily in class as you discuss various geometries. The students are always impressed that these are the natural geometries assumed.

In addition, large scale molecular models are very useful. Styrofoam orbital models can be purchased from Aldrich Chemical Company. The smaller models, such as those used in the photos in Table 10.6, can be purchased from Allyn and Bacon, Inc.

The program **KC? Discoverer** is useful here when discussing electronegativity. The MS-DOS version of **KC?** will allow you to plot electronegativity against atomic number or plot the position of the element in a group or period against electronegativity. The Macintosh Hypercard version of **KC?**, called the **Periodic Table Stack**, can create graphs such as that in Figure 10.10, but only if you use Cricket Graph® to construct the graphs. The graph in Figure 10.10 was actually done using another, similar program for the Macintosh called **MacMendeleev**. This was purchased from Kinko's Academic Courseware Exchange, but it is uncertain if this is still in business. One can try Intellimation, Department XA, 130 Cremona Drive, PO Box 1922, Santa Barbara, CA 93116-1922.

# SOLUTIONS TO SELECTED STUDY QUESTIONS

## VALENCE ELECTRONS

10-28. Number of valence electrons

| *Element* | *Group Number* | *Number of Valence Electrons* |
|---|---|---|
| C | 4A | 4 |
| Cl | 7A | 7 |
| P | 5A | 5 |
| Ne | 8A | 8 |
| Si | 4A | 4 |

## IONIC BONDING

10-30. (a) NaCl (because the ion-ion distance is smaller than in NaI)

(b) CaO (because the ion charges are larger than in KCl)

(c) LiF (because the ion-ion distances are smaller than in CsF)

10-32. (a) RbCl should be more soluble, since the enthalpy of dissociation of RbCl should be smaller than that for NaCl owing to the larger ion-ion distance in RbCl.

(b) CsI should be more soluble than CsBr.

(c) NaI should be more soluble than KF, since the ion-ion distance in NaI is larger, making its dissociation enthalpy smaller.

10-34. (a) KCl should have a higher melting point than CsI.

(b) CaO has a higher melting point than KCl.

(c) $CaCl_2$ has a higher melting point than KCl.

## THE OCTET RULE AND EXPANDED VALENCE

10-36. (a) P; (e) Cl; (g) Se, and (h) Sn.

10-38. (a) $Cl_2CO$

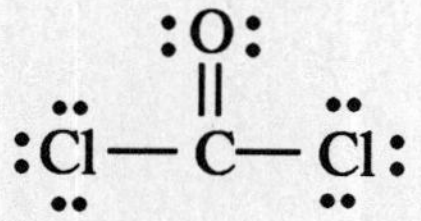

(b) $BF_4^-$

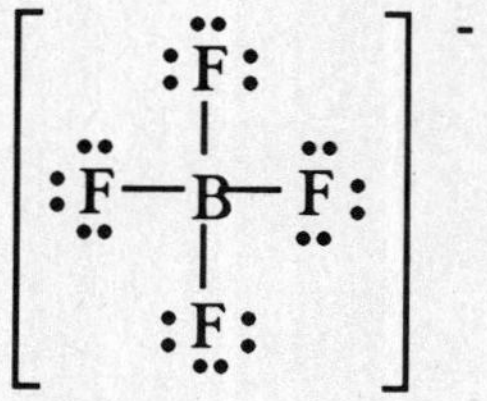

(c) $NO_2^+$

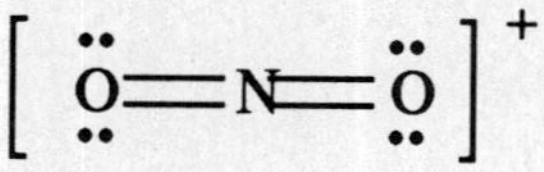

(d) $PO_4^{3-}$

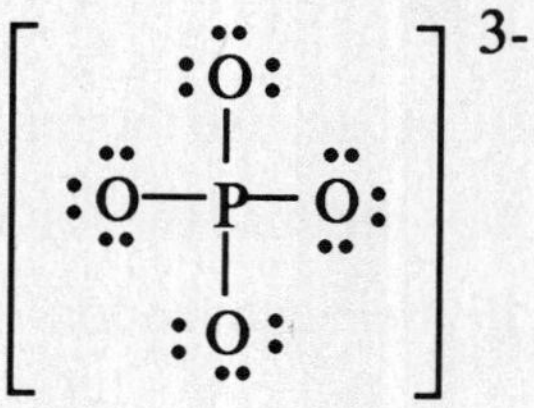

10-40. (a) $F_2CCF_2$

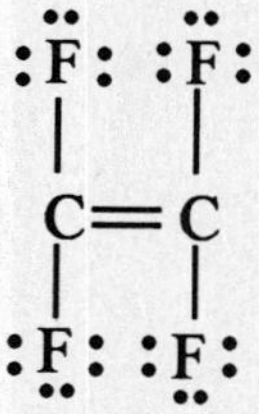

(b) $H_2CCHCl$

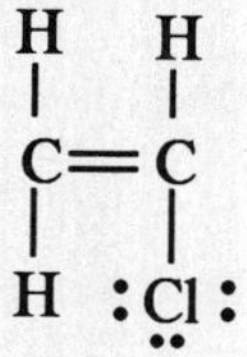

(c) $H_2CCHCN$

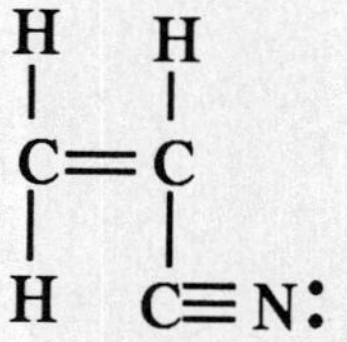

(d) $H_3CSiCl_3$

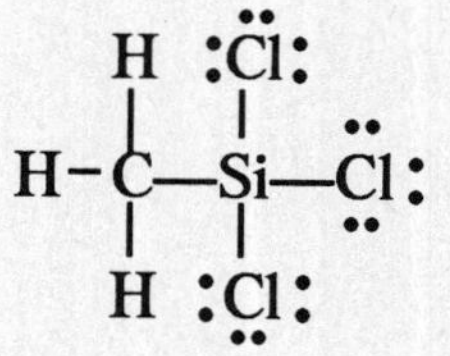

10-42. (a) $NO_3^-$

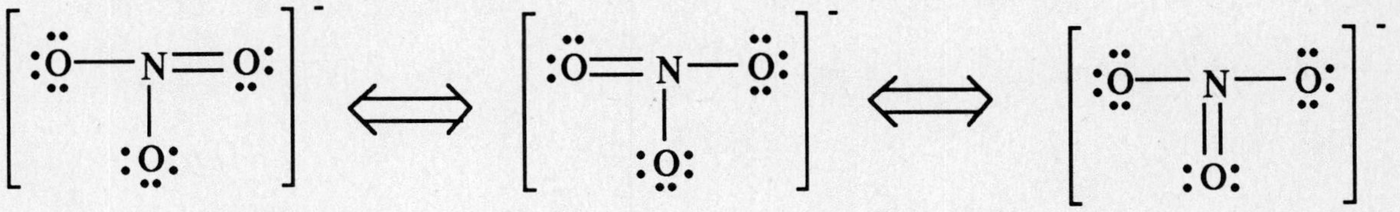

(b) $HNO_3$

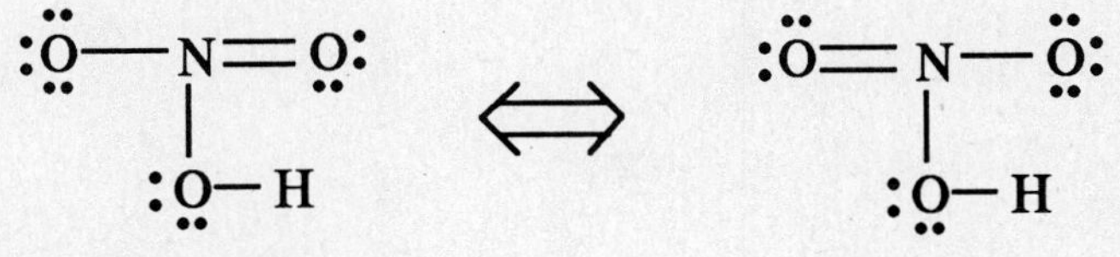

(c) NNO

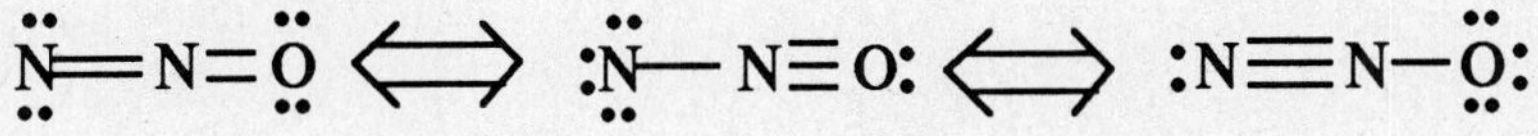

10-44. (a) $BrF_5$

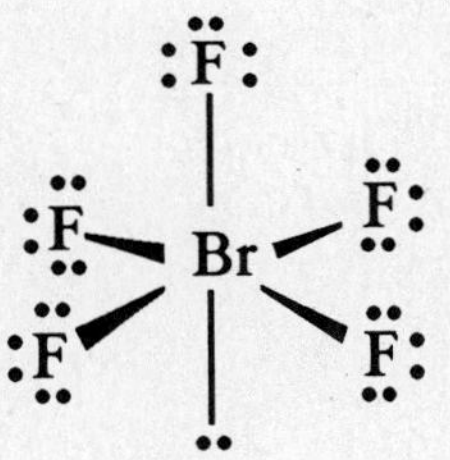

(b) $IF_3$

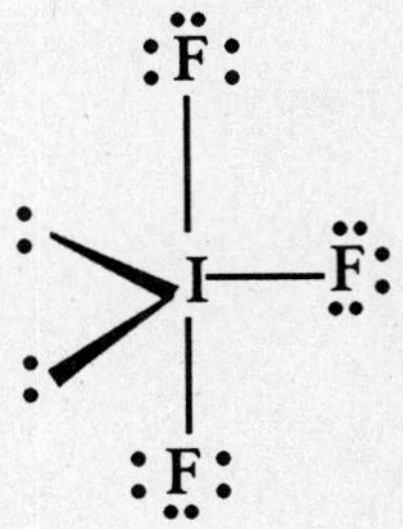

(c) $IBr_2^-$

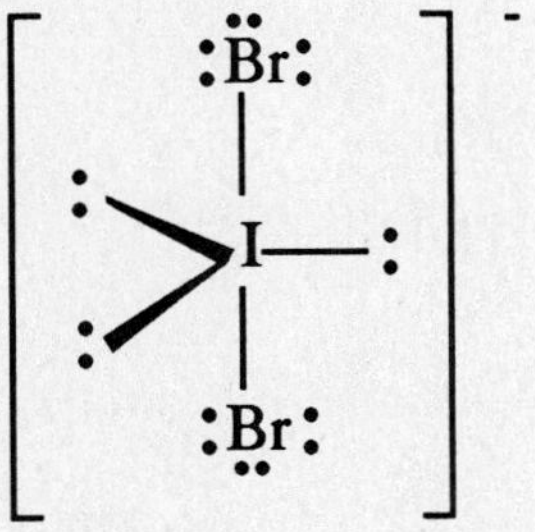

## BOND PROPERTIES

10-46. (a) $SCN^-$ Two of the three resonance structures of shown. In all cases there are 2 π bonds and 2 σ bonds.

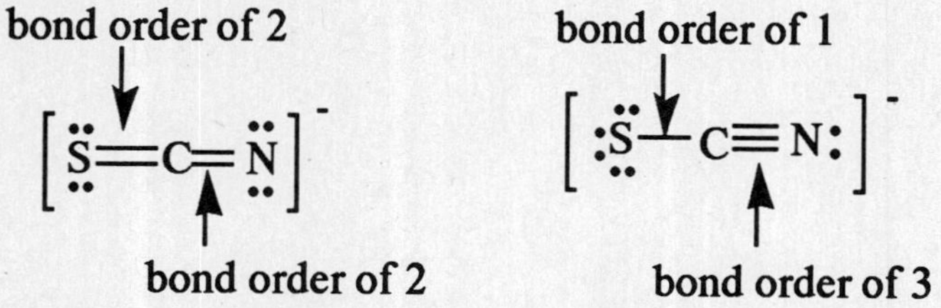


(b) $H_3CCN$. There are five σ bonds and 2 π bonds.

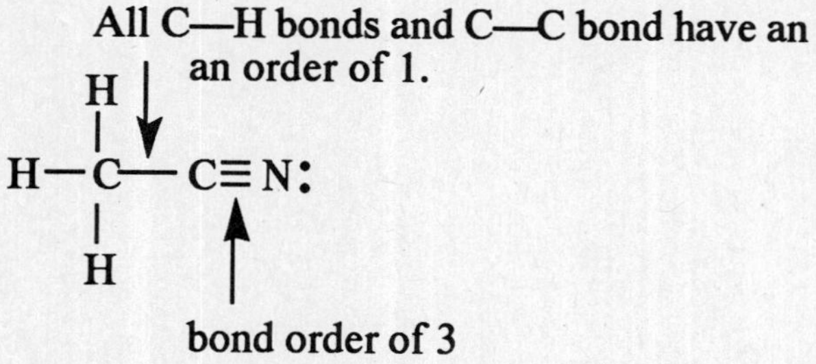


(c) $SO_3$. There are three resonance structures, all having 1 π bond and three σ bonds.

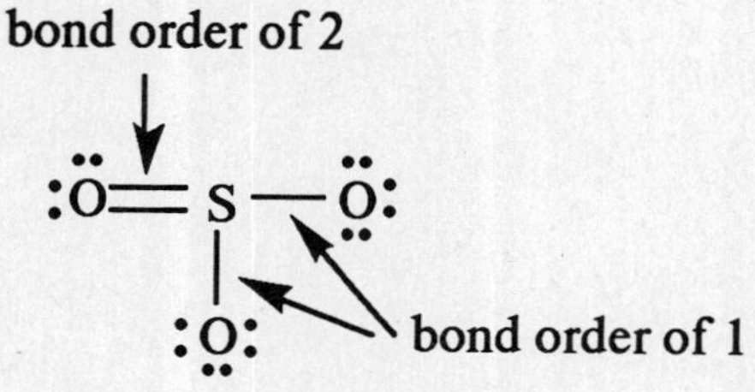


10-48. (a) P—O (176 pm) should be shorter than Si—N (187 pm) because both P and O are smaller than Si and N, respectively.

(b) C—O (143 pm) should be shorter than Si—O (183 pm) because C is smaller than Si.

(c) C—F (141 pm) is shorter than C—Br (191 pm) because F is smaller than Br.

(d) The C=C double bond (134 pm) is longer than the C≡N triple bond (115 pm).

10-50. In hydrazine the N—N bond is a single bond and so is expected to be longer and weaker than the N=N double bond in $N_2O$. (Only one resonance structure of dinitrogen oxide is shown. However, taking all three resonance structures into account, the average NN bond order is still a double bond. See Question 10-42.)

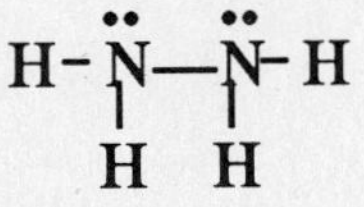

hydrazine

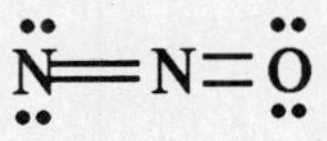

dinitrogen oxide

10-52. The CO bonds in the formate ion have a bond order of 1.5. In contrast, the CO bond in the carbonate ion has an average bond order of 1.33.

Lewis structure for the formate ion

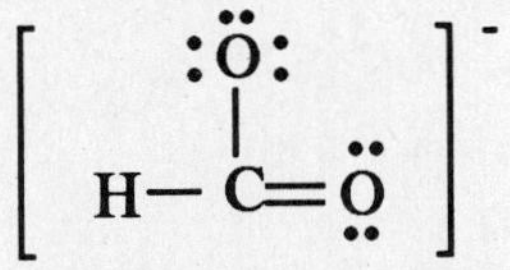

Resonance structures for the carbonate ion

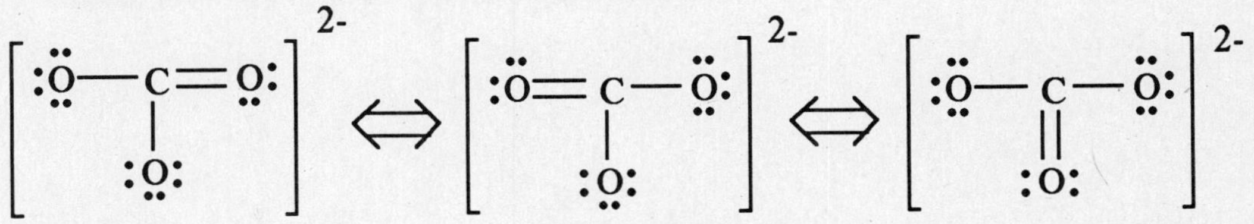

## BOND ENERGIES AND REACTION ENTHALPIES

10-54. (a) Enthalpy of formation of chloramine

$H_2(g) + \frac{1}{2} N_2(g) + \frac{1}{2} Cl_2(g) \rightarrow$

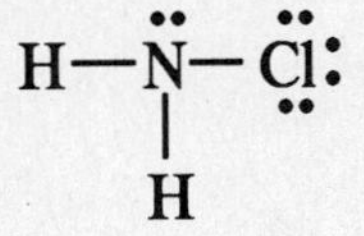

$$\Delta H_f^\circ = D_{H\text{-}H} + \tfrac{1}{2}(D_{N\equiv N}) + \tfrac{1}{2}(D_{Cl\text{-}Cl}) - 2\,D_{N\text{-}H} - D_{N\text{-}Cl}$$

$$= 1\text{ mol }(436\text{ kJ/mol}) + \tfrac{1}{2}\text{ mol }(946\text{ kJ/mol}) + \tfrac{1}{2}\text{ mol }(243\text{ kJ/mol}) - 2\text{ mol }(389\text{ kJ/mol}) - 1\text{ mol }(201\text{ kJ/mol})$$

$$= 52\text{ kJ per mole of chloramine formed}$$

(b) Enthalpy of formation of hydrogen peroxide

$H_2(g) + O_2(g) \rightarrow H\text{—}O\text{—}O\text{—}H(g)$

$$\Delta H_f^\circ = D_{H\text{-}H} + D_{O=O} - 2\,D_{O\text{-}H} - D_{O\text{-}O}$$

$= 1\text{ mol }(436\text{ kJ/mol}) + 1\text{ mol }(498\text{ kJ/mol})$

$- 2\text{ mol }(464\text{ kJ/mol}) - 1\text{ mol}(138\text{ kJ/mol})$

$= -132$ kJ per mole of hydrogen peroxide formed

10-56. Enthalpy of formation of ethylene

$2\,H_2(g) + 2\,C(\text{graphite}) \rightarrow H_2C{=}CH_2(g)$

$\Delta H^\circ_f = 2\,D_{H\text{-}H} + 2\,\Delta H^\circ_{vap}[C(\text{graphite})] - 4\,D_{C\text{-}H} - D_{C=C}$

$= 2\text{ mol }(436\text{ kJ/mol}) + 2\text{ mol }(717\text{ kJ/mol})$

$- 4\text{ mol }(414\text{ kJ/mol}) - 1\text{ mol }(611\text{ kJ/mol})$

$= 39$ kJ per mole of $C_2H_4$ formed

The enthalpy of formation of ethylene is much more negative than that for acetylene owing to the fact that the formation of two additional C—H bonds partly offsets the energy required to break an additional mole of H—H bonds.

10-58. Formation of urea

$2\,H_2(g) + C(\text{graphite}) + N_2(g) + \frac{1}{2}\,O_2(g) \rightarrow H_2N\text{—}C(O)\text{—}NH_2(g)$

$\Delta H^\circ_f = 2\,D_{H\text{-}H} + \Delta H^\circ_{vap}[C(\text{graphite})] + D_{N\equiv N} + \frac{1}{2}\,D_{O=O}$

$- 4\,D_{N\text{-}H} - 2\,D_{C\text{-}N} - D_{C=O}$

$= 2\text{ mol }(436\text{ kJ/mol}) + 1\text{ mol }(717\text{ kJ/mol}) + 1\text{ mol }(946\text{ kJ/mol})$

$+ 0.5\text{ mol }(498\text{ kJ/mol}) - 4\text{ mol }(389\text{ kJ/mol}) - 2\text{ mol }(293\text{ kJ/mol})$

$- 1\text{ mol }(745\text{ kJ/mol})$

$= -103$ kJ per mole of urea formed

10-60. $H_2(g) + S(s) \rightarrow H_2S(g)$

$\Delta H^\circ_f[H_2S(g)] = -20.63\text{ kJ/mol} = D_{H\text{-}H} + \Delta H^\circ_{vap}[S(s)] - 2\,D_{H\text{—}S}$

$-20.63\text{ kJ} = 1\text{ mol }(436\text{ kJ/mol}) + 1\text{ mol }(278.8\text{ kJ/mol}) - 2\,D_{H\text{—}S}$

$D_{H\text{—}S} = 368\text{ kJ/mol}$

The H—O bond energy is 464 kJ/mol, almost 100 kJ/mol larger than that for H—S. From this we may conclude, tentatively, that bonds between hydrogen and Group 6A elements decrease in energy as the atomic weight of the Group 6A element increases.

10-62. Estimate the bond dissociation energy for the B—F bond.

$B(solid) + 3/2\ F_2(g) \rightarrow BF_3(g) \qquad \Delta H^\circ_f[BF_3(g)] = -1137\ kJ/mol$

$\Delta H^\circ_f[BF_3(g)] = -1137\ kJ/mol = \Delta H^\circ_{vap}[B(s)] + 3/2\ D_{F\text{-}F} - 3\ D_{B\text{-}F}$

$-1137\ kJ/mol = 1\ mol\ (563\ kJ/mol) + 3/2\ mol\ (159\ kJ/mol) - 3\ D_{B\text{-}F}$

$D_{B\text{-}F} = 646\ kJ/mol$

10-64. $2\ H_3COH(g) + 3\ O_2(g) \rightarrow 2\ CO_2(g) + 4\ H_2O(g)$

$\Delta H_{rxn} = 6\ D_{C\text{-}H} + D_{C\text{-}O} + D_{O\text{-}H} + 3\ D_{O=O} - 4\ D_{C=O} + 8\ D_{O\text{-}H}$

$= 6\ mol\ (414\ kJ/mol) + 1\ mol\ (351\ kJ/mol) + 1\ mol\ (464\ kJ/mol)$
$+ 3\ mol\ (498\ kJ/mol) - 4\ mol\ (803\ kJ/mol) - 8\ mol\ (464\ kJ/mol)$

$= -2131\ kJ$

10-66. Estimate the enthalpy of the reaction

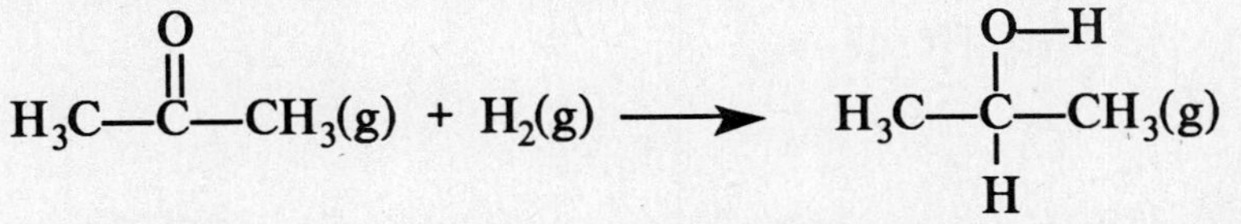

To find the enthalpy change for this reaction we need only see what happens to several of the bonds. That is, the reaction progresses by breaking a CO double bond and a H—H bond and the making a C—O single bond, a C—H bond, and a O—H bond.

$\Delta H_{rxn} = D_{C=O} + D_{H\text{-}H} - D_{C\text{-}O} - D_{O\text{-}H} - D_{C\text{-}H}$

$= 1\ mol\ (745\ kJ/mol) + 1\ mol\ (436\ kJ/mol) - 1\ mol\ (351\ kJ/mol)$
$- 1\ mol\ (464\ kJ/mol) - 1\ mol\ (414\ kJ/mol)$

$= -48\ kJ$

## ELECTRONEGATIVITY AND BOND POLARITY

10-68. (a)

| Bond | Atom More Negatively Charged | Δχ |
|---|---|---|
| C—N | N | 0.5 |
| C—H | C | 0.4 |
| C—Br | Br | 0.3 |
| S—O | O | 1.0 |

(b) The most polar bond is the one with the largest difference in electronegativity. Therefore, the S—O bond is the most polar.

10-70. Bonding in urea

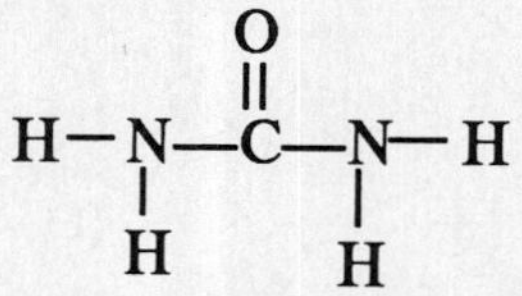

(a) All bonds in the molecule are polar.

(b) The C=O bond is the most polar, with O the negative end of the bond dipole.

## OXIDATION NUMBERS AND ATOM FORMAL CHARGES

10-72.

| | Atom | Oxidation Number |
|---|---|---|
| (a) | Cl | +3 |
| | F | -1 |
| (b) | Xe | +2 |
| | F | -1 |
| (c) | O | +2 |
| | F | -1 |
| (d) | H | +1 |
| | C | +2 |
| | N | -3 |

As an example of the way the oxidation numbers were derived, consider HCN.

H—C≡N: H :C :::N:

+1 +2 -3

We assign the bond electrons to the more electronegative atom. Since C is more electronegative than H, the H—C bonding electrons are assigned to C. In the same way, the 6 electrons of the CN triple bond are assigned to N. This leads to the indicated oxidation numbers.

10-74.

| | Atom | Oxidation Number | Formal Charge |
|---|---|---|---|
| (a) | I | +1 | -1 |
| | Cl | -1 | 0 |
| (b) | N | -3 | 0 |
| | H | +1 | 0 |
| (c) | O | -2 | -1 |
| | Cl | +1 | 0 |
| (d) | S | +4 | 0 |
| | F | -1 | 0 |

***Note*** that in the first printing of the book the charge on $ICl_2^-$ was not included. Also note that there is an interesting inversion of oxidation number and formal charge on the central I atom. In $OCl^-$, the charges were derived assuming the bonding electrons are evenly distributed since O and Cl have the same electronegativities.

10-76. Resonance structures of the formate ion. The formal charges are indicated on one of the resonance structures.

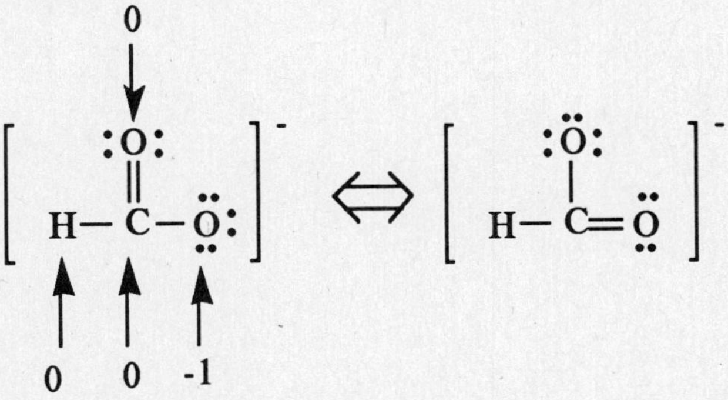

10-78. (a) Resonance structures for $CNO^-$ with formal charges.

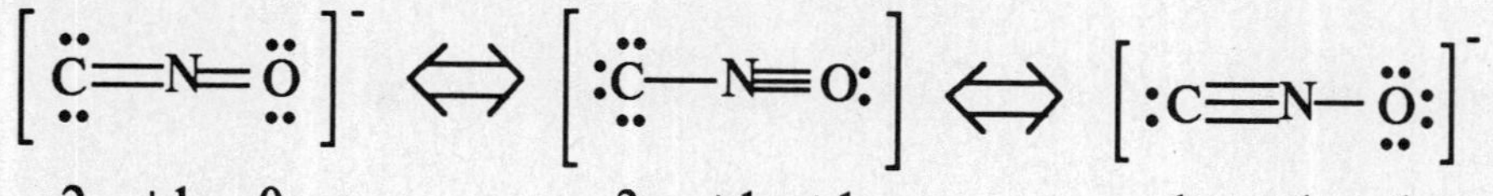

-2 +1 0 -3 +1 +1 -1 +1 -1

(b) The most reasonable structure is the one at the far right because it is the one in which the formal charges on the atoms are at a minimum.

(c) The instability of the ion could be attributed to the fact that the least electronegative atom in the ion bears a negative charge in all resonance structures.

## MOLECULAR SHAPE

10-80. NOTE: In this question, and many that follow, only the lone pairs of electrons around the central atom are indicated.

(a) $ClF_2^+$ Structural pair geometry = tetrahedral

Molecular geometry = bent

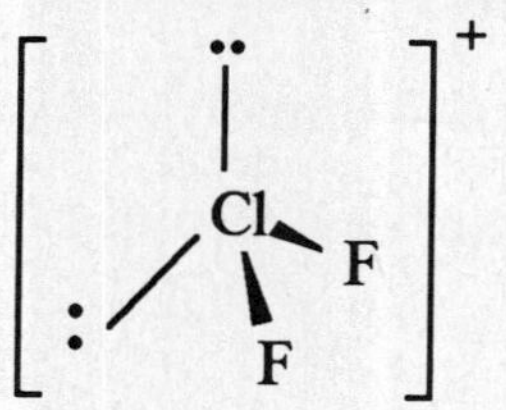

(b) $SnCl_3^-$ Structural pair geometry = tetrahedral

Molecular geometry = pyramidal

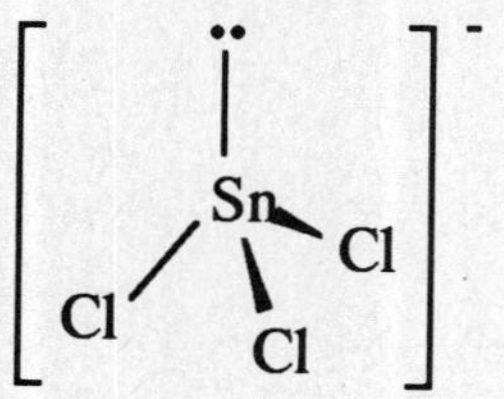

(c) $PO_4^{3-}$ Structural pair geometry = tetrahedral

Molecular geometry = tetrahedral

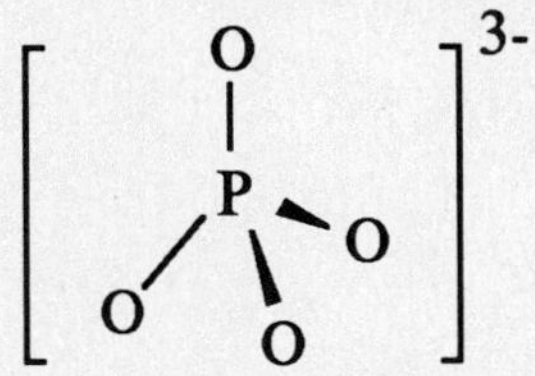

(d) $CS_2$ Structural pair geometry = linear

Molecular geometry = linear

S=C=S

10-82. (a) $BO_3^{3-}$, borate, 24 valence electrons

Structural pair geometry = trigonal planar

Molecular geometry = trigonal planar

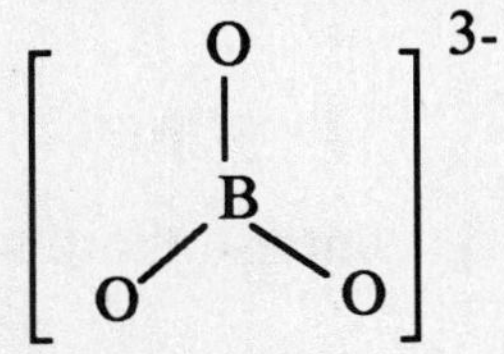

(b) $CO_3^{2-}$, carbonate, 24 valence electrons

Structural pair geometry = trigonal planar

Molecular geometry = trigonal planar

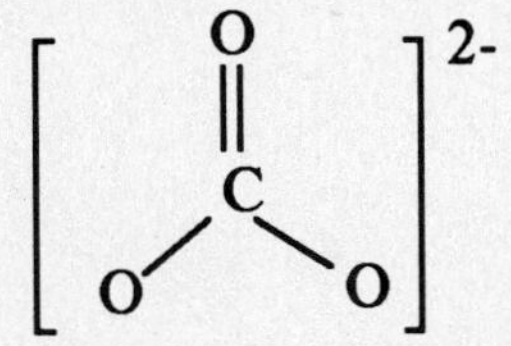

(only one of three resonance structures shown)

(c) $NO_3^-$. nitrate, 24 valence electrons

Structural pair geometry = trigonal planar

Molecular geometry = trigonal planar

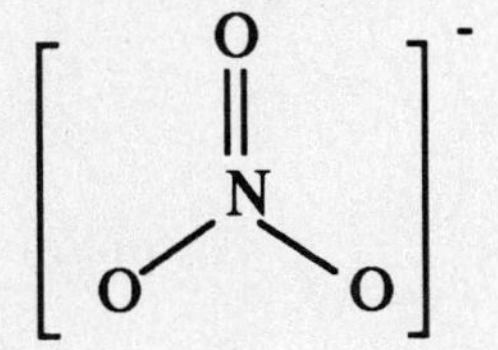

(only one of three resonance structures shown)

(d) $SO_3^{2-}$, sulfite, 26 valence electrons

Structural pair geometry = tetrahedral
Molecular geometry = pyramidal

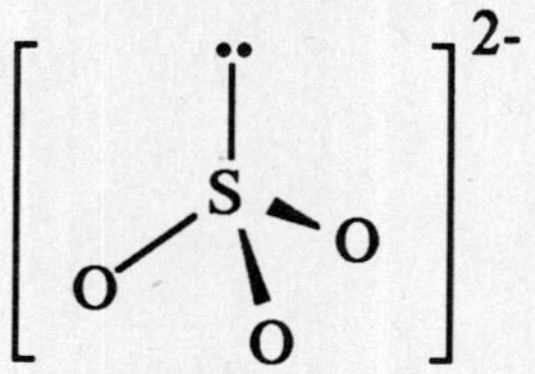


(e) $ClO_3^-$, chlorate, 26 valence electrons

Structural pair geometry = tetrahedral
Molecular geometry = pyramidal

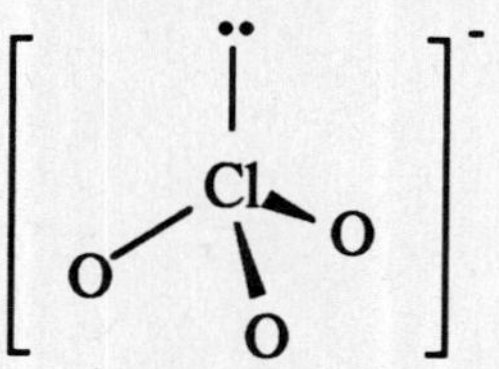


The structures above illustrate that, for ions of the type $MO_3^{n-}$, those with 24 valence electrons have a trigonal planar structural pair geometry. In contrast, those with 26 valence electrons have a tetrahedral structural pair geometry.

10-84. As in the question above, we have again only indicated the lone pairs of electrons on the central atom.

(a) $SiF_6^{2-}$, hexafluorosilicate, 48 valence electrons

Structural pair geometry = octahedral
Molecular geometry = octahedral

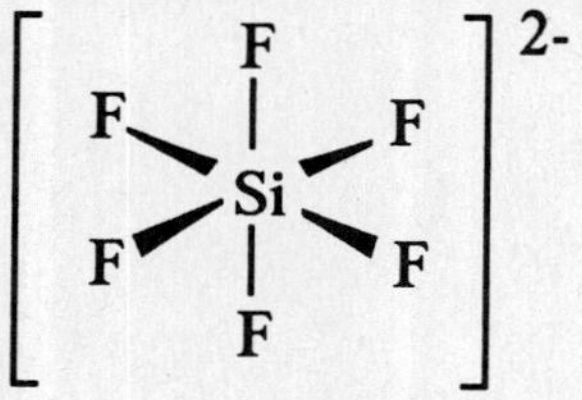

(b) $PF_5$, phosphorus pentafluoride, 40 valence electrons

Structural pair geometry = trigonal bipyramidal

Molecular geometry = trigonal bipyramidal

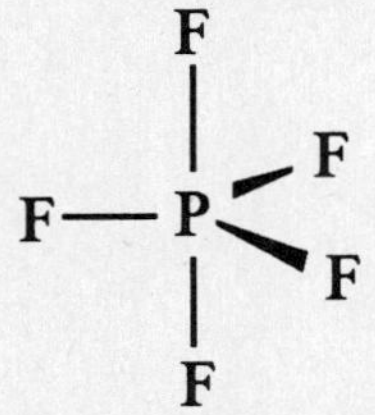


(c) $SF_4$, sulfur tetrafluoride, 34 valence electrons

Structural pair geometry = trigonal bipyramidal

Molecular geometry = see-saw shape

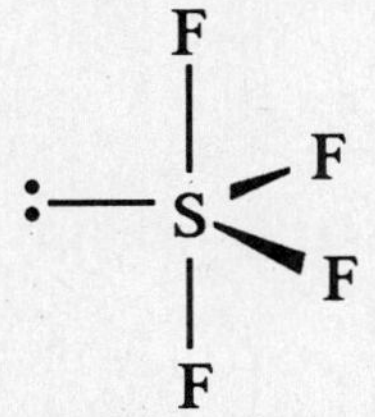


(d) $XeF_4$, xenon tetrafluoride, 36 valence electrons

Structural pair geometry = octahedral

Molecular geometry = square planar

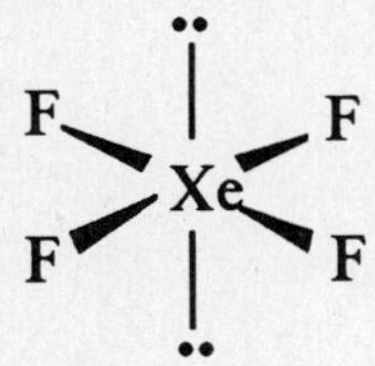


10-86. (a) Cl—S—Cl in $SCl_2$ ≈ 109°

(b) N—N—O in $N_2O$ = 180°

(c) Angle 1 = 120° and angle 2 = 109°

(d) Angle 1 = 109° and angle 2 = 109°

10-88. (a) 13 σ bonds and 1 π bond.

(b) Angle 1 = 109°, angle 2 = 109°, angle 3 = 109°, angle 4 = 109°, and angle 3 = 120°.

10-90. (a) The structural pair geometry of $SF_6$ is octahedral. Therefore, the F—S—F angles are all 90°.

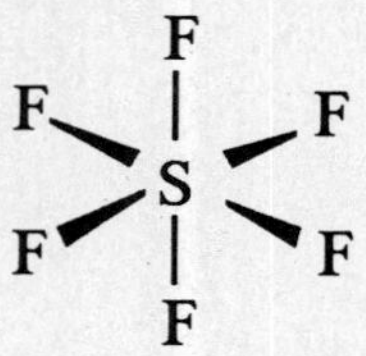


(b) The structural pair geometry of $XeF_2$ is trigonal bipyramidal. Therefore, the F—Xe—F angles are 180°.

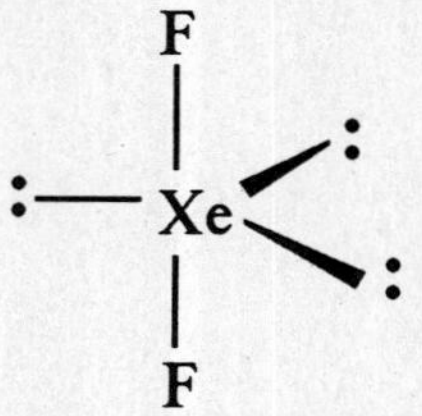


(c) The structural pair of $ClF_2^-$ is like that of $XeF_2$, that is, trigonal bipyramidal. Therefore, the F—Cl—F angles are 180°.

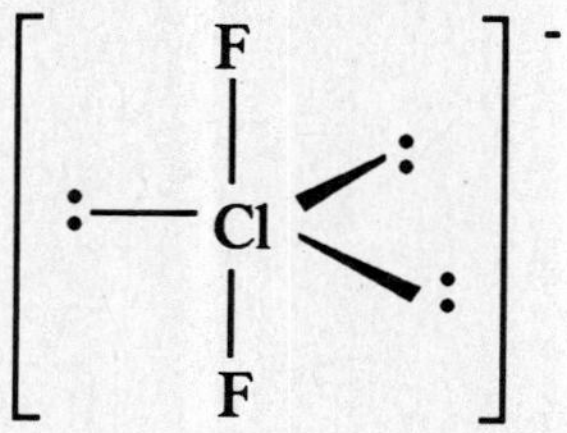


10-92. The F—Cl—F angle in $ClF_2^-$ is 180° (see above), whereas the same angle in $ClF_2^+$ is 109°.

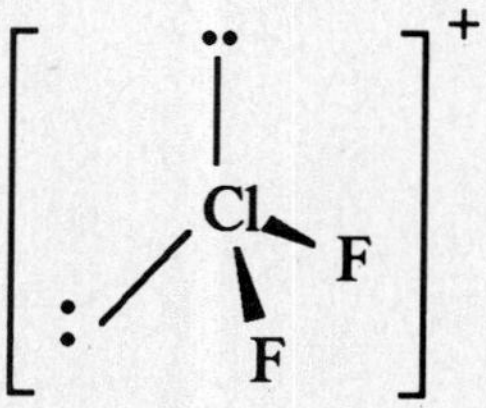

## MOLECULAR POLARITY

10-94.

| **Molecule** | **Δχ for bond** |
|---|---|
| $CH_4$ | H - C = 2.1 - 2.5 = -0.4 |
| $NCl_3$ | Cl - N = 3.0 - 3.0 = 0.0 |
| $BF_3$ | F - B = 4.0 - 2.0 = 2.0 |
| $CS_2$ | S - C = 2.5 - 2.5 = 0.0 |

The results above show the B—F bonds are the most polar of this group.

The only molecule in the group above that is polar is $NCl_3$ (molecular geometry = pyramidal) The remainder are nonpolar.

10-96.

| **Molecule** | **Δχ for bond** | **Molecular Geometry** | **Molecular Polarity** |
|---|---|---|---|
| CO | C - O = 2.1 - 3.5 = 1.4 | linear | polar |
| $BCl_3$ | Cl - B = 3.0 - 2.0 = 1.0 | trigonal planar | not polar |
| $CF_4$ | F - C = 4.0 - 2.1 = 1.9 | tetrahedral | not polar |
| $PCl_3$ | Cl - P = 3.0 - 2.1 = 0.9 | pyramidal | polar |
| $GeH_4$ | H - Ge = 2.1 - 1.9 = 0.2 | tetrahedral | not polar |

The molecule with the most polar bonds is $CF_4$. Both $GeH_4$ and $CF_4$ are tetrahedral and so are not polar, while $BCl_3$ is trigonal planar and so is not polar. See Figure 10.14 on page 401.

## GENERAL QUESTIONS

10-98. No. It is not possible to have two atoms bonded by a $\pi$ bond only.

10-100. $NO_2^-$ is a 18 electron ion, while $NO_2^+$ is a 16 electron ion.

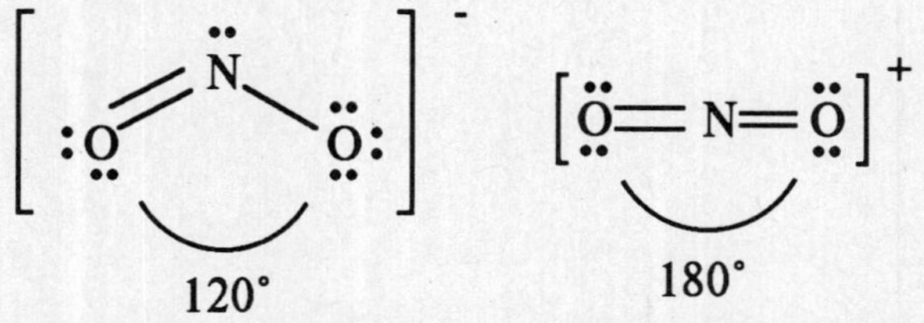

This means that the former has a trigonal planar structural pair geometry and so has a bent molecular geometry with a 120° O—N—O angle. In contrast, $NO_2^+$ has a linear structural pair and molecular geometry and has a 180° bond angle.

10-102. $1/4\ P_4(s) + 3/2\ H_2(g) \rightarrow PH_3(g)$

$\Delta H^\circ_f[PCl_3(g)] = 1/4\ \Delta H^\circ_f[P(g)] + 3/2\ D_{H\text{-}H} - 3\ D_{P\text{-}H}$

$= 1/4$ mol (1259 kJ/4 mol P) + 3/2 mol (436 kJ/mol) - 3 mol (318 kJ/mol)

$= 15$ kJ

10-104. (a) Angle 1 = 120° and angle 2 = 180°

(b) The C=C double bond is shorter than the C—C single bond.

(c) The C=C double bond is stronger than the C—C single bond.

(d) The C≡N bond is the most polar bond, with the N atom being the negative end of the bond.

10-106. The three resonance structures are given below.

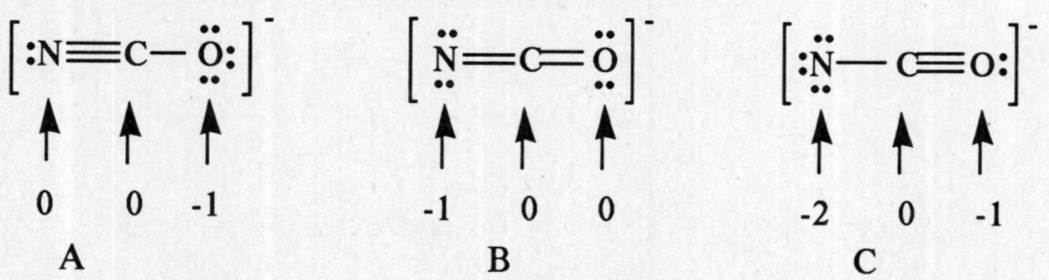

Structure C contributes very little to the overall structure of the ion. Further, since O is more electronegative than N, this means that structure A is slightly favored. Therefore, $H^+$ should attack the O end of the ion to give H—OCN.

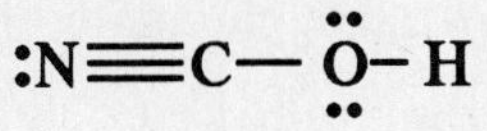

10-108. The resonance structures of nitryl chloride are given below.

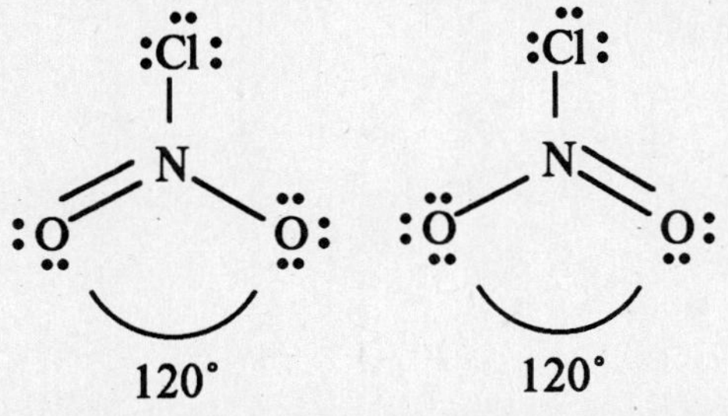


The molecule has trigonal planar structural pair and molecular geometries, and all angles are 120°.

10-110. Only the *trans* isomer has a dipole moment.

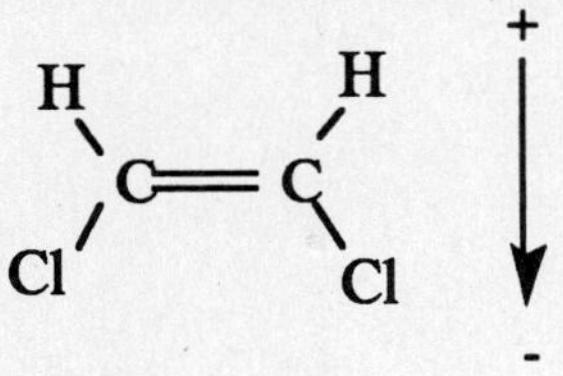

10-112. (a) There are 15 σ bonds and 3 π bonds.

(b) Angle 1 = 120°; angle 2 = 109°; and angle 3 = 109°

(c) The C=O bonds.

10-114. (a) Calculate the enthalpy change for the reaction from enthalpies of formation

$$\Delta H^\circ_{rxn} = \Delta H^\circ_f[C_6H_{12}(g)] - \Delta H^\circ_f[C_6H_6(g)]$$

$$= 1 \text{ mol } (-123.1 \text{ kJ/mol}) - 1 \text{ mol } (+82.8 \text{ kJ/mol})$$

$$= -205.9 \text{ kJ}$$

(b) Calculate the enthalpy change for the reaction using bond energies.

$$\Delta H^\circ_{rxn} = 3\,D_{C=C} + 3\,D_{H\text{-}H} - 3\,D_{C\text{-}C} - 6\,D_{C\text{-}H}$$

$$= 3 \text{ mol } (611 \text{ kJ/mol}) + 3 \text{ mol } (436 \text{ kJ/mol}) - 3 \text{ mol } (347 \text{ kJ/mol}) - 6 \text{ mol } (414 \text{ kJ/mol})$$

$$= -384 \text{ kJ/mol}$$

(c) The smaller energy calculated in part (a) compared with that calculated from bond energies in part (b) suggests that resonance stabilizes a compound (by 178 kJ/mol in this case).

# CHAPTER 11

# FURTHER CONCEPTS OF CHEMICAL BONDING

## INSTRUCTOR'S NOTES

About three or four lectures are scheduled for this chapter. The emphasis is on valence bond theory, with only about one and a half lectures on molecular orbital theory. Metallic bonding is mentioned only briefly.

The following computer programs may be used for this material.

- Chem Draw®, a drawing program for the Macintosh® was used to draw the figures in this chapter and in the previous one. The program is produced by Cambridge Scientific Computing, Inc., 875 Massachusetts Avenue, Suite 41, Cambridge, MA 02139.
- Molecular Editor for the Macintosh® can be used to construct scale models of molecules as ball-and stick models, wire-frame, or space filling models. It is available from Intellimation, Department XA, 130 Cremona Drive, PO Box 1922, Santa Barbara, CA 93116-1922.
- Alchemy II is available for the IBM-compatible computers and the Macintosh. This is a very sophisticated molecular modeling program with a reasonable academic price. It is available from TRIPOS Associates, 1699 S. Hanley Road, Suite 303, St. Louis, MO 63144.

## LIST OF TRANSPARENCIES FOR THIS CHAPTER

| No. | Figure | Page | Subject |
|---|---|---|---|
| 42 | 11.3 | 415 | Sigma bond formation |
| 43 | 11.5 | 417 | Hybrid orbital formation for $sp^3$ |
| 44 | 11.6 | 419 | Formation of $sp$ and $sp^2$ hybrid orbitals |
| 45 | 11.7 | 423 | Geometry of hybrid orbital sets |

| | | | |
|---|---|---|---|
| 46 | 11.8 | 425 | Formation of a $\pi$ bond |
| 47 | 11.9 | 425 | Bonding in ethene, $C_2H_4$ |
| 48 | 11.10 | 427 | Bonding in acetylene, $C_2H_2$ |
| 49 | 11.12 | 430 | Formation of bonding and antibonding molecular orbitals |
| 50 | 11.13 | 432 | Molecular orbital energy level diagram for $H_2$ |
| 51 | 11.15 | 433 | Molecular orbital energy level diagram for combination of two atoms with 1*s* and 2*s* atomic orbitals. |
| 52 | 11.16 | 435 | Bonding and antibonding molecular orbitals that come from *s* and *p* orbital overlap. |
| 53 | 11.17 | 436 | Molecular orbital energy level diagram for homonuclear diatomic molecule. |
| 54 | Table 11.2 | 437 | Molecular orbital occupations and data for homonuclear diatomic molecules. |
| 55 | 11.18 | 439 | Bands of molecular orbitals in a metal |
| 56 | 11.20 | 442 | Diamond structure |

## SUGGESTED DEMONSTRATIONS

In these lectures it is most useful to show models of molecules with the hybrid orbitals in place. Suitable models can be obtained from Aldrich Chemical Company, PO Box 2060, Milwaukee, WI 53201.

# SOLUTIONS TO SELECTED STUDY QUESTIONS

## HYBRID ORBITALS

11-16. The electron dot structure for $OF_2$, a 20-electron molecule, is

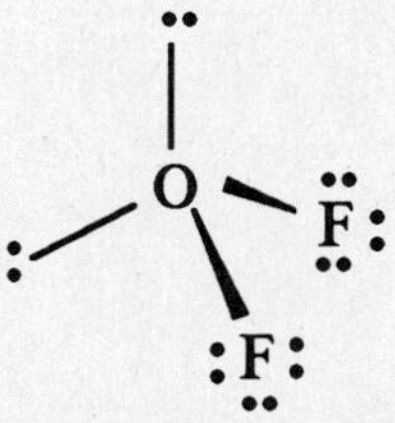


The structural pair geometry is tetrahedral, and the molecule geometry is bent. The O atom is $sp^3$ hybridized. Two of these hybrid orbitals are assigned to lone pairs of electrons, and the other two are each assigned a single electron. Each of these latter orbitals overlaps a fluorine 2p orbital, which has been assigned a single electron.

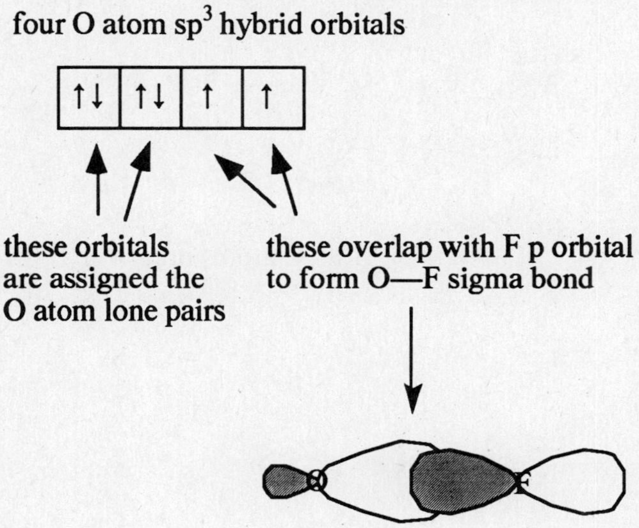


11-18. The electron dot structure of $BF_4^-$ is illustrated below. (The three lone pairs of electrons on each F atom are not included.)

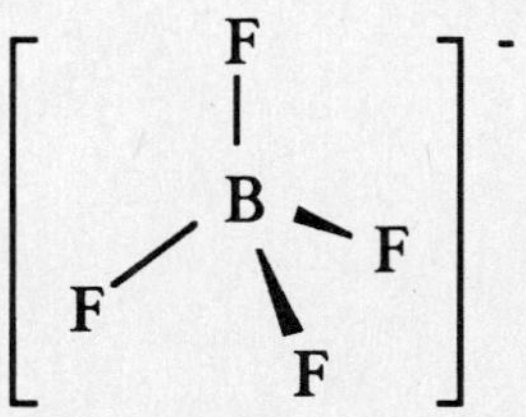


The B atom is $sp^3$ hybridized, with one electron assigned to each hybrid orbital (the three

valence electrons of B plus the electron that accounts for the negative charge of the ion).

four B atom $sp^3$ hybrid orbitals

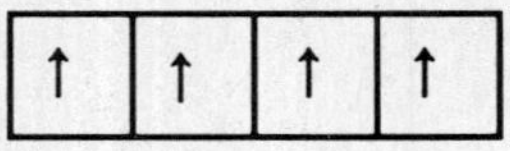

Each of these orbitals forms a bond with an F atom as in $OF_2$ above.

11-20.

| **Molecule** | **Hybrid Orbital Set** |
|---|---|
| $CS_2$ | C atom is sp hybridized |
| $NO_2^-$ | N atom is $sp^2$ hybridized |
| $NO_3^-$ | N atom is $sp^2$ hybridized |
| $BH_4^-$ | B atom is $sp^3$ hybridized |

11-22.

| **Question** | **Underlined atom** | **Hybrid Orbital Set** |
|---|---|---|
| (a) | N | $sp^3$ |
| | C | $sp^2$ |
| (b) | C | Both C atoms are $sp^2$ hybridized |
| (c) | C of C=C | $sp^2$ |
| | C of the CN group | sp |

11-24. The electron dot structure for the $ICl_4^-$ ion is

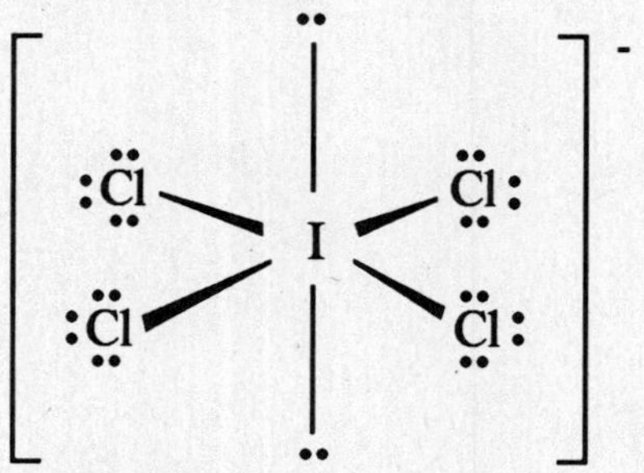

Structural pair geometry = octahedral

Molecular geometry = square planar

The I atom must be $sp^3d^2$ hybridized, and the six orbitals accommodate the seven valence electrons of I plus the electron accounting for the negative charge.

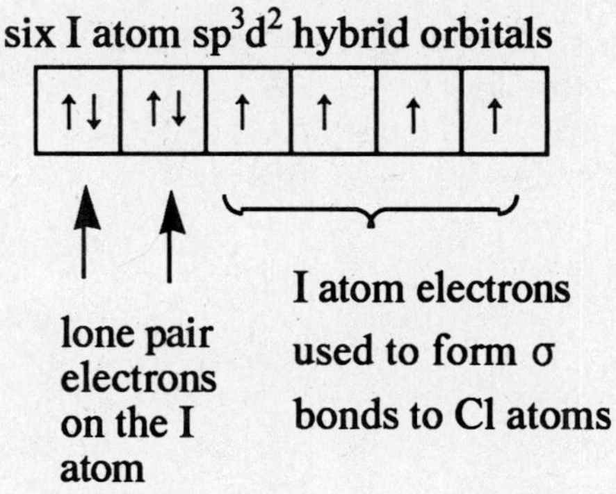


The I—Cl σ bond is formed by overlap of an I atom hybrid orbital and a Cl atom p orbital, each orbital containing a single unpaired electron. See $OF_2$ above.

11-26.

| Molecule Or Ion | Structural Pair Geometry About Underlined Atom | Hybrid Orbital Set For Underlined Atom |
|---|---|---|
| $OXeF_4$ | octahedral (see p. 399) | $sp^3d^2$ |
| $ClF_3$ | trigonal bipyramid (Table 10.8) | $sp^3d$ |
| $Br_3^-$ | trigonal bipyramid | $sp^3d$ |

11-28. The bonding in the $NCO^-$ ion can be explained by noting that the ion is linear with C=O and C=N double bonds.

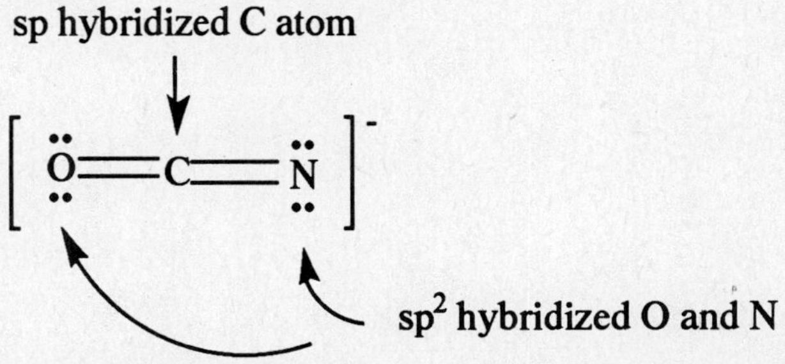


This tells us the C atom is sp hybridized, with one electron assigned to each of the two hybrid orbitals. The other two C atom valence electrons are assigned, one each, to unhybridized C atom p orbitals.

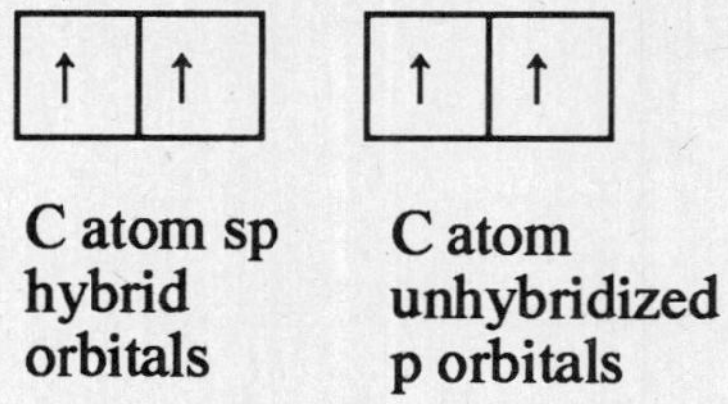


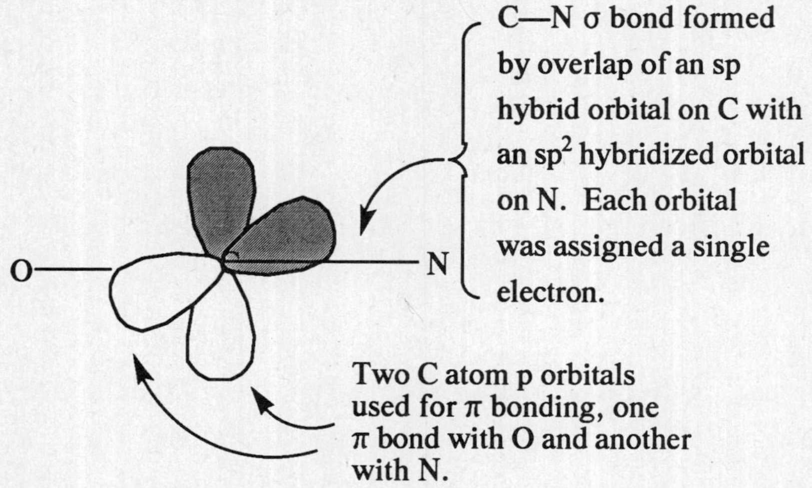


The O and N are both $sp^2$ hybridized. One of the $sp^2$ hybrid orbitals is used to bond to C, and the remaining two $sp^2$ hybrid orbitals are assigned lone pairs. The unhybridized p orbital on N (or O) is used to form the $\pi$ bonds with the central C atom.

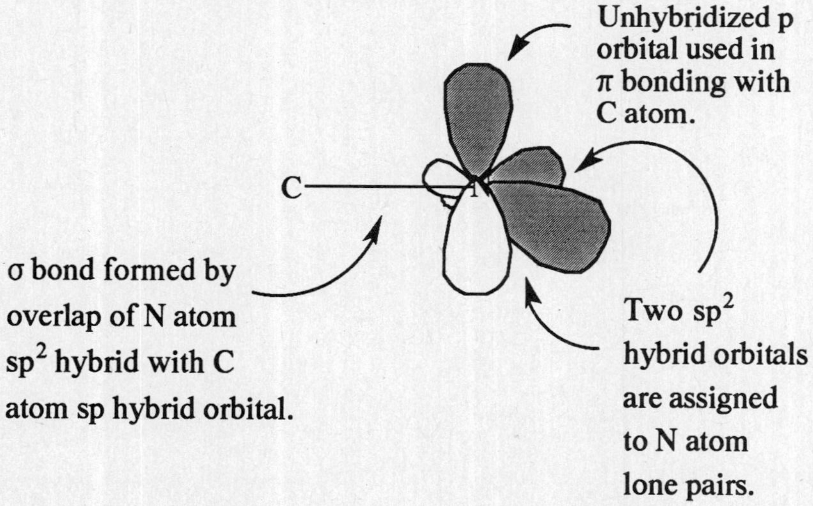


11-30. The dot structure for $SO_2$ is given on page 396, and the bonding description for the molecule is identical with that for $O_3$ on pages 428-429.

## MOLECULAR ORBITAL THEORY

11-32. $Li_2^+$: $(\sigma_{1s})^2(\sigma^*_{1s})^2(\sigma_{2s})^1(\sigma^*_{2s})^0$ Bond order = ½

$Li_2^-$: $(\sigma_{1s})^2(\sigma^*_{1s})^2(\sigma_{2s})^2(\sigma^*_{2s})^1$ Bond order = ½

The bond order of $Li_2$ ( = 1) is greater than that of either of its ions.

11-34. $O_2$, oxygen molecule

[core electrons] $(\sigma_{2s})^2(\sigma^*_{2s})^2(\pi_{2p})^4(\sigma_{2p})^2(\pi^*_{2p})^2$

$O_2^-$, superoxide ion

[core electrons] $(\sigma_{2s})^2(\sigma^*_{2s})^2(\pi_{2p})^4(\sigma_{2p})^2(\pi^*_{2p})^3$

$O_2^{2-}$: peroxide ion

[core electrons] $(\sigma_{2s})^2(\sigma^*_{2s})^2(\pi_{2p})^4(\sigma_{2p})^2(\pi^*_{2p})^4$

| Property | $O_2$ | $O_2^-$ | $O_2^{2-}$ |
|---|---|---|---|
| Magnetism | paramagnetic | paramagnetic | diamagnetic |
| Net σ bonds | 1 | 1 | 1 |
| Net π bonds | 1 | ½ | 0 |
| Bond order | 2 | 1.5 | 1 |
| Bond length | short | medium | long |
| | 121 pm | 132 pm | 149 pm |

11-36. (a) The $NO^+$ ion has an even number of valence electrons (10 electrons) and so is predicted to be diamagnetic.

(b) Assuming that $NO^+$ behaves like a homonuclear, the configuration for the ion is [core electrons] $(\sigma_{2s})^2(\sigma^*_{2s})^2(\pi_{2p})^4(\sigma_{2p})^2(\pi^*_{2p})^0$, a configuration that confirms that the ion is diamagnetic. The highest energy orbital assigned electrons is a σ bonding mo that arises from 2p overlap.

(c) The N—O bond order is 3, a bond order also suggested by the dot structure.

(d) The bond order of the NO molecule is 2.5, whereas that of $NO^+$ is 3. Therefore, ionizing the molecule increases the bond order because the electron is removed from an antibonding molecular orbital.

## GENERAL QUESTIONS

11-38. $ClF_2^+$ $ClF_2^-$

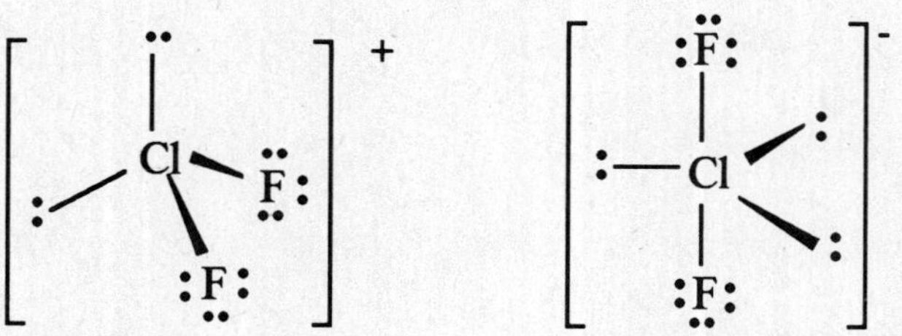

| | $ClF_2^+$ | $ClF_2^-$ |
|---|---|---|
| Structural-pair geometry | tetrahedral | trigonal bipyramidal |
| Molecular geometry | bent | linear |
| Cl hybrid orbitals | $sp^3$ | $sp^3d$ |

11-40. There are three resonance structures for the $SO_3$ molecule.

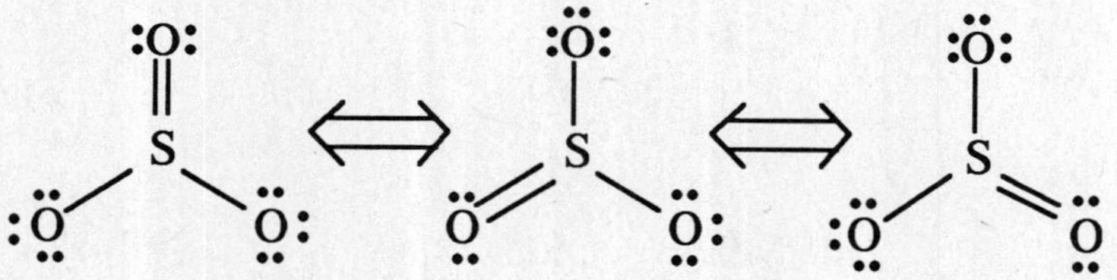

The trigonal planar structural pair geometry implies that S is $sp^2$ hybridized. This means there is an unhybridized p orbital on the S atom to be used for $\pi$ bond formation with a p orbital on an O atom.

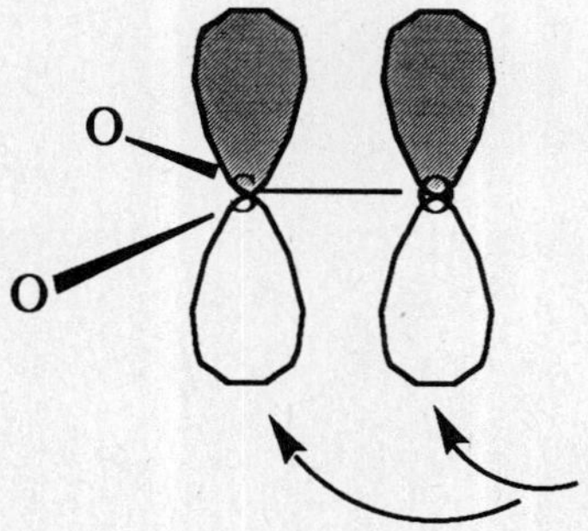

11-42. Hybridizations: Both the designated C atoms are $sp^2$ hybridized.
All three angles are approximately 120°.

11-44. (a) There are 3 $\pi$ bonds and 20 $\sigma$ bonds.

(b) Atom 1 = $sp^3$, atom 2 = $sp^3$, atom 3 = sp, and atom 4 = $sp^3$.

(c) The C≡N triple bond is much shorter than the C—N single bonds.

(d) Angle A = 109°, angle B = 180°, and angle C = 109°

11-46. $BF_3$ has a trigonal planar structural pair geometry and the B atom is $sp^2$ hybridized. In the complex $H_3N$—$BF_3$ the structural pair geometry is tetrahedral, and the B atom is $sp^3$ hybridized.

11-48. $N_2^+$ ion: [core electrons] $(\sigma_{2s})^2(\sigma^*_{2s})^2(\pi_{2p})^4(\sigma_{2p})^1(\pi^*_{2p})^0$

$N_2^-$ ion: [core electrons] $(\sigma_{2s})^2(\sigma^*_{2s})^2(\pi_{2p})^4(\sigma_{2p})^2(\pi^*_{2p})^1$

| **Property** | $N_2^+$ | $N_2^-$ |
|---|---|---|
| Magnetism | paramagnetic | paramagnetic |
| Net $\sigma$ bonds | $\frac{1}{2}$ | 1 |
| Net $\pi$ bonds | 2 | 1.5 |
| Bond order | 2.5 | 2.5 |
| Bond length | These ions should have similar bond lengths; both should be greater than that of N≡N. | |
| Bond strength | These ions should have similar bond strengths; both should be less than that of N≡N. | |

11-50. (a) $SbF_5$, antimony pentafluoride, has 40 valence electrons and so has the dot structure below (where the three lone pairs on each F atom are not shown). Since the structural

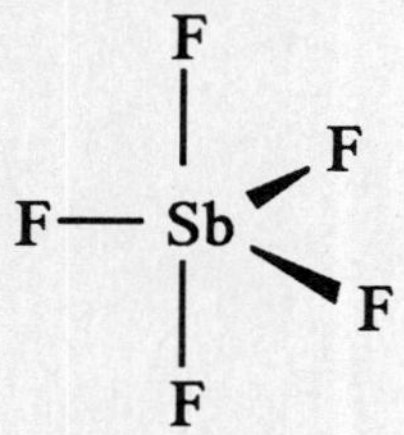

SbF5

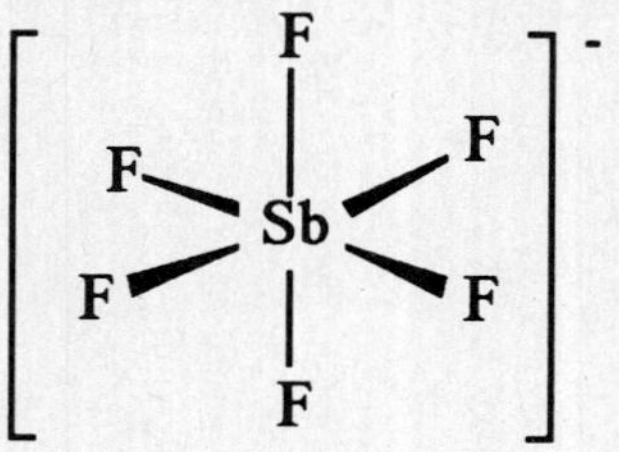

$SbF_6^-$

pair geometry is trigonal bipyramidal, the Sb atom hybridization is $sp^3d$. In $SbF_6^-$, which has 48 valence electrons, the octahedral structural pair geometry implies that Sb has $sp^3d^2$ hybridization.

(b) The $H_2F^+$ ion has 8 valence electrons (like water), and so it has a tetrahedral structural pair geometry with an $sp^3$ hybridized F atom.

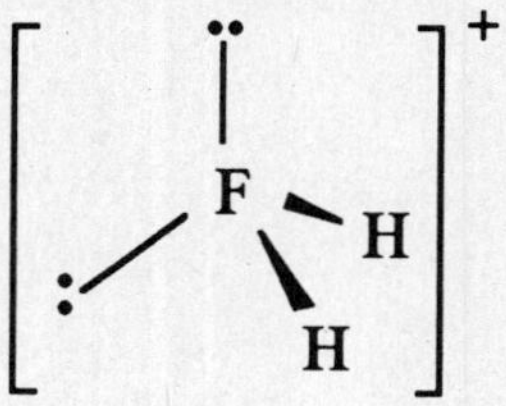

11-52.

| | $XeO_3$ | $XeO_4$ |
|---|---|---|
| | 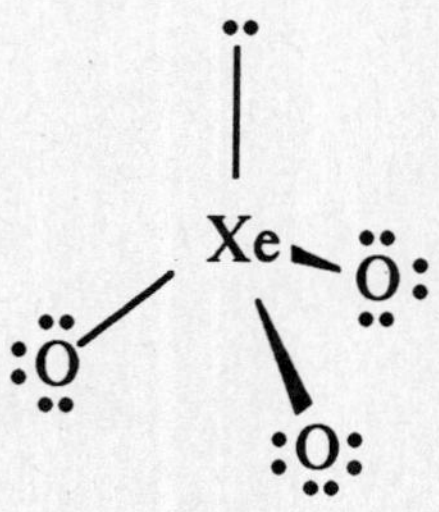 | 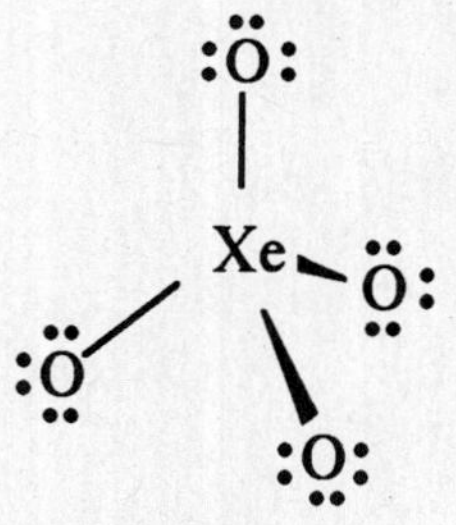 |
| Structural-pair geometry | tetrahedral | tetrahedral |
| Molecular geometry | tetrahedral | tetrahedral |
| Xe hybrid orbitals | $sp^3$ | $sp^3$ |

11-54.

| Molecule or Ion | Number of Valence Electrons | Magnetic Behavior |
|---|---|---|
| NO | 5 + 6 = 11 | paramagnetic |
| $OF^-$ | 6 + 7 + 1 = 14 | diamagnetic |
| $O_2^{2-}$ | (2 x 6) + 2 = 14 | diamagnetic (see question 34 above) |
| $Ne_2^+$ | (2 x 8) - 1 = 15 | paramagnetic |
| CN | 4 + 5 = 9 | paramagnetic |

11-56. $SO_3$ reacts with $NH_2^-$ by forming a bond between S and N where one of the lone pairs of electrons of $NH_2^-$ attacks the unhybridized p orbital of the S atom. Both S and N have a tetrahedral structural pair geometry in the product, so both a $sp^3$ hybridized and both have bond angles of 109°.

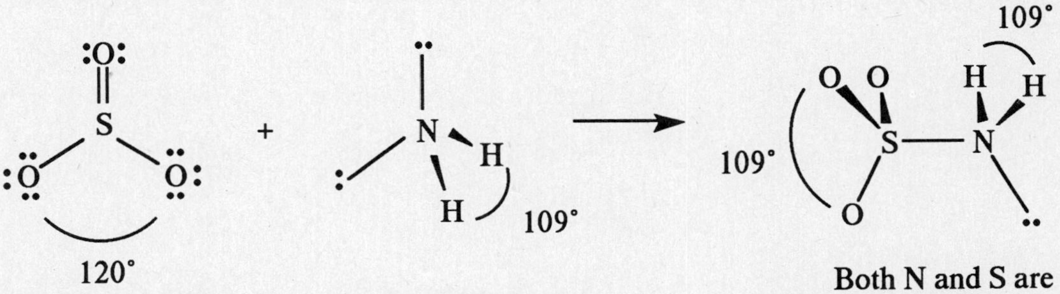


$sp^2$ hybridized S

$sp^3$ hybridized N

Both N and S are $sp^3$ hybridized.

11-58. The CN molecule has (4 + 5) 9 valence electrons. This means its molecular orbital electron configuration is the same as that of $N_2^+$ in question 48:

$$[\text{core electrons}]\ (\sigma_{2s})^2(\sigma^*_{2s})^2(\pi_{2p})^4(\sigma_{2p})^1(\pi^*_{2p})^0$$

Therefore, the highest energy mo (called a HOMO) to which an electron is assigned is the $\sigma_{2p}$ orbital. The bond order is 2.5, since there are 2 net $\pi$ bonds and $\frac{1}{2}$ a $\sigma$ bond. Like $N_2^+$, the CN molecule is paramagnetic.

## SUMMARY QUESTION

11-60. (a) The *enol*→ *keto* transformation

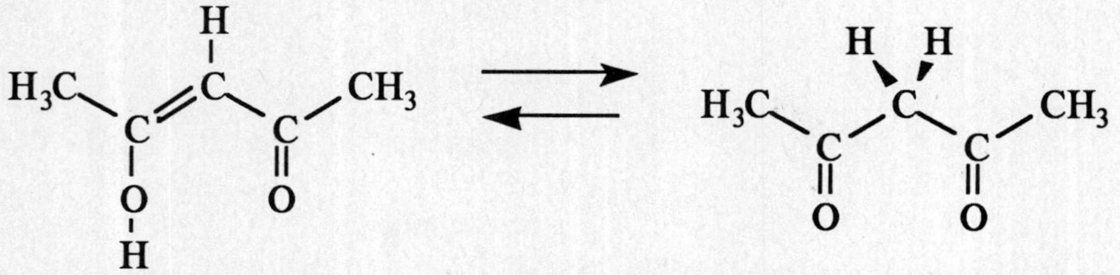


occurs by breaking a C—O single bond, an O—H bond, and a C=C double bond and by then making a C=O double bond, a C—C single bond, and a C—H bond. Therefore, we can estimate the reaction enthalpy change using only these bond energies:

$$\Delta H = D_{C\text{-}O} + D_{O\text{-}H} + D_{C=C} - D_{C=O} - D_{C\text{-}C} - D_{C\text{-}H}$$

$$= 351\text{ kJ} + 464\text{ kJ} + 611\text{ kJ} - 745\text{ kJ} - 347\text{ kJ} - 414\text{ kJ}$$

$$= -80\text{ kJ}$$

This result indicates the reaction is expected to be exothermic.

(b) Hybridization in the *enol* form

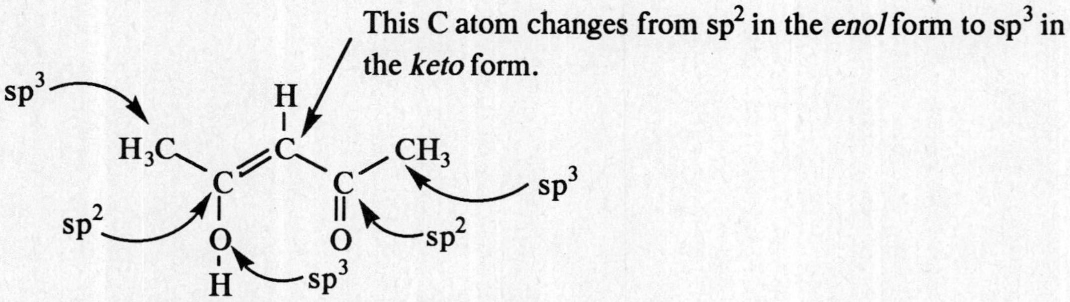


(c) C atom geometries in *enol* and *keto* forms.

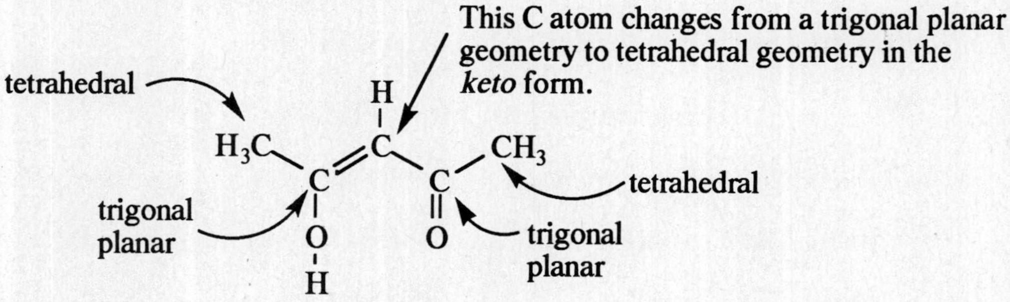

(d) 15.0 g Cr compd • 1 mol/349.33 g = $4.29 \times 10^{-2}$ mol Cr compd

$4.29 \times 10^{-2}$ mol Cr compd • 1 mol $CrCl_3$/1 mol Cr compd • 158.36 g/mol

= 6.80 g $CrCl_3$ required

$4.29 \times 10^{-2}$ mol Cr compd • 3 mol $H_3CC(OH)CHCOCH_3$/1 mol Cr compd

• 100.12 g/mol = 12.9 g $H_3CC(OH)CHCOCH_3$ required

$4.29 \times 10^{-2}$ mol Cr compd • 3 mol NaOH • 40.0 g/mol = 5.15 g NaOH required

# PART III

# STATES OF MATTER

# CHAPTER 12
# GASES AND THEIR BEHAVIOR

## INSTRUCTOR'S NOTES

In the first edition this chapter was placed in the first quarter of the book. However, it now seems more logical to place it somewhat later in the year, since we can then make a very nice connection between the phases of matter by covering the chapters on gases, liquids and solids, and solutions in that order. When discussing nonideal gases, we make the point that intermolecular forces play an important role, so it is a natural transition to move from there to the discussion of liquids and solids.

In spite of the fact that the chapter has been moved, it is important to emphasize that *this chapter can be used almost anywhere in the sequence of topics, as long as the basic ideas of stoichiometry have been covered.*

Three lectures to four lectures are normally given on this material. At the beginning of this series we take the opportunity to emphasize two reasons for studying gases and their behavior. (i) There are gases in the atmosphere and many of these and others are commercially important. We must understand how to deal with them. (ii) The behavior of gases is well understood and can be modeled mathematically. This has two benefits: (a) This has led to a better understanding of other aspects of the physical world (such as the energy distribution in liquids). (b) It is important to recognize that nature and some aspects of social behavior can be subjected to a mathematical models, an approach that is important in business and industry today.

Another very useful point to make is that one can clearly see the difference between a law (the ideal gas law and the laws on which it is based) and a theory (the kinetic molecular theory).

## LIST OF TRANSPARENCIES FOR THIS CHAPTER

| No. | Figure | Page | Subject |
|---|---|---|---|
| 57 | 12.1 | 457 | Barometer |
| 58 | 12.7 | 465 | Summary of gas laws |
| 59 | --- | 471 | Scheme for stoichiometry calculations involving gases |
| 60 | 12.12 | 478 | Collecting a gas over water. |
| 61 | 12.14 | 482 | Boltzmann distribution of molecular speeds for gases at two different temperatures. |
| 62 | 12.15 | 483 | Effect of molecular mass on Boltzmann distribution curve. |
| 63 | 12.17 | 484 | Effusion of gases |

## SUGGESTED DEMONSTRATIONS

Volume 2 of Shakhashiri's *Chemical Demonstrations* lists 26 demonstrations on "The Physical Behavior of Gases."

Boyle's law has been demonstrated in Figure 12.2 using a large syringe. Instead of lead shot in a beaker, one can also use a pile of textbooks to provide the pressure as described in Experiment 5.5 of volume 2 of *Chemical Demonstrations.* (For more on gas law demonstrations using syringes, see D. Davenport, *Journal of Chemical Education,* **1962**, *39*, 252.)

Yet another demonstration, popular with our students in illustrated in Figure 12.3. Simply place a large number of marshmallows in a large filter flask and evacuate the flask with a vacuum pump. The marshmallows swell in size appreciably, but the effect does not last long.

Finally, we have also illustrated Boyle's law using a standard bicycle pump, an object with which students are more familiar.

A favorite demonstration with the students in shown in Figure 12.4 (page 461). Perhaps 15-20 balloons are passed out to students at the beginning of the lecture, and they are asked to blow them up and tie them off. When the demonstration begins, each of these students comes to the front of the room and places the balloon he or she has blown up in a large beaker of liquid nitrogen. After all the balloons have been placed in the beaker, the beaker is turned over, and the balloons slowly reinflate as they warm again to room temperature. (However, a few may break, as they become twisted when frozen.)

Figure 12.8 illustrates the relative densities of two gases and is an easy demonstration. This can also be used to illustrate diffusion, since He atoms diffuse from a balloon much faster than Ar or air ($O_2$ and $N_2$).

Effusion and diffusion can be illustrated in several ways. One method we have used is to release a *small* quantity of a concentrated scent (rose, lavender, or lemon oil) at the beginning of the lecture. The odor is soon noticed throughout the lecture room.

The following is a list of some demonstrations obtained through the Project SERAPHIM database (see page 1 of this *Manual*):


- Carney, G.D.; Kern, C.W. "Balloon Balance Thermometer. A Lecture Demonstration of Charles's Law" *J. Chem. Educ.* **1979**, *56*, 823 .
- Davenport, D.L "Boyle's Law" *ibid.* **1979**, *56*, 322.
- Moeller, M.B. "An Exercise with Boyle's Law" *ibid.* **1978**, *55*, 584.
- Murdock, H.D.; Hawthorne, R.M. Jr. "A Charles's Law Experiment using a Balloon or a Plastic Bag" *ibid.* **1973**, *50*, 528.
- Zaborowski, L.M. "Determination of the Molar Volume of a Gas at Standard Temperature and Pressure" *ibid.* **1972**, *49*, 361.

# SOLUTIONS TO SELECTED STUDY QUESTIONS

## UNITS OF MEASUREMENT

12-12. (a) 781 mmHg • 1 atm/760 mmHg = 1.03 atm

(b) 356 mmHg • 1 atm/760 mmHg • 101.325 kPa/atm = 47.5 kPa

(c) 35 torr • 1 atm/760 torr = 0.046 atm

(d) 210 kPa • 1 atm/101.325 kPa • 760 mmHg/atm = 1600 mmHg

(e) 0.015 kPa • 1 atm/101.325 kPa = $1.5 \times 10^{-4}$ atm

(f) 95.0 kPa • 1 atm/101.325 kPa • 760 mmHg/atm = 713 mmHg

## PRESSURE MEASUREMENT

12-14. 713 mmHg

## THE GAS LAWS

12-16. $V_2 = P_1V_1/P_2$ = (67.5 mmHg • 256 mL) / 23.6 mmHg = 732 mL

12-18. $P_2 = P_1(V_1/V_2)$ = 735 mmHg • (3.50 L/15.0 L) = 172 mmHg

12-20. $P_2 = P_1(T_2/T_1)$ = (55 lb/in$^2$) • (308 K/288 K) = 59 lb/in$^2$

12-22. $SiH_4(g) + 2\,O_2(g) \rightarrow SiO_2(s) + 2\,H_2O(g)$

5.2 L    10.4 L    10.4 L

## THE GENERAL GAS LAW

12-24. $P_1$ = 165 mmHg    $P_2$ = 265 mmHg

$V_1$ = 754 mL    $V_2$ = ?

$T_1$ = 295 K    $T_2$ = 315 K

$V_2 = V_1 \cdot (T_2/T_1) \cdot (P_1/P_2)$

= 754 mL • (315 K/295 K) • (165 mmHg/265 mmHg) = 501 mL

12-26. $P_1$ = 136.5 mmHg $P_2$ = 94.3 mmHg

$V_1$ = 265 mL $V_2$ = ?

$T_1$ = 295.7 K $T_2$ = 297.7 K

$V_2 = V_1 \cdot (T_2/T_1) \cdot (P_1/P_2)$

= 265 mL • (297.7 K/295.7 K) • (136.5 mmHg/94.3 mmHg) = 386 mL

12-28. $P_1$ = 745 mmHg $P_2$ = 600. mmHg

$V_1$ = 1.05 x $10^5$ L $V_2$ = ?

$T_1$ = 293.2 K $T_2$ = 240. K

$V_2 = V_1 \cdot (T_2/T_1) \cdot (P_1/P_2)$

= (1.05 x $10^5$ L) • (240. K/293.2 K) • (745 mmHg/600. Hg) = 1.07 x $10^5$ L

Therefore, the balloon displaces the required volume of air.

## THE IDEAL GAS LAW

12-30. 30.0 x $10^3$ g He • (1 mol/4.003 g) = 7.50 x $10^3$ mol He

$$V = nRT/P = \frac{(7.50 \times 10^3 \text{ mol He})(0.082057 \text{ L} \cdot \text{atm/K} \cdot \text{mol})(295 \text{ K})}{1.20 \text{ atm}} = 1.51 \times 10^5 \text{ L}$$

12-32. 1.50 g • (1 mol/86.18 g) = 0.0174 mol

$$P = nRT/V = \frac{(0.0174 \text{ mol})(0.082057 \text{ L} \cdot \text{atm/K} \cdot \text{mol})(523 \text{ K})}{0.250 \text{ L}} = 2.99 \text{ atm}$$

12-34.

$$n = PV/RT = \frac{(1.1 \text{ atm})(5.0 \text{ L})}{(0.082057 \text{ L} \cdot \text{atm/K} \cdot \text{mol})(298 \text{ K})} = 0.23 \text{ mol He}$$

0.23 mol He • (4.003 g/mol) = 0.90 g He

12-36. 700. mmHg • 1 atm/760 mmHg = 0.921 atm

$$n = PV/RT = \frac{(0.921\ \text{atm})(0.450\ \text{L})}{(0.082057\ \text{L} \cdot \text{atm/K} \cdot \text{mol})(296\ \text{K})} = 0.0171\ \text{mol unknown}$$

Molar mass = 0.982 g / 0.0171 mol = 57.5 g/mol

12-38. *Step 1:* Determine the empirical formula of the compound.

The weight percentages of the elements in the compound are: 11.79 % C, 69.57 % Cl, and, by difference, 18.64 % F. Therefore, we first convert the percentages into moles of each element.

11.79 g C • 1 mol/12.011 g = 0.9816 mol C

69.57 g Cl • 1 mol/35.453 g = 1.962 mol Cl

18.64 g F • 1 mol/18.998 g = 0.9812 mol F

This leads to an empirical formula of $CCl_2F$.

*Step 2:* Calculate the molar mass of the compound.

21.3 mmHg • 1 atm/760 mmHg = 0.0280 atm

$$n = PV/RT = \frac{(0.0280\ \text{atm})(0.458\ \text{L})}{(0.082057\ \text{L} \cdot \text{atm/K} \cdot \text{mol})(298\ \text{K})} = 5.25 \times 10^{-4}\ \text{mol}$$

Molar mass = 0.107 g / $5.25 \times 10^{-4}$ mol = 204 g/mol

*Step 3:* Determine the molecular formula

204 g/mol • 1 unit/102 g = 2 units of $CCl_2F$ per mol

Therefore, the molecular formula is $C_2Cl_4F_2$

12-40. *Step 1:* Determine the empirical formula of the compound

The weight percentages of the elements in the compound are: 25.23 % S and, by difference, 74.77 % F. Therefore, we first convert the percentages into moles of each element.

25.23 g S • 1 mol/32.066 g = 0.7868 mol S

74.77 g F • 1 mol/18.998 g = 3.936 mol F

This leads to an empirical formula of $SF_5$.

*Step 2:* Calculate the molar mass of the compound.

83.8 mmHg • 1 atm/760 mmHg = 0.110 atm

$$n = PV/RT = \frac{(0.110\text{ atm})(0.089\text{ L})}{(0.082057\text{ L}\cdot\text{atm/K}\cdot\text{mol})(318\text{ K})} = 3.75 \times 10^{-4}\text{ mol}$$

Molar mass = 0.0955 g /3.75 x $10^{-4}$ mol = 255 g/mol

*Step 3:* Determine the molecular formula

255 g/mol • 1 unit/ 127 g = 2 units of $SF_5$ per mol

Therefore, the molecular formula is $S_2F_{10}$

12-42. This problem can be solved using equation 12.6 (page 469), or in the following steps:

*Step 1:* Calculate the volume occupied by 1.00 mol of ether under the specified conditions.

233 mmHg • 1 atm/760 mmHg = 0.307 atm

$$V = nRT/P = \frac{(1.00\text{ mol})(0.082057\text{ L}\cdot\text{atm/K}\cdot\text{mol})(298\text{ K})}{0.307\text{ atm}} = 79.7\text{ L}$$

*Step 2:* Determine the density. 1.00 mole of the ether has a mass of 74.12 g. Therefore,

$$\text{Density} = \frac{74.12\text{ g}}{79.7\text{ L}} = 0.931\text{ g/L}$$

12-44. This problem can be solved using equation 12.6 (page 469), or in the following steps:

*Step 1:* Find the number of moles of gas that occupy a volume of 1.00 L under the specified conditions.

715 mmHg • 1 atm/760 mmHg = 0.941 atm

$$n = PV/RT = \frac{(0.941\text{ atm})(1.00\text{ L})}{(0.082057\text{ L}\cdot\text{atm/K}\cdot\text{mol})(296\text{ K})} = 0.0387\text{ mol}$$

*Step 2:* Calculate the molar mass of the gas.

$$\text{Molar mass} = \frac{2.39\text{ g}}{0.0387\text{ mol}} = 61.7\text{ g/mol}$$

12-46. (a) This question is most easily answered if the information is set out in table form.

| | **Helium Balloon** | **Hydrogen Balloon** |
|---|---|---|
| Volume | V | 2V |
| Pressure | 2 atm | 1 atm |
| Temperature | 70 °F (21 °C) | 30 °F (-1.1 °C) |

The moles of either gas can be found from n = PV/RT. Let us find the *ratio* of the number of moles of each gas.

$$\frac{n(He)}{n(H_2)} = \frac{(2\text{ atm}) \cdot (V) / R \cdot 294\text{ K}}{(1\text{ atm}) \cdot 2V / R \cdot 272\text{ K}} = 272/294 = 0.93$$

Therefore, n(He) = 0.93 $n(H_2)$. With this result we see that the number of moles of He is *less* than the number of moles of $H_2$.

(b) Assume that the hydrogen balloon contains 1.00 mol of $H_2$, that is, 2.0 g of $H_2$. This must mean that the helium balloon contains only 0.93 mol of He, or 3.7 g of He. Therefore, while there are more moles of $H_2$ present, the *mass of helium present is greater than that of the $H_2$.*

12-48. (a) False. The molar masses of $O_2$ (32 g/mol) and $N_2$ (28 g/mol) are different. Therefore, equal masses of the two compounds represent different numbers of moles and therefore different numbers of molecules.

(b) True. Since the masses of $O_2$ and $N_2$ are the same, say 1 g each, this means there are more moles of the lighter compound, $N_2$, than of the heavier compound, $O_2$. Therefore, there are more molecules of $N_2$ than of $O_2$, and so the pressure of $N_2$ is greater than that of $O_2$.

(c) False. See (b) above.

12-50. (a) Since P = nRT/V, and since T and V are the same for both cylinders, the cylinder with the greater pressure will be the one with the greater number of moles of gas. Since each cylinder contains 1 kg of gas, the cylinder containing molecules having the smaller molar mass will contain a greater number of moles of gas and so have the larger pressure. In this case, the molar mass of acetylene ($C_2H_2$, 26 g/mol) is smaller than that of CO (28 g/mol), so the pressure in the acetylene cylinder is greater than that in the CO cylinder.

(b) Begin by defining the parameters for each gas.

$P_A$ = 10 atm $P_B$ = 9 atm

$V_A = V_B$ $V_B = V_A$

$T_A$ = 25 °C = 298 K $T_B$ = 0 °C = 273 K

$n_A$ = ? $n_B$ = ?

Now define the ratio of moles of gas.

$$\frac{n_A}{n_B} = \frac{P_AV_A/RT_A}{P_BV_B/RT_B} = \frac{10\text{ atm}/298\text{ K}}{9\text{ atm}/273\text{ K}} = 1.02 \text{ or 1 to one significant figure.}$$

Under the conditions specified, there are essentially the same number of moles of gas (and therefore the same number of molecules) in each cylinder.

## GAS LAWS AND STOICHIOMETRY

12-52. *Step 1:* Calculate moles of $B_4H_{10}$

$0.050\text{ g} \cdot 1\text{ mol}/53.3\text{ g} = 9.4 \times 10^{-4}\text{ mol}$

*Step 2:* Find moles of water after reaction.

$9.4 \times 10^{-4}\text{ mol } B_4H_{10} \cdot (10\text{ mol } H_2O/2\text{ mol } B_4H_{10}) = 4.7 \times 10^{-3}\text{ mol } H_2O$

*Step 3:* Calculate the pressure of water.

$$P = nRT/V = \frac{(4.7 \times 10^{-3}\text{ mol})(0.082057\text{ L} \cdot \text{atm/K} \cdot \text{mol})(303\text{ K})}{4.25\text{ L}} =$$

P = 0.027 atm (or 21 mmHg)

12-54. *Step 1:* From the data given, we can find the moles of butane that must have been present.

$2.45\text{ g } O_2 \cdot 1\text{ mol}/32.00\text{ g} = 0.0766\text{ mol } O_2$

$0.0766\text{ mol } O_2 \cdot 2\text{ mol } C_4H_{10} / 13\text{ mol } O_2 = 0.0118\text{ mol } C_4H_{10}$

*Step 2:* Find the moles of butane required under the new conditions using the general gas law.

$P_1$ = 1.5 atm $P_2$ = 2.0 atm

$T_1$ = 500 °C = 773 K $T_2$ = 773 K

$V_1 = V_2$ $V_2 = V_1$

$n_1$ = 0.0118 mol $C_4H_{10}$ $n_2$ = ?

$$P_1V_1/n_1T_1 = P_2V_2/n_2T_2$$

$$\frac{1.5 \text{ atm}}{0.0118 \text{ mol}} = \frac{2.0 \text{ atm}}{n_2}$$

$$n_2 = 0.016 \text{ mol}$$

*Step 3:* Find the number of grams of $O_2$

0.016 mol butane • 13 mol $O_2$/2 mol butane • 32.00 g/mol = 3.33 g $O_2$

12-56. *Step 1:* Write a balanced equation

$$Ni(s) + 4\ CO(g) \rightarrow Ni(CO)_4(\ell)$$

*Step 2:* Find the number of moles of CO present

$$n = PV/RT = \frac{(0.550 \text{ atm})(1.50 \text{ L})}{(0.082057 \text{ L} \cdot \text{atm/K} \cdot \text{mol})(298.2 \text{ K})} = 0.0337 \text{ mol CO}$$

*Step 3:* Find the number of moles of product that can be formed

0.0337 mol CO • 1 mol $Ni(CO)_4$ / 4 mol CO = 0.00843 mol $Ni(CO)_4$

*Step 4:* Find grams of $Ni(CO)_4$

0.00843 mol $Ni(CO)_4$ • 170.74 g/mol = 1.44 g $Ni(CO)_4$

12-58. *Step 1:* Calculate the number of moles of $CO_2$

$$n = PV/RT = \frac{(0.0918 \text{ atm})(0.285 \text{ L})}{(0.082057 \text{ L} \cdot \text{atm/K} \cdot \text{mol})(298 \text{ K})} = 0.00107 \text{ mol CO}$$

*Step 2:* Find moles of $MCO_3$

Since M is a Group 2A metal, the metal carbonate must have the formula $MCO_3$. Therefore, there must have been 1 mole of $MCO_3$ present for each mole of $CO_2$ generated. This means there are 0.00107 moles of $MCO_3$.

*Step 3:* Find the molar mass of $MCO_3$

0.158 g / 0.00107 mol = 148 g/mol

*Step 4:* Find the molar mass of M

148 g/mol = M g/mol + 60. g/mol for $CO_3$

M = 88 g/mol

Therefore, M is Sr, strontium.

12-60. *Step 1:* Calculate the number of moles of $CH_4$

$$n = PV/RT = \frac{(0.196 \text{ atm})(0.250 \text{ L})}{(0.082057 \text{ L} \cdot \text{atm/K} \cdot \text{mol})(298 \text{ K})} = 0.00200 \text{ mol } CH_4$$

*Step 2:* Find the mass of $CH_3$. Each mole of $CH_4$ evolved originated as a $CH_3$ group in the unknown compound. Therefore, we need to find the grams of $CH_3$ so that we can eventually relate moles of $CH_3$ to moles of AuBr.

0.00200 mol $CH_4$ • 1 mol $CH_3$ / 1 mol $CH_4$ • 15.04 g $CH_3$ / mol = 0.0301 g $CH_3$

*Step 3:* Find the mass of AuBr in the sample

0.307 g $Au(CH_3)_xBr$ - 0.0301 g $CH_3$ = 0.277 g AuBr

*Step 4:* Find the moles of AuBr

0.277 g AuBr • 1 mol/ 276.9 = 0.00100 mol AuBr

*Step 5:* Determine the formula of the compound.

Our calculations indicate that there are 0.00200 mole of $CH_3$ per 0.00100 mole of AuBr. Therefore, the formula must be $Au(CH_3)_2Br$.

12-62. *Step 1:* Calculate moles of each reactant

Moles of Si = 1.95 g Si • 1 mol/28.09 g = 0.0694 mol Si

$$\text{Moles } CH_3Cl = n = PV/RT = \frac{(0.638 \text{ atm})(6.56 \text{ L})}{(0.082057 \text{ L} \cdot \text{atm/K} \cdot \text{mol})(298 \text{ K})}$$

Moles $CH_3Cl$ = 0.171 mol

*Step 2:* Determine the limiting reagent

0.171 mol $CH_3Cl$ / 0.0694 mol Si = 2.47 mol $CH_3Cl$ / 1 mol Si

The reaction requires 2 moles of $CH_3Cl$ per mole of Si. However, there is clearly an excess of $CH_3Cl$. The limiting reagent is Si.

*Step 3:* Determine the number of grams of product.

0.0694 mol Si • 1 mol $(CH_3)_2SiCl_2$ / 1 mol Si • 129.1 g/mol

= 8.96 g $(CH_3)_2SiCl_2$

*Step 4:* Determine the pressure of the product, $CH_3)_2SiCl_2$.

$$P = nRT/V = \frac{(0.0694 \text{ mol})(0.082057 \text{ L} \cdot \text{atm/K} \cdot \text{mol})(368 \text{ K})}{6.56 \text{ L}} = 0.319 \text{ atm}$$

## GAS MIXTURES

12-64. *Step 1:* Calculate the partial pressure of $O_2$

0.0870 g $O_2$ • 1 mol/32.00 g $O_2$ = 0.00272 mol $O_2$

$$P = nRT/V = \frac{(0.00272 \text{ mol})(0.082057 \text{ L} \cdot \text{atm/K} \cdot \text{mol})(297 \text{ K})}{0.550 \text{ L}} = 0.120 \text{ atm}$$

*Step 2:* Calculate the partial pressure of CO

When $CO_2$ is removed, $P_{total}$ = 1.34 atm = P(CO) + P($O_2$). Since P($O_2$) = 0.120 atm, this means that P(CO) = 1.22 atm.

*Step 3:* Calculate the partial pressure of $CO_2$

Total pressure = 1.56 atm = P(CO) + P($CO_2$) + P($O_2$)

1.56 atm = (1.22 atm) + P($CO_2$) + (0.120 atm) and P($CO_2$) = 0.22 atm

12-66. (a) *Grams of He*

$$n = PV/RT = \frac{(1.0 \text{ atm})(10. \text{ L})}{(0.082057 \text{ L} \cdot \text{atm/K} \cdot \text{mol})(293 \text{ K})} = 0.42 \text{ mol He}$$

0.42 mol He • 4.003 g/mol = 1.7 g

(b) Final *partial pressure of He* (after adding $O_2$)

$$P = nRT/V = \frac{(0.42 \text{ mol})(0.082057 \text{ L} \cdot \text{atm/K} \cdot \text{mol})(293 \text{ K})}{30. \text{ L}} = 0.33 \text{ atm}$$

(c) *Partial pressure of* $O_2$

$P_{total} = P(He) + P(O_2) = 1.0$ atm

Since P(He) = 0.33 atm, this means $P(O_2) = 0.7$ atm

(d) *Mole fraction of each gas*

To find the mole fraction, we first need the number of moles of $O_2$.

$$n = PV/RT = \frac{(0.7 \text{ atm})(30. \text{ L})}{(0.082057 \text{ L} \cdot \text{atm/K} \cdot \text{mol})(293 \text{ K})} = 0.8 \text{ mol } O_2$$

$$X_{He} = \frac{0.42 \text{ mol He}}{0.42 \text{ mol He} + 0.8 \text{ mol } O_2} = 0.3$$

Therefore, mole fraction of $O_2$ = 0.7

12-68. The mole ratio of $SO_2$ to $Cl_2$ must be 1 to 2. Therefore, the ratios of pressures of $SO_2$ to $Cl_2$ must also be 1 to 2. Since the pressure of $SO_2$ is 125 mmHg, the pressure of $Cl_2$ must be 250. mmHg.

12-70. *Step 1:* Calculate moles of $SO_2$

0.235 g $SO_2$ • 1 mol/64.06 g = $3.67 \times 10^{-3}$ mol $SO_2$

*Step 2:* Calculate moles of product

$3.67 \times 10^{-3}$ mol $SO_2$ • 2 mol product/1 mol $SO_2$ = $7.34 \times 10^{-3}$ mol product

*Step 3:* Total pressure of products

$$P = nRT/V = \frac{(7.34 \times 10^{-3}\ \text{mol})(0.082057\ \text{L} \cdot \text{atm/K} \cdot \text{mol})(358\ \text{K})}{0.250\ \text{L}} = 0.862\ \text{atm}$$

12-72. *Step 1:* Calculate moles of $(NH_4)_2Cr_2O_7$

$0.50$ g $(NH_4)_2Cr_2O_7 \cdot 1$ mol/252 g $= 0.0020$ mol $(NH_4)_2Cr_2O_7$

*Step 2:* Calculate moles of each product

$0.020$ mol $(NH_4)_2Cr_2O_7 \cdot 1$ mol $N_2 / 1$ mol $(NH_4)_2Cr_2O_7 = 0.0020$ mol $N_2$

$0.020$ mol $(NH_4)_2Cr_2O_7 \cdot 4$ mol $H_2O / 1$ mol $(NH_4)_2Cr_2O_7 = 0.0080$ mol $H_2O$

*Step 3:* Calculate the partial pressure of each product

$$P(N_2) = nRT/V = \frac{(0.0020\ \text{mol})(0.082057\ \text{L} \cdot \text{atm/K} \cdot \text{mol})(298\ \text{K})}{6.0\ \text{L}} = 0.0081\ \text{atm}$$

Since there are 4 times as many moles of $H_2O$ as there are of $N_2$, this means the partial pressure of $H_2O$ is four times that of $N_2$. Therefore, $P(N_2) = 4 \cdot 0.0081$ atm $= 0.032$ atm.

*Step 4:* Calculate the total pressure of products

$P_{total} = P(N_2) + P(H_2O) = 0.0081\ \text{atm} + 0.32\ \text{atm} = 0.040\ \text{atm}$

12-74. *Step 1:* Calculate pressure of dry $O_2$

$P_{total} = P(O_2) + P(H_2O)$

$735\ \text{mmHg} = P(O_2) + 16.5\ \text{mmHg}$ (where the vapor pressure of water is taken from Appendix D)

$P(O_2) = 719$ mmHg or $0.946$ atm

*Step 2:* Calculate moles of $O_2$

$$n = PV/RT = \frac{(0.946\ \text{atm})(0.327\ \text{L})}{(0.082057\ \text{L} \cdot \text{atm/K} \cdot \text{mol})(292\ \text{K})} = 0.0129\ \text{mol}\ O_2$$

*Step 3:* Find number of moles of $KClO_3$ and then grams of $KClO_3$

$0.0129$ mol $O_2 \cdot 2$ mol $KClO_3 / 3$ mol $O_2 = 0.00861$ mol $KClO_3$

$0.00861 \text{ mol } KClO_3 \cdot 122.5 \text{ g/mol} = 1.05 \text{ g } KClO_3$

*Step 4:* Calculate weight percent of $KClO_3$

$(1.05 \text{ g } KClO_3/1.56 \text{ g})100 = 67.3\ \%\ KClO_3$

12-76. *Step 1:* Determine the quantity of excess $CH_3Cl$

$0.0694 \text{ mol Si} \cdot 2 \text{ mol } CH_3Cl / 1 \text{ mol Si} = 0.139 \text{ mol } CH_3Cl$ required

$CH_3Cl$ left after reaction = 0.171 mol available - 0.139 mol req'd = 0.032 mol

*Step 2:* Calculate the partial pressure of the excess $CH_3Cl$

$$P = nRT/V = \frac{(0.032 \text{ mol})(0.082057 \text{ L} \cdot \text{atm/K} \cdot \text{mol})(368 \text{ K})}{6.56 \text{ L}} = 0.15 \text{ atm}$$

*Step 3:* Calculate the total pressure

$$P_{total} = P(\text{excess } CH_3Cl) + P[(CH_3)_2SiCl_2]$$

$$= 0.15 \text{ atm} + 0.319 \text{ atm} = 0.47 \text{ atm}$$

## KINETIC MOLECULAR THEORY

12-78. (a) False. There are equal masses of Ar and $N_2$ in the flask. There is a greater number of moles of the gas with the smaller molar mass, so there are more moles — and so more molecules — of $N_2$ in the flask.

(b) False. At the same temperature all molecules have the same average energy. The lighter molecules will have the higher rms speed. This means the $N_2$ molecules have a greater average speed than the Ar atoms.

(c) True. Since the $N_2$ molecules have a greater velocity than Ar atoms, they collide more frequently with the container wall.

12-80. Here we use equation 12.17 (page 482)

$$\text{rms speed for } CCl_4 = \frac{\sqrt{3(8.314510 \text{ J/K} \cdot \text{mol})(298 \text{ K})}}{\sqrt{154 \times 10^{-3} \text{ kg/mol}}} = 220.\ \text{m/s}$$

The relative rms speeds of $CCl_4$ and $H_2$ can be found from equation 12.18 (page 483).

$$\frac{\text{rms speed of } H_2}{\text{rms speed of } CCl_4} = \frac{\sqrt{\text{molar mass of } CCl_4}}{\sqrt{\text{molar mass of } H_2}}$$

$$\frac{\text{rms speed of } H_2}{220.\text{ m/s}} = \frac{\sqrt{154 \text{ g/mol}}}{\sqrt{2.02 \text{ g/mol}}}$$

rms speed of $H_2$ = 1920 m/s

12-82. $OSCl_2$ (119 g/mol) > $Cl_2O$ (87 g/mol) > $Cl_2$ (71 g/mol) > $SO_2$ (64 g/mol)

*—increasing average molecular speed→*

## GRAHAM'S LAW

12-84. (a) $F_2$

(b) $N_2$

(c) $B_2H_6$ (27.67 g/mol) > $C_2H_4$ (28.05 g/mol)

(d) $CFCl_3$ (137 g/mol) > $C_2Cl_2F_4$ (171 g/mol)

12-86. The usual way of stating Graham's law is as in equation 12.18 on page 483.

$$\text{rate}_1/\text{rate}_2 = \sqrt{M_2/M_1}$$

where the rates are given in molecules per unit time or moles per unit time. However, when the rate is expressed in mass per time, then the law should be written

$$\text{mass}_2/\text{mass}_1 = \sqrt{M_2/M_1}$$

This expression was derived explicitly by T. Kenney, *J. Chem. Educ.* **1990**, *67*, 871. Therefore, here the appropriate expression would be

$$\frac{17.7 \text{ mg/hr for unknown}}{15.0 \text{ mg/hr for } I_2} = \frac{\sqrt{M \text{ for unk}}}{\sqrt{253.8}}$$

and *M* for unknown = 353 g/mol, a molar mass corresponding to $UF_6$.

## NONIDEAL GASES

12-88. (a) Pressure of gas calculated using the ideal gas law

165 g $CO_2$ • 1 mol / 44.01 g = 3.75 mol

$$P = nRT/V = \frac{(3.75\ \text{mol})(0.082057\ \text{L} \cdot \text{atm/K} \cdot \text{mol})(298\ \text{K})}{12.5\ \text{L}} = 7.33\ \text{atm}$$

(b) Pressure of the gas calculated from van der Waals's equation

a = 1.36 atm • $L^2/mol^2$   b = 0.0318 L/mol

$$[P + (3.59\ \text{atm} \cdot \text{L}^2/\text{mol}^2)(3.75\ \text{mol}/12.5)^2]\ [12.5\ \text{L} - (0.0427\ \text{L/mol})(3.75\ \text{mol})]$$
$$= (3.75\ \text{mol})(0.082057\ \text{L} \cdot \text{atm/K} \cdot \text{mol})(298\ \text{K})$$

P = 7.11 atm

## GENERAL QUESTIONS

12-90. $P_1$ = 1.00 atm   $P_2$ = ?

$V_1$ = 1.50 L   $V_2$ = 1.50 L

$T_1$ = 273 K   $T_2$ = 346 K

$P_2 = P_1 \cdot (T_2/T_1)$ = 1.00 atm • (346 K/273 K) = 1.27 atm

12-92. $P_1$ = 575 mmHg   $P_2$ = 385 mmHg

$V_1$ = 615 mL   $V_2$ = 455 mL

$T_1$ = 300.2 K   $T_2$ = ? K

$T_2 = T_1 \cdot (P_2/P_1) \cdot (V_2/V_1)$

= 300.2 K • (385 mmHg/575 mmHg) • (455 mL/615 mL) = 149 K

12-94. 2000. psi • 1 atm/14.7 psi = 136.1 atm

Temperature = 70 °F → 21 °C = 294 K

$$n = PV/RT = \frac{(136.1\ \text{atm})(75\ \text{L})}{(0.082057\ \text{L} \cdot \text{atm/K} \cdot \text{mol})(294\ \text{K})} = 420\ \text{mol}$$

$420 \text{ mol} \cdot 26.0 \text{ g/mol} = 1.1 \times 10^4 \text{ g acetylene}$

12-96. With all gases measured under the same conditions or T and P, the number of molecules is proportional to the volume of the gas.

$$1.0 \text{ L compound} \rightarrow 2.0 \text{ L } CO_2 + 3.5 \text{ L } H_2O + 0.50 \text{ L } N_2$$

This tells us, essentially, that 1 mole of the compound contains 2 moles of C, 7 moles of H, and 1 mole of N. Therefore, the empirical formula is $C_2H_7N$.

12-98. The partial pressure of each gas can be calculated from

$$P_2 = P_1 \cdot (V_1/V_2)$$

where $V_2$ in each case is 5.0 L.

Partial pressure of He = 145 mmHg • (3.0 L/5.0 L) = 87 mmHg

Partial pressure of Ar = 355 mmHg • (2.0 L/5.0 L) = 140 mmHg

$P_{total} = P_{He} = P_{Ar} = 87 \text{ mmHg} + 140 \text{ mmHg} = 227 \text{ mmHg}$

12-100. *Step 1:* Calculate moles of $N_2$ produced.

$250 \text{ L } N_2H_4 \cdot 0.010 \text{ mol/L} \cdot 1 \text{ mol } N_2/1 \text{ mol } N_2H_4 = 2.5 \text{ mol } N_2$

*Step 2:* Calculate the volume of $N_2$.

$$V = nRT/P = \frac{(2.5 \text{ mol})(0.082057 \text{ L} \cdot \text{atm/K} \cdot \text{mol})(298 \text{ K})}{0.96 \text{ atm}} = 64 \text{ L}$$

12-102. *Step 1:* Calculate the partial pressures before reaction.

For methane, $CH_4$

$0.450 \text{ g} \cdot 1 \text{ mol}/16.04 \text{ g} = 0.0281 \text{ mol } CH_4$

$$P(CH_4) = nRT/V = \frac{(0.0281 \text{ mol})(0.082057 \text{ L} \cdot \text{atm/K} \cdot \text{mol})(298 \text{ K})}{1.25 \text{ L}}$$

$P(CH_4) = 0.549 \text{ atm } (417 \text{ mmHg})$

For oxygen, $O_2$

$0.516\ g \cdot 1\ mol/32.00\ g = 0.0161\ mol\ O_2$

$$P(O_2) = nRT/V = \frac{(0.0161\ mol)(0.082057\ L \cdot atm/K \cdot mol)(298\ K)}{1.25\ L}$$

$P(O_2) = 0.315\ atm\ (239\ mmHg)$

*Step 2:* Determine the limiting reagent

The ratio of moles — and therefore the ratio of pressures — should be $2\ O_2 / 1\ CH_4$. The actual ratio is $0.315\ mol\ O_2 / 0.549\ atm\ CH_4$, must less than required. Therefore, $O_2$ is the limiting reagent.

*Step 3:* Determine the partial pressures of the product gases.

$0.315\ atm\ O_2 \cdot 1\ atm\ CO_2 / 2\ atm\ O_2 = 0.158\ atm\ CO_2$

$0.315\ atm\ O_2 \cdot 2\ atm\ H_2O / 2\ atm\ O_2 = 0.315\ atm\ H_2O$

*Step 4:* Determine the pressure of excess $CH_4$

$0.315\ atm\ O_2 \cdot 1\ atm\ CH_4 / 2\ atm\ O_2 = 0.158\ atm\ CH_4$

$0.549\ atm - 0.158\ atm = 0.391\ atm$ excess $CH_4$

*Step 5:* Calculate the total pressure after reaction.

$P_{total} = P(CH_4) + P(CO_2) + P(H_2O)$

$= 0.391\ atm + 0.158\ atm + 0.315\ atm = 0.864\ atm\ (656\ mmHg)$

12-104. (a) $O_2$ (32 g/mol) < $B_2H_6$ (27.7 g/mol) < $H_2O$ (18 g/mol)

*—increasing speed→*

(b) Both reacting gases are contained in tanks of equal volume at the same temperature. Therefore, their pressures are proportional to the quantity of gas present. Stoichiometry demands that there must be 1/3 as many moles of $B_2H_6$ as there are of $O_2$. Alternatively, the pressure of $B_2H_6$ must be 1/3 as large as that of $O_2$. Since the pressure of $O_2$ is 45 atm, this means that the pressure of $B_2H_6$ should be 15 atm.

12-106. Once again all the gases here are measured in the same container at the same temperature. Therefore, their pressures are proportional to the quantity of gas present in moles.

*Step 1:* Calculate the quantity of $F_2$ used.

$F_2$ used = total pressure of gas present initially - initial P(Xe) - $P(F_2)$ at end of reaction

= 0.72 atm - 0.12 atm - 0.36 atm

$F_2$ used = 0.24 atm

*Step 2:* Find the ratio of F to Xe

0.12 atm of Xe were used in reacting with 0.24 atm of $F_2$. Therefore, a balanced equation for the reaction must have been

$Xe(g) + 2\ F_2(g) \rightarrow XeF_4(s)$

and the product is seen to have been $XeF_4$.

12-108. (a) $P_{total}$ = 740. mmHg = $P(N_2) + P(CO_2) + P(H_2O) + P(O_2)$

740. mmHg = 557 mmHg + 23 mmHg + 40. mmHg + $P(O_2)$

$P(O_2)$ = 120. mmHg

(b) $CO_2$ (44 g/mol) < $O_2$ (32 g/mol) < $N_2$ (28 g/mol) < $H_2O$ (18 g/mol)

(c) Probably $N_2$ and $O_2$

12-110. Mole fraction $N_2$ = 500 mmHg/750 mmHg = 0.67

Mole fraction $C_2H_2$ = 250 mmHg/750 mmHg = 0.33

The mole fraction of $N_2$ is greater than that of $C_2H_2$, and the mass of $N_2$ is greater than that of $C_2H_2$. The latter means that the rms speed of $C_2H_2$ should be greater than that of $N_2$. Therefore, graph (a) is the correct one.

12-112.

$$\frac{\text{rate of diffusion of } NH_3}{\text{rate of diffusion of HCl}} = \frac{\sqrt{M(\text{HCl})}}{\sqrt{M(NH_3)}} = \frac{\sqrt{36.46 \text{ g/mol}}}{\sqrt{17.03 \text{ g/mol}}} = 1.463$$

The calculation above tells us that, if a HCl molecule travels a distance x in a given time, then a $NH_3$ molecule travels a distance $1.463x$. Therefore,

$x + 1.463\,x = 50.$ cm

$x = 20.3$ cm = distance traveled by HCl in a given time

$1.463\,x = 29.7$ cm = distance traveled by $NH_3$ in a given time

12-114. *Step 1:* Calculate the theoretical yield of acetylene

$2.65 \text{ g } CaC_2 \cdot 1 \text{ mol}/64.10 \text{ g/mol} \cdot 1 \text{ mol } C_2H_2/1 \text{ mol } CaC_2 \cdot 26.04 \text{ g/mol}$

$= 1.08$ g

*Step 2:* Calculate the partial pressure of acetylene

$P_{total} = P(C_2H_2) + P(H_2O)$

$735.2 \text{ mmHg} = P(C_2H_2) + 24.1 \text{ mmHg}$

$P(C_2H_2) = 711.1 \text{ mmHg } (0.936 \text{ atm})$

*Step 3:* Calculate the moles and then grams of $C_2H_2$ actually isolated.

$$n = PV/RT = \frac{(0.936 \text{ atm})(0.795 \text{ L})}{(0.082057 \text{ L} \cdot \text{atm/K} \cdot \text{mol})(298.4 \text{ K})} = 0.0304 \text{ mol}$$

$0.0304 \text{ mol} \cdot 26.04 \text{ g/mol} = 0.791 \text{ g}$

*Step 4:* Calculate the percentage yield

$(0.791 \text{ g}/1.08 \text{ g})100 = 73.3\ \%$

12-116. Part (a): EMPIRICAL FORMULA

*Step 1:* Find the moles of B and H in the sample of $B_xH_y$

$0.540 \text{ g } H_2O \cdot 1 \text{ mol } H_2O/18.02 \text{ g} \cdot 2 \text{ mol H}/1 \text{ mol } H_2O = 0.0599 \text{ mol H}$

$0.0599 \text{ mol H} \cdot 1.008 \text{ g/mol} = 0.0604 \text{ g H}$

$0.492 \text{ g sample} - 0.0604 \text{ g H} = 0.432 \text{ g B}$

$0.432 \text{ g B} \cdot 1 \text{ mol}/10.81 \text{ g} = 0.0400 \text{ mol B}$

*Step 2:* Find mole ratio of H to B

$0.0599 \text{ mol H}/0.0400 \text{ mol B} = 3 \text{ H}/2 \text{ B}$

Empirical formula = $B_2H_3$

Part (b): MOLECULAR FORMULA

*Step 1:* Find the molar mass

$$n = PV/RT = \frac{(0.130 \text{ atm})(0.120 \text{ L})}{(0.082057 \text{ L} \cdot \text{atm/K} \cdot \text{mol})(296 \text{ K})} = 6.41 \times 10^{-4} \text{ mol}$$

$0.0631 \text{ g}/ 6.41 \times 10^{-4} \text{ mol} = 98.4 \text{ g/mol}$

*Step 2:* Compare the empirical formula mass with the molar mass.

$$\frac{98.4 \text{ g/mol}}{24.65 \text{ g/unit}} = 3.99 \text{ units per mole}$$

Therefore, the molecular formula is $(B_2H_3)_4 = B_8H_{12}$

12-118. *Step 1:* Calculate the moles of $H_2$ obtained

$$n = PV/RT = \frac{(1.00 \text{ atm})(1.69 \text{ L})}{(0.082057 \text{ L} \cdot \text{atm/K} \cdot \text{mol})(273 \text{ K})} = 0.0755 \text{ mol}$$

*Step 2:* This problem involves finding two unknowns, so we need two equations.

(i) X g Al + Y g Zn = sample mass = 1.67 g

(ii) Moles of $H_2$ from Al + moles of $H_2$ from Zn = 0.0755 mol

In equation (i), we can say that

X = A mol Al • 26.98 g/mol

Y = Z mol Zn • 65.38 g/mol

and so

A(26.98) + Z(65.38) = 1.67

Solving for A we find A = 0.0619 - 2.42Z.

We can now modify equation (ii) as follows:

A mol Al (3 mol $H_2$/1 mol Al) + Z mol Zn (1 mol $H_2$/1 mol Zn) = 0.0755 mol

and then substitute for A.

(0.0619 - 2.42Z)(1.5) + Z = 0.0755

Solving we find that Z = 0.00659 mol Zn

0.00659 mol Zn • 65.38 g/mol = 0.431 g Zn

Mass of Al = 1.67 g - 0.431 g = 1.29 g

## SUMMARY QUESTION

12-120. (a) Balance the equation

$Co(s) \rightarrow Co^{2+}(aq) + 2e^-$

$NO_3^-(aq) + 2\,H^+(aq) + e^- \rightarrow NO_2(g) + H_2O(\ell)$

---

$Co(s) + 2\,NO_3^-(aq) + 4\,H^+(aq) \rightarrow Co^{2+}(aq) + 2\,NO_2(g) + 2\,H_2O(\ell)$

$Co(s) + 4\,HNO_3(aq) \rightarrow Co(NO_3)_2(aq) + 2\,NO_2(g) + 2\,H_2O(\ell)$

(b) The oxidizing agent is nitric acid and Co metal is the reducing agent.

(c) *Step 1:* Determine the limiting reagent

2.115 g Co • 1 mol/58.933 g = 0.03589 mol Co

0.025 L • 6.00 mol/L = 0.150 mol $HNO_3$

The ratio of available moles is 4.18 mol $HNO_3$/1 mol Co. The required ratio is 4/1, so Co is the limiting reagent.

*Step 2:* Find the number of grams of $Co(NO_3)_2$ isolated.

0.03589 mol Co • 1 mol $Co(NO_3)_2$/1 mol Co • 182.94 g/mol = 6.565 g

*Step 3:* Find the number of moles and then grams of $NO_2$ isolated.

0.03589 mol Co • 2 mol $NO_2$/1 mol Co = 0.07178 mol $NO_2$

0.07178 mol $NO_2$ • 46.0055 g/mol = 3.302 g $NO_2$

(d) Percent yield of $NO_2$

*Step 1:* Find the actual quantity of $NO_2$ isolated.

$$n = PV/RT = \frac{(0.493\ \text{atm})(2.56\ \text{L})}{(0.082057\ \text{L} \cdot \text{atm/K} \cdot \text{mol})(298\ \text{K})} = 0.0517\ \text{mol}$$

0.0517 mol • 46.0055 g/mol = 2.38 g $NO_2$

Step 2: Calculate the percent yield of $NO_2$

(2.38 g/3.302 g)100 = 72.0 %

# CHAPTER 13
# INTERMOLECULAR FORCES, LIQUIDS, AND SOLIDS

## INSTRUCTOR'S NOTES

We believe the material in this chapter has been neglected in introductory courses, particularly the matter of intermolecular forces. Therefore, we set aside 5-6 lecture periods for this chapter.

One of the difficulties our students have with this chapter is keeping the material organized. Therefore, an outline of the chapter is given in Figure 13.1. Other, similar organizational tools are found on pages 509 and 535.

Users of the first edition of this text will find that this chapter has been extensively revised. First, a discussion of ionic bonding has been moved to Chapter 10. Second, the section on intermolecular forces has been rewritten to improve its clarity. Most importantly, we have added the box on page 509 ("Deciphering Intermolecular Forces") as a way to show students the organization of this subject.

The section on solids has also been extensively revised. One of the main changes has been to introduce solid state structures by looking first at the simple cubic, face-centered cubic, and body-centered cubic structures of some of the elements. The structures of ionic solids are introduced after this discussion. Second, we have dropped an extensive discussion of lattice energies, largely because such a discussion seemed beyond the usual first-year course.

## LIST OF TRANSPARENCIES FOR THIS CHAPTER

| No. | Figure | Page | Topic |
|---|---|---|---|
| 64 | 13.1 | 502 | Organization of the chapter on solids, liquids, and intermolecular forces |
| 65 | Table 13.1 | 504 | Summary of intermolecular forces |
| 66 | --- | 509 | Types of intermolecular forces and their interrelation |
| 67 | 13.8 | 511 | Boiling points of nonmetal hydrogen-containing compounds |

| No. | Figure | Page | Topic |
|---|---|---|---|
| 68 | 13.12 | 517 | Vapor pressure |
| 69 | 13.14 | 521 | Vapor pressure curves |
| 70 | 13.26 | 528 | Cubic unit cells |
| 71 | 13.28 | 529 | Cubic close packing and hexagonal-close packing |
| 72 | 13.29 | 531 | Atom sharing in cube faces and at corners |
| 73 | 13.30 | 532 | NaCl unit cell and ion packing |
| 74 | 13.31 | 533 | Octahedral holes in a face-centered cubic lattice |
| 75 | 13.34 | 536 | Graphite |
| 76 | 13.37 | 538 | Trends in lattice energies |
| 77 | 13.40 | 544 | Ice |
| 78 | 13.43 | 545 | Temperature dependence of the densities of ice and water. |
| 79 | 13.45 | 547 | Phase diagram for water |
| 80 | 13.48 | 549 | Phase diagram for carbon dioxide |

## SUGGESTED DEMONSTRATIONS

Molecular models are invaluable to show intermolecular forces and solid state structures. Various kinds of kits are available from Aldrich Chemical Company. In addition, the Klinger Educational Products Corp. (112-19 14th Road, College Point, New York, NY 11356-1453) sells a very large number of models of various solids.

The close packing of spheres (such as the photos in Figure 13.27) is easy to do on an overhead projector. We have also used oranges and grapefruits, as well as styrofoam balls.

We have searched for good demonstrations of vapor pressure for some time. One of the demonstrations we use is a so-called "Love Meter", a toy that can be obtained in novelty and gift shops. The closed, thin-walled glass vessel contains methyl chloride (bp. -23.7 °C). Warming the vessel in your hand raises the vapor pressure of the liquid and forces the liquid to another compartment of the vessel.

Finally, *KC? Discoverer* or the *Periodic Table Stack* has excellent diagrams of the structures of the elements. In addition, one can search the database for all elements that have a given type of solid structure. Some excellent discussions of periodic trends can result.

# SOLUTIONS TO SELECTED STUDY QUESTIONS

## INTERMOLECULAR FORCES

13-19. Hydrogen bonding between water molecules must be broken down, and ion-ion forces between $Na^+$ and $Cl^-$ in the NaCl crystal must also be broken down. The $Na^+$ and $Cl^-$ ions bind to water molecules by ion-dipole forces.

13-21. $I_2$ molecules interact with one another by weak induced dipole forces, while molecules of methanol are held together by hydrogen bonding (and dipole-dipole forces). The solution process occurs because the polar methanol molecules can induce a dipole in the nonpolar $I_2$ molecules.

13-23. (a) liquid $N_2$: induced dipole (dispersion)

(b) mercury: induced dipole (dispersion)

(c) $CCl_3F$ (a chlorofluorocarbon): dipole-dipole

(d) $CH_3CH_2OH$ (ethyl alcohol): hydrogen bonding

13-25. He < $H_3C—CH_2—CH_2—CH_2—CH_3$ < $SO_2$ < $H_3C—OH$

—predicted order of increasing strength of IM forces→

It is difficult to decide on the order of the relative forces in pentane and $SO_2$. Pentane is a very volatile liquid at room temperature (bp, 36 °C), whereas $SO_2$ is a gas, observations that would lead to the conclusion that the forces are in the order $SO_2$ < pentane. However, while $SO_2$ is polar while pentane is not, it is logical for students to arrive at the order above.

13-27. $H_2S$ molecules interact through dipole-dipole forces as well as very weak hydrogen bonds. Water forms much stronger hydrogen bonds.

13-29. The normal boiling point of $CH_3Cl$ (chloromethane) is 3.6 °C, whereas that for $CH_3I$ is 42.4 °C. Therefore, $CH_3I$ has the stronger intermolecular forces in the liquid phase, presumably owing to its high molar mass.

13-31. Which of the following are expected to form strong hydrogen bonds with water?

(a) $H_2Te$: The molecule is polar, but H-bonding is expected to be very weak.

(b) $H_3C\text{-}O\text{-}H$: H-bonding expected and observed.

(c) $H_3C\text{-}O\text{-}CH_3$: No H-bonding expected.

(d) HI: Weak H-bonding only.

(e) $H_3C{-}C({=}O){-}OH$ : H-bonding expected and indeed strongly observed.

13-33. Which compound in each of the following pairs of salts is more likely found with a hydrated cation?

(a) $Li_2SO_4$ should be more strongly hydrated than $Cs_2SO_4$, owing to the fact that the $Li^+$ more strongly attracts water than does $Cs^+$.

(b) $Mg(NO_3)_2$ should be more strongly hydrated than $NH_4NO_3$, since the doubly charged $Mg^{2+}$ ion strongly attracts water molecules.

(c) $NiCl_2$ should be more strongly hydrated than RbCl, since the doubly charged $Ni^{2+}$ ion more strongly attracts water molecules than the $Rb^+$ ion.

## LIQUIDS

13-35. Methyl alcohol, $H_3COH$, has a normal boiling point of 64.7 °C and a vapor pressure of 100 mmHg at 21.2 °C, while formaldehyde ($H_2C{=}O$), has a normal boiling of -19.5 °C and a vapor pressure of 100 mmHg at -57.3 °C. The reason that methyl alcohol requires a much higher temperature to achieve the same vapor pressure as formaldehyde is that $H_3COH$ molecules interact strongly through hydrogen bonding, whereas $H_2C{=}O$ molecules interact only through somewhat weaker dipole-dipole forces.

13-37. $1.00 \times 10^3$ g $CCl_2F_2$ • 1 mol $CCl_2F_2$ / 120.9 g • 20.0 kJ/mol = 165 kJ

13-39. $5.50 \times 10^6$ L • 1000 mL/L • 1.00 g/mL • 2260 J/g • 1 kJ/1000 J = $1.24 \times 10^{10}$ kJ

13-41. (a) The equilibrium vapor pressure of diethyl ether at room temperature (approximately 20 °C) is about 400 mmHg.

(b) Order of increasing intermolecular forces: ether < alcohol < water

(c) At 40 °C and 400 mmHg, ether is a gas, while the alcohol and water would be predominantly in the liquid state.

13-43. (a) You put some water at 60 °C in a plastic milk carton and seal the top very tightly so gas cannot enter or leave the carton. When the water cools the air pressure on the outside of the container is larger than the pressure inside the container, so the container collapses.

(b) If you put a few drops of liquid diethyl ether on your hand, it will evaporate completely, since the normal boiling point of the ether is lower than the temperature of your body (37 °C).

13-45. ***NOTE: In the first printing of the book, the column labels on the data in this question were reversed. The correct data should be:***

| Vapor Pressure (mmHg) | Temperature (°C) |
|---|---|
| 13.6 | 25 |
| 45.3 | 50. |
| 127.2 | 75 |
| 310.8 | 100. |

Using the Clausius-Clapeyron equation on page 522, and the data for 50. °C and 100. °C., we have

$T_1$ = 323 K $T_2$ = 373 K

$P_1$ = 45.3 mmHg $P_2$ = 310.8 mmHg

$\ln (310.8/45.3) = (\Delta H_{vap} / 8.3145 \times 10^{-3}\ \text{kJ/K} \cdot \text{mol}) [(1/323) - (1/373)]$

$\Delta H_{vap} = 38.6$ kJ/mol

Now that we have a value of $\Delta H_{vap}$ we can calculate the normal boiling point using one of the known vapor pressures in the table above.

$$\ln(760./45.3) = (38.6 \text{ kJ/mol} / 8.3145 \times 10^{-3} \text{ kJ/K} \cdot \text{mol}) [(1/323) - (1/T_2)]$$

$$T_2 = 402 \text{ K}$$

13-47. Since the critical temperature for propane is well above ordinary room temperatures, or even outdoor temperatures, propane can be liquefied. A common use for liquefied propane is for home heating and cooking.

## SOLID STATE STRUCTURES

13-49. See the outline on page 535.

*Step 1:* Calculate the mass of one Ca atom

$$1 \text{ mol}/6.022 \times 10^{23} \text{ atoms} \cdot 40.078 \text{ g/mol} = 6.655 \times 10^{-23} \text{ g/atom}$$

*Step 2:* Calculate the mass of the unit cell.

$$4 \text{ Ca atoms/unit cell} \cdot 6.655 \times 10^{-23} \text{ g/atom} = 2.662 \times 10^{-22} \text{ g/unit cell}$$

*Step 3:* Calculate the volume and then the length of the edge of the unit cell.

$$2.662 \times 10^{-22} \text{ g/unit cell} \cdot 1 \text{ cm}^3/1.54 \text{ g} = 1.73 \times 10^{-22} \text{ cm}^3$$

$$[1.729 \times 10^{-22} \text{ cm}^3]^{1/3} = 5.58 \times 10^{-8} \text{ cm}$$

*Step 4:* Calculate the size of the Ca atom

diagonal distance of unit cell = $d = \sqrt{2} \cdot$ edge of unit cell

$$d = 7.89 \times 10^{-8} \text{ cm}$$

Diagonal distance = $4 \cdot$ Ca atom radius

Ca atom radius = $1.97 \times 10^{-8}$ cm (or 197 pm)

*KC? Discoverer* lists 197.0 pm as the Ca atom radius.

13-51. See Example 13.5.

*Step 1:* Calculate the mass of one Cu atom

$$1 \text{ mol}/6.022 \times 10^{23} \text{ atoms} \cdot 63.546 \text{ g/mol} = 1.055 \times 10^{-22} \text{ g/atom}$$

*Step 2:* Assuming that the unit cell is face-centered cubic, calculate the volume of the cell.

Cu atom radius = $127.8 \text{ pm} \cdot 1 \text{ m}/1 \times 10^{12} \text{ pm} \cdot 100 \text{ cm}/1 \text{ m} = 1.278 \times 10^{-8}$ cm

Cell diagonal = $4 \cdot$ radius of Cu = $4 \cdot 1.278 \times 10^{-8} \text{ cm} = 5.112 \times 10^{-8}$ cm

Since cell diagonal = $\sqrt{2} \cdot$ edge, the length of the edge = $3.615 \times 10^{-8}$ cm

Volume of cell = $[3.615 \times 10^{-8}\,cm]^3 = 4.723 \times 10^{-23}\,cm^3$

*Step 3:* Calculate the mass of the unit cell

$4.723 \times 10^{-23}\,cm^3/\text{unit cell} \cdot 8.95\,g/cm^3 = 4.23 \times 10^{-22}\,g/\text{unit cell}$

*Step 4:* Calculate the number of atoms per unit cell

$4.23 \times 10^{-22}\,g/\text{unit cell} \cdot 1\,\text{atom}/1.055 \times 10^{-22}\,g = 4.01\,\text{atoms/unit cell}$

A face-centered cubic unit cell should indeed have 4 atom per unit cell.

13-53. If M cations occupy all of the octahedral holes in a face-centered cubic lattice of X anions, the formula of the salt can *only* be MX, since the ratio of M to X can *only* be 4:4 or 1:1.

13-55. *Step 1:* Calculate the length of an edge of the unit cell (see Examples 13.7 and 13.8).

$Na^+$ radius = 116 pm and $I^-$ radius = 206 pm

Therefore, the length of the cube edge = $2 \cdot 116\,pm + 2 \cdot 206\,pm = 644\,pm$

$644\,pm \cdot 1 \times 10^{-12}\,m/1\,pm \cdot 100\,cm/1\,m = 6.44 \times 10^{-8}\,cm$

*Step 2:* Calculate the volume of the unit cell.

$[6.44 \times 10^{-8}\,cm]^3 = 2.67 \times 10^{-22}\,cm^3$

*Step 3:* Calculate the mass of the unit cell.

$4\,NaI/\text{unit cell} \cdot 1\,mol/6.022 \times 10^{23}\,NaI \cdot 149.89\,g/mol = 9.96 \times 10^{-22}\,g$

*Step 4:* Calculate the density of the solid.

Mass/Volume = $9.96 \times 10^{-22}\,g/2.67 \times 10^{-22}\,cm^3 = 3.73\,g/cm^3$

(The literature density of NaI = $3.667\,g/cm^3$)

13-57. *Step 1:* Calculate the volume of the unit cell.

$[3.85 \times 10^{-8}\,cm]^3 = 5.71 \times 10^{-23}\,cm^3/\text{unit cell}$

*Step 2:* Calculate the mass of the unit cell

$5.71 \times 10^{-23}\,cm^3/\text{unit cell} \cdot 7.00\,g/cm^3 = 4.00 \times 10^{-22}\,g/\text{unit cell}$

*Step 3:* Calculate the mass of one TlCl unit

$239.8\,g/mol \cdot 1\,mol/6.022 \times 10^{23}\,TlCl = 3.98 \times 10^{-22}\,g/1\,TlCl$

*Step 4:* Calculate the number of atoms per unit cell

$4.00 \times 10^{-22}\,g/\text{unit cell} \cdot 1\,TlCl/3.98 \times 10^{-22}\,g = 1.00\,TlCl/\text{unit cell}$

Therefore, the unit cell must be a simple cubic lattice of $Cl^-$ ions with a $Tl^+$ ion in the center of the cube. (See CsCl structure on page 533.)

## PHYSICAL PROPERTIES OF SOLIDS

13-59. The charges on the cation and anion in MgO are +2 and -2, respectively. In contrast, those for NaF are only +1 and -1, respectively. This means the force of attraction between anion and cation in MgO is much greater (roughly by a factor of four) than in NaF. This increased force of attraction means a much higher melting point.

13-61. The lattice energy of a salt depends on the ion-ion forces of attraction, which are in turn inversely proportional to the distance between the ions. Since the $K^+$ ion is smaller than the $Rb^+$ ion (Figure 9.15), the lattice energy for KBr is expected to be larger than that of RbBr (see Figure 13.37), making the melting point and boiling point of KBr higher than that of RbBr.

13-63. 0.500 mL • 13.6 g/mL • 294 J/g • 1 kJ/1000 J = 2.00 kJ

## HEATING AND COOLING CURVES AND PHASE DIAGRAMS

13-65. *Step 1:* Heating 10. kg of ammonia from -50.0 °C to -33.6 °C ($\Delta T$ = 16.4 K) requires

$10. \times 10^3$ g $NH_3$ • (4.7 J/g • K) • 16.4 K = $7.7 \times 10^5$ J

*Step 2:* Converting liquid ammonia to gaseous ammonia requires

$10. \times 10^3$ g $NH_3$ • 1 mol $NH_3$/ 17.0 g $NH_3$ • 23.5 kJ/mol • 1000 J/kJ

= $1.4 \times 10^7$ J

*Step 3:* Heating the $NH_3$ vapor from -33.6 °C to 0 °C requires

$10. \times 10^3$ g $NH_3$ • (2.2 J/g • K) • 33.6 K = $7.4 \times 10^5$ J

*Step 4:* Calculate the total energy input

Total energy = $7.7 \times 10^5$ J + $1.4 \times 10^7$ J + $7.4 \times 10^5$ J = $1.5 \times 10^7$ J

13-67. Approximate phase diagram for oxygen.

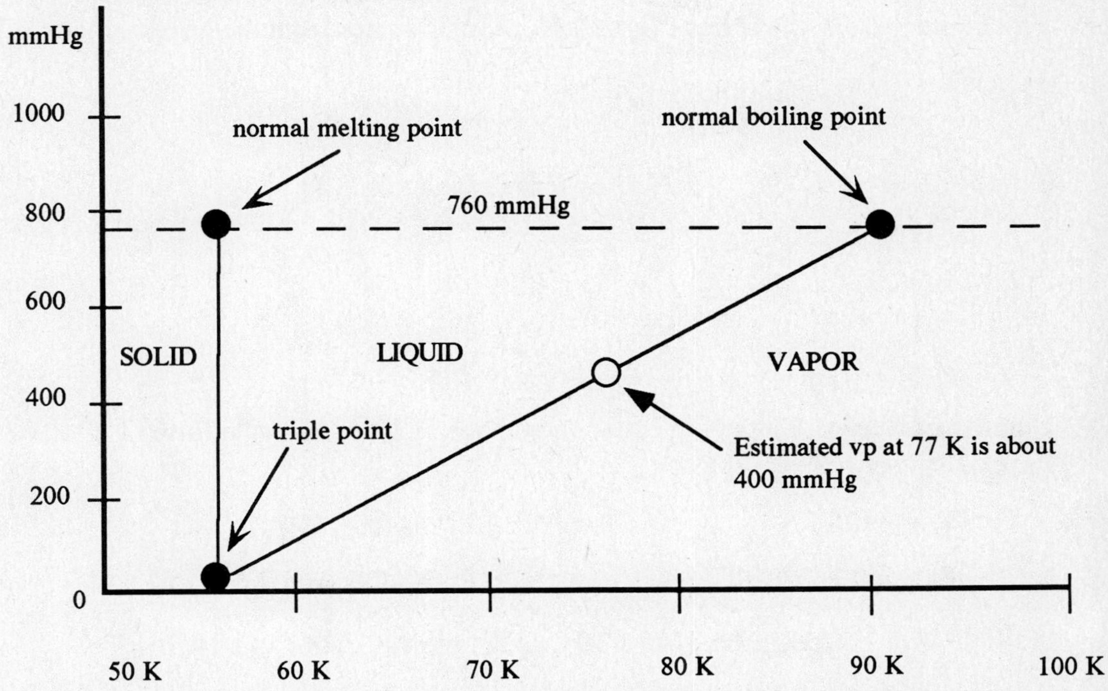


The very slight positive slope of the solid/liquid equilibrium line indicates the solid is more dense the liquid.

## GENERAL QUESTIONS

13-69. Acetone readily absorbs water owing to hydrogen bonding between the C=O oxygen atom and the O—H bonds of water.

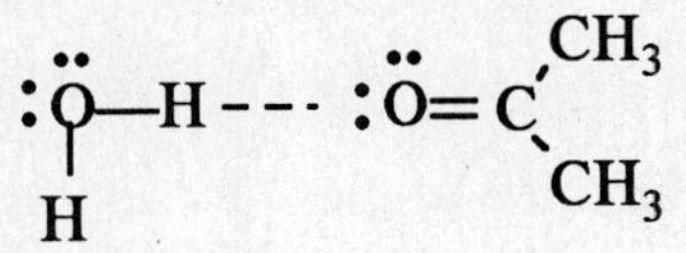

13-71.

| **Compound** | **Boiling Point (°C)** |
|---|---|
| $NH_3$ | -34.5 |
| $PH_3$ | -87.5 |
| $AsH_3$ | -62.4 |
| $SbH_3$ | -18.4 |

These data reflect the data plotted in Figure 13.8 on page 511. The boiling point of $NH_3$ is higher than that of $PH_3$ owing to the relatively strong hydrogen bonding in $NH_3$. The general increase in the series $PH_3 < AsH_3 < SbH_3$ comes from the general increase in dispersion forces.

13-73.

| **Temperature (°C)** | **Vapor Pressure (mmHg)** |
|---|---|
| 549 | 10. |
| 633 | 40. |
| 701 | 100. |
| 823 | 400. |

Using the Clausius-Clapeyron equation on page 522, and the data for 633 °C and 823 °C., we have

$T_1$ = 906 K $T_2$ = 1096 K

$P_1$ = 40. mmHg $P_2$ = 400. mmHg

$\ln(400./40.) = (\Delta H_{vap} / 8.3145 \times 10^{-3}\ \text{kJ/K} \cdot \text{mol})\ [(1/906) - (1/1096)]$

$\Delta H_{vap} = 1.0 \times 10^{2}\ \text{kJ/mol}$

Now that we have a value of $\Delta H_{vap}$ we can calculate the normal boiling point using one of the known vapor pressures in the table above.

$\ln(760./40.) = (1.0 \times 10^{2}\ \text{kJ/mol} / 8.3145 \times 10^{-3}\ \text{kJ/K} \cdot \text{mol})\ [(1/906) - (1/T_2)]$

$T_2 = 1170\ \text{K}$

13-75. The polar liquid methyl alcohol has strong cohesive forces. The polar molecules will also adhere to the glass surface (also polar). The combination of these forces results in a *concave surface* for the alcohol.

13-77. *Step 1:* Calculate the volume of the unit cell

$[5.4 \times 10^{-8}\ \text{cm}]^3 = 1.6 \times 10^{-22}\ \text{cm}^3\text{/unit cell}$

*Step 2:* Calculate the mass of a unit cell

$4\ CaF_2\text{/unit cell} \cdot 1\ \text{mol}/6.022 \times 10^{23}\ CaF_2 \cdot 78.08\ \text{g/mol}\ CaF_2$

$= 5.2 \times 10^{-22}\ \text{g/unit cell}$

*Step 3:* Calculate the density of the solid

$5.2 \times 10^{-22}$ g/unit cell • 1 unit cell / $1.6 \times 10^{-22}$ $cm^3$ = 3.3 g/$cm^3$

13-79. *Step 1:* Calculate the mass of the unit cell.

1 CsI/unit cell • 1 mol/$6.022 \times 10^{23}$ CsI • 259.8 g/mol

= $4.314 \times 10^{-22}$ g/unit cell

*Step 2:* Calculate the volume of the unit cell

$4.314 \times 10^{-22}$ g/unit cell • 1 $cm^3$ / 4.56 g = $9.46 \times 10^{-23}$ $cm^3$

*Step 3:* Calculate the length of one edge

$[9.46 \times 10^{-23}\ cm^3]^{1/3}$ = $4.56 \times 10^{-8}$ cm

13-81. *Step 1:* Calculate the volume and then the mass of water in the lake

Area of the lake = 1.00 $mile^2$ • [1 km/0.62137 mile]$^2$

• [1000m/1 km]$^2$ • [100 cm/1 m]$^2$ = $2.59 \times 10^{10}$ $cm^2$

Depth of the lake = 100. ft • 12.0 inches/ft • 2.54 cm/inch = $3.05 \times 10^3$ cm

Volume of the lake = ($2.59 \times 10^{10}$ $cm^2$) • ($3.05 \times 10^3$ cm) = $7.89 \times 10^{13}$ $cm^3$

Mass of water = $7.89 \times 10^{13}$ $cm^3$ • 1.0 g/$cm^3$ = $7.9 \times 10^{13}$ g

*Step 2:* Calculate the heat transferred from the lake as it cools from 25 °C to 4 °C.

heat transferred = ($7.9 \times 10^{13}$ g) • (4.184 J/g • K) (4 °C - 25 °C)

= $6.9 \times 10^{15}$ J (or $6.9 \times 10^{12}$ kJ)

13-83. (a) 190. g • 1 mol / 58.12 g $C_4H_{10}$ • 24.3 kJ/mol = 79.4 kJ

(b) Calculate the vapor pressure at 0 °F.

First, 0 °F = -17.8 °C or 255.2 K

From Table 13.5, $\Delta H_{vap}$

= 24.3 kJ/mol at -0.5 °C where the vapor pressure is 760 mmHg

Therefore, we use the following data in the Clausius-Clapeyron equation:

$T_1$ = 272.7 K $T_2$ = 255.2 K

$P_1$ = 760. mmHg $P_2$ = ?

$\ln (P_2/P_1)$ = (24.3 kJ / $8.3145 \times 10^{-3}$ kJ/K • mol) [(1/272.7) - (1/255.2)]

$\ln (P_2/P_1) = -0.727$

$P_2/P_1 = 0.483$

Since $P_1 = 760.$ mmHg, $P_2$ must be 367 mmHg

The stove does not work well at low temperatures because there is not enough butane in the vapor phase.

13-85. *For packing scheme A:*

Area of the square = $(\text{side})^2 = (2 \cdot r)^2$ where r = radius of circles making up the pattern

Since 1/4 of each circle is within the square, the total area of the square occupied by the circles is equal to the area of one circle ($\pi \cdot r^2$). Therefore, the percentage of the area of the square covered is

$$(\pi \cdot r^2) / 4r^2 = 3.14/4 = 0.785 \text{ or approximately } 79\%$$

*For packing scheme B:*

The area of any triangle with angles A, B, and C, and sides opposite a, b, and c, respectively, is

$$\text{Area} = \tfrac{1}{2}\, a \cdot b \cdot \sin C$$

Here the sides are all the same length ($2 \cdot r$), and the angles are all 60°. Therefore,

$$\text{Area of triangle} = \tfrac{1}{2}\,(2 \cdot r)^2(\sin 60°) = 1.732\, r^2$$

One-sixth of each of the circles is taken up by a portion of the blue triangle. Therefore, the blue triangle occupies, overall, 3 x 1/6 or 1/2 of the three circles. Since the area of a circle is $\pi \cdot r^2$, the percentage of the area occupied is

$$\tfrac{1}{2}\,\pi \cdot r^2 / 1.732\, r^2 = 1.57 / 1.73 = 0.907 \text{ or about } 90\%$$

13-87. *Step 1:* Calculate the length of the cube diagonal from the length of an edge of the unit cell.

$$\text{Cube diagonal} = \sqrt{3} \cdot \text{edge}$$

$$\text{Cube diagonal} = \sqrt{3} \cdot 0.445 \text{ nm} = 0.771 \text{ nm}$$

*Step 2:* Calculate the radius of the Cs+ ion

$$\text{Cube diagonal} = \text{radius of } I^- + 2 \cdot \text{radius of } Cs^+ + \text{radius of } I^-$$

Since an I- ion has a radius of 206 pm (or 0.206 nm), this means the radius of $Cs^+$

is 0.180 nm or 180. pm. (*KC? Discoverer* gives the radius of Cs+ as 181 pm.)

13-89. (a) Heat transferred by cooling 1.00 mole of ammonia

$$q = (17.03 \text{ g}) \cdot (4.70 \text{ J/g} \cdot \text{K}) \cdot (10. \text{ K}) = 800. \text{ J}$$

(b) Heat transferred by cooling 1.0 mole of water 10 °C.

$$q = (18.02 \text{ g}) \cdot (4.184 \text{ J/g} \cdot \text{K}) \cdot (10. \text{ K}) = 754 \text{ J}$$

## SUMMARY QUESTION

13-91. (a) Calculate the vapor pressure of Hg at 25 °C.

$T_1$ = 25 °C = 298 K    $T_2$ = 357 °C = 630. K

$P_1$ = ? mmHg    $P_2$ = 760 mmHg

$$\ln (P_2/P_1) = (59.0 \text{ kJ} / 8.3145 \times 10^{-3} \text{ kJ/K} \cdot \text{mol}) [(1/298) - (1/630.)]$$

$$\ln (P_2/P_1) = 12.6$$

$$P_2/P_1 = 2.82 \times 10^5$$

Since $P_2$ = 760. mmHg, $P_1$ must be $2.7 \times 10^{-3}$ mmHg

(b) Calculate the number of atoms of mercury in the gas phase in a room.

*Step 1:* Calculate the volume of the room in liters

$$(12 \cdot 12 \cdot 8)\text{ft}^3 \cdot (12 \text{ inches/ft})^3 \cdot (2.54 \text{ cm/in})^3 \cdot 1 \text{ L}/1000 \text{ cm}^3 = 3.26 \times 10^4 \text{ L}$$

*Step 2:* Calculate the number of moles of Hg needed to create a pressure of $2.7 \times 10^{-3}$ mmHg ( = $3.6 \times 10^{-6}$ atm) in a room with a volume of $3.26 \times 10^4$ L.

$$n = PV/RT = \frac{(3.6 \times 10^{-6} \text{ atm}) \cdot (3.26 \times 10^4 \text{ L})}{(0.082057 \text{ L} \cdot \text{atm/K} \cdot \text{mol}) \cdot (298 \text{ K})} = 4.7 \times 10^{-3} \text{ mol Hg}$$

$4.7 \times 10^{-3}$ mol Hg • $6.022 \times 10^{23}$ atoms/mol = $2.9 \times 10^{21}$ atoms Hg in the room

$2.9 \times 10^{21}$ atoms / room • 1 room / $1.15 \times 10^3$ ft$^3$ = $2.5 \times 10^{18}$ atoms / ft$^3$

# CHAPTER 14

# SOLUTIONS AND THEIR BEHAVIOR

## INSTRUCTOR'S NOTES

We generally cover all the topics in this chapter (except for Section 14.4) in 2-3 lectures. Students are able to deal with straightforward problems and discuss solution phenomena with little difficulty.

## LIST OF TRANSPARENCIES FOR THIS CHAPTER

| No. | Figure | Page | Topic |
|---|---|---|---|
| 81 | 14.16 | 584 | Osmosis |
| 82 | 14.17 | 586 | Osmosis and living cells |

## SUGGESTED DEMONSTRATIONS

Several demonstrations in Volume 1 of *Chemical Demonstrations* by Shakhashiri are suitable. Suggestions include "Crystallization from Supersaturated Solutions of Sodium Acetate," "Chemical Hot Packs," and "Endothermic Reactions of Hydrated Barium Chloride and Ammonium Salts."

Several of the photos in the chapter represent demonstrations we do in class. For example, Figure 14.1 shows the difference between molality and molarity, and Figure 14.3 shows a "hot pack" we purchased at a local camping center. Figure 14.6, showing the preferential solubility of iodine in $CCl_4$, is easy to do. One of our favorites is the frozen solution in Figure 14.15. We place an indicator in water and then allow the solution to freeze *slowly* in the freezer of a lab refrigerator. The result clearly shows that the solute concentrates in the solution and that the solid is pure solvent.

# SOLUTIONS TO SELECTED STUDY QUESTIONS

## CONCENTRATIONS

14-14. 45.0 g $C_4H_8N_2O_2$ • 1 mol/116.12 g = 0.388 mol DMG

395 g $C_2H_5OH$ • 1 mol/46.07 g = 8.57 mol

(a) *Molality* of the solution

$m$ = 0.388 mol/0.395 kg = 0.981 molal

(b) *Mole fraction*

$$X_{DMG} = \frac{\text{mol DMG}}{\text{mol DMG + mol alcohol}} = \frac{0.388 \text{ mol DMG}}{0.388 \text{ mol DMG} + 8.57 \text{ mol alc}}$$

$X_{DMG}$ = 0.0433

(c) Weight percent

$$\frac{45.0 \text{ g DMG}}{45.0 \text{ g DMG} + 395 \text{ g alc}} \cdot 100 = 10.2\ \%$$

14-16.

| **Compound** | **Molality** | **Weight Percent** | **Mole Fraction** |
|---|---|---|---|
| $NH_3$ | *25.2* | 30.0 | *0.312* |
| $CH_3COOH$ | 0.0083 | *0.050* | *0.00015* |
| $C_2H_4(OH)_2$ | *2.84* | 15.0 | *0.0487* |

Calculations for the case of $NH_3$

(a) *Molality:* There are 30.0 g of $NH_3$ and 70.0 g of $H_2O$ in every 100.0 g of solution.

Moles of $NH_3$ = 30.0 g $NH_3$ • 1 mol/17.03 g = 1.76 mol

$m$ = 1.76 mol/0.070 kg water = 25.2 molal

(b) *Mole fraction:*

Moles water = 70.0 g • 1 mol/18.02 g = 3.89 mol

$$X_{ammonia} = \frac{1.76 \text{ mol}}{1.76 \text{ mol} + 3.89 \text{ mol}} = 0.312$$

Calculations for $CH_3COOH$ (acetic acid)

From the molality you know that there are 0.0083 mol of acetic acid in 1.0 kg of water. Therefore, there are 0.0083 mol • 60.0 g/mol = 0.50 g of acetic acid and

$1.0 \times 10^3$ g water • 1 mol/18.02 g = 55 mol water.

(a) *Percent:*

$$\frac{0.50 \text{ g acid}}{1.0 \times 10^3 \text{ g water}} \cdot 100 = 0.050\ \%$$

(b) *Mole fraction:*

$$X_{acid} = \frac{0.0083 \text{ mol}}{0.0083 \text{ mol} + 55 \text{ mol}} = 0.00015$$

14-18. (a) Grams $K_2CO_3$ required

0.0512 molal = X mol $K_2CO_3$ / 0.250 kg solvent

X mol = 0.0128 mol $K_2CO_3$

0.0128 mol $K_2CO_3$ • 138.2 g/mol = 1.77 g $K_2CO_3$

(b) Mole fraction $K_2CO_3$

$$X_{salt} = \frac{0.0128 \text{ mol}}{0.0128 \text{ mol} + (250.\text{ g} \cdot 1 \text{ mol}/18.02 \text{ g})} = 9.21 \times 10^{-4}$$

14-20. (a) Calculate grams of glycol required

$$X_{glycol} = 0.125 = \frac{? \text{ mol glycol}}{? \text{ mol glycol} + \text{mol } H_2O}$$

Since there are 52.7 moles of $H_2O$ in 950. g of water, this means ? mol glycol = 7.53.

7.53 mol glycol • 62.07 g/mol = 468 g glycol

(b) Calculate the molality of the solution.

$m$ = 7.53 mol glycol / 0.950 kg water = 7.93

14-22. 5.12 g $MgCl_2$ • 1 mol/95.21 g = 0.0538 mol

652 g $H_2O$ • 1 mol/18.02 g = 36.2 mol $H_2O$

(a) *Molality:*

m = 0.0538 mol/0.652 kg = 0.0825 molal

(b) *Mole fraction of water:*

$$X_{water} = \frac{36.2 \text{ mol } H_2O}{36.2 \text{ mol } H_2O + 0.0538 \text{ mol } MgCl_2} = 0.999$$

14-24.

| Compound | Grams Compound | Grams Water | Molality of Compound | Mole Fraction of Compound |
|---|---|---|---|---|
| $Pb(NO_3)_2$ | *6.49* | 800. | 0.0245 | *4.41 x $10^{-4}$* |
| $C_2H_4(OH)_2$ | *49.6* | 250. | *3.20* | 0.0545 |
| $Pt(NH_3)_2Cl_2$ | 0.0075 | 200. | *1.3 x $10^{-4}$* | *2.3 x $10^{-6}$* |

(a) Calculations for $Pb(NO_3)_2$

*m* = 0.0245 = X mol/0.800 kg

X mol = 0.0196 mol

0.0196 mol • 331.2 g/mol = 6.49 g $Pb(NO_3)_2$

$$X_{lead\ salt} = \frac{0.0196 \text{ mol}}{0.0196 \text{ mol} + 44.4 \text{ mol}} = 4.41 \times 10^{-4}$$

(b) Calculations for $C_2H_4(OH)_2$ (ethylene glycol)

Moles water = 250. g • 1 mol/18.02 g = 13.9 mol

$$X_{glycol} = \frac{? \text{ mol glycol}}{? \text{ mol glycol} + 13.9 \text{ mol } H_2O} = 0.0545$$

? mol glycol = 0.800 mol

0.800 mol • 62.07 g/mol = 49.6 g

*m* = 0.800 mol/0.250 kg = 3.20 molal

(c) Calculations for $Pt(NH_3)_2Cl_2$

$0.0075\ g \cdot 1\ mol\ Pt(NH_3)_2Cl_2 / 300.\ g = 2.5 \times 10^{-5}\ mol\ Pt(NH_3)_2Cl_2$

$m = 2.5 \times 10^{-5}\ mol\ Pt(NH_3)_2Cl_2 / 0.200\ kg = 1.3 \times 10^{-4}$ molal

$$X_{Pt\ salt} = \frac{2.5 \times 10^{-5}\ mol}{2.5 \times 10^{-5}\ mol + 11.1\ mol} = 2.3 \times 10^{-6}$$

14-26. (a) Calculate *molality*. If the solution is 95.0% acid, it contains 95.0 g of $H_2SO_4$ and 5.0 g of water in every 100.0 g of solution.

$95.0\ g \cdot 1\ mol/98.07\ g = 0.969\ mol\ H_2SO_4$

$m = 0.969\ mol\ H_2SO_4 / 0.0050\ kg = 194$ molal

(b) Calculate *molarity.*

$$\frac{95.0\ g\ H_2SO_4}{100.0\ g\ solution} \cdot \frac{1.84\ g\ solution}{1.00\ cm^3\ solution} \cdot \frac{1000\ cm^3\ solution}{1.00\ L\ solution} \cdot \frac{1\ mol\ H_2SO_4}{98.07\ g}$$

$= 17.8$ molar in $H_2SO_4$

14-28. (a) Calculate the *molality* of the solution. From the molarity and density, we can first find the number of moles of $NH_3$ per gram of solution.

$$\frac{14.8\ mol\ NH_3}{1000\ cm^3\ solution} \cdot \frac{1\ cm^3\ solution}{0.90\ g\ solution} = 0.0164\ mol\ NH_3/g\ solution$$

$(0.0164\ mol\ NH_3/g\ solution) \cdot (17.03\ g\ NH_3/mol) = 0.280\ g\ NH_3 / 1.00\ g\ solution$

This means that 1.00 g of solution contains 0.28 g of $NH_3$ and 0.72 g $H_2O$.

Therefore, $m = 0.0164\ mol\ NH_3 / 0.72 \times 10^{-3}\ kg = 22.8$ molal

(b) Calculate the *mole fraction* of the solution.

$$X_{ammonia} = \frac{0.0164 \text{ mol}}{0.0164 \text{ mol} + 0.040} = 0.29$$

(c) Calculate the *weight percent* of ammonia.

$(0.280 \text{ g } NH_3 / 1.00 \text{ g solution}) \cdot 100 = 28.0\ \%$

14-30. $MgCl_2(aq) \rightarrow Mg^{2+}(aq) + 2\ Cl^-(aq)$

A solution 0.100 molal in ions is 0.100/3 = 0.0333 molal in $MgCl_2$. Therefore,

0.0333 mol • 95.21 g/mol = 3.17 g/kg water

3.17 g/kg water • 0.150 kg water = 0.476 g $MgCl_2$

14-32. (a) 28 ppb = 28 g $Ag^+$ / $1.0 \times 10^9$ g water ≡ $2.8 \times 10^{-5}$ g $Ag^+$ / kg $H_2O$

$m$ = ($2.8 \times 10^{-5}$ g $Ag^+$ / kg $H_2O$) • (1 mol $Ag^+$ / 108 g) = $2.6 \times 10^{-7}$ molal

(b) $1.0 \times 10^2$ g • (1 kg $H_2O$ / $2.8 \times 10^{-5}$ g $Ag^+$) • (1 L/kg) = $3.6 \times 10^6$ L

## THE SOLUTION PROCESS

14-34. As the carbon chain length increases, this interferes with hydrogen bonding between the alcohol —OH group and $H_2O$.

14-36. $\Delta H_{solution} = \Delta H_f^\circ[CaCl_2(aq)] - \Delta H_f^\circ[CaCl_2(s)]$

$= -877.89 \text{ kJ} - (-795 \text{ kJ}) = -83 \text{ kJ}$

14-38. $\Delta H_{solution} = \Delta H_f^\circ[OC(NH_2)_2(aq, \text{dilute})] - \Delta H_f^\circ[OC(NH_2)_2(s)]$

$= -318 \text{ kJ} - (-333.5 \text{ kJ}) = 16 \text{ kJ}$

The enthalpy of solution is positive, so the process is endothermic.

urea(s) + heat → urea(aq)

This means the compound should become *more soluble* as the temperature increases.

## HENRY'S LAW

14-40. Molality of $O_2$ = $k \cdot P(O_2)$

$$= (1.7 \times 10^{-6} \text{ molal/mmHg}) \cdot (40.\text{ mmHg}) = 6.8 \times 10^{-5} \text{ molal}$$

## VAPOR PRESSURE CHANGES

14-42. We want to find $P_A$ in the equation $P_A = X_A P^\circ_A$ where A refers to water in a mixture of water (A) and urea (B).

*Step 1:* Calculate moles of urea and water

9.00 g urea • 1 mol/60.06 g = 0.150 mol urea

10.0 g water • 1 mol/18.02 g = 0.555 mol water

*Step 2:* Calculate the mole fraction of water ($X_A$)

$$X_A = \frac{0.555 \text{ mol}}{0.150 \text{ mol} + 0.555 \text{ mol}} = 0.787$$

*Step 3:* Fined the vapor pressure of water over the solution

$$P_A = X_A P^\circ_A = (0.787)(22.4 \text{ mmHg}) = 17.6 \text{ mmHg}$$

14-44. *Step 1:* Calculate the moles of benzene

100.0 g • 1 mol/78.115 g = 1.280 mol

*Step 2:* Calculate the mole fraction of benzene

$$X_{benzene} = \frac{1.280 \text{ mol}}{0.200 \text{ mol} + 1.280 \text{ mol}} = 0.865$$

*Step 3:* Calculate the vapor pressure of benzene over the solution

$$P_{benzene} = X_{benzene} P^\circ_{benzene} = (0.865)(760.\text{ mmHg}) = 657 \text{ mmHg}$$

14-46. *Step 1:* Use Raoult's law to find the mole fraction of solvent (A = $CCl_4$)

$$P_A = X_A P^\circ_A$$

$$504 \text{ mmHg} = X_A \cdot 531 \text{ mmHg}$$

$$X_A = 0.949$$

*Step 2:* Calculate the number of moles of solute in the solution

Moles of $CCl_4$ = 154 g • 1 mol/153.8 g = 1.00 mol

$$X_{solvent} = X_A = \frac{1.00 \text{ mol}}{? \text{ mol solute} + 1.00 \text{ mol}} = 0.949$$

? mol solute = 0.054 mol

*Step 3:* Calculate the molar mass of the unknown

20.0 g / 0.054 mol = 370 g/mol

## BOILING POINT ELEVATION

14-48. *Step 1:* Calculate the moles of urea

15.0 g urea • 1 mol/60.06 g = 0.250 mol urea

*Step 2:* Calculate the molality of the solution

$m$ = 0.250 mol / 0.500 kg = 0.500 molal

*Step 3:* Calculate the boiling point elevation

$\Delta t_{bp} = k_{bp} \cdot m$ = (+0.512 K/molal) • 0.500 molal = 0.256 °C

Boiling point = 100.256 °C

14-50. *Step 1:* Calculate the moles of wintergreen

0.755 g urea • 1 mol/152.2 g = 0.00496 mol oil of wintergreen

*Step 2:* Calculate the molality of the solution

$m$ = 0.00496 mol / 0.100 kg = 0.0496 molal

*Step 3:* Calculate the boiling point elevation

$\Delta t_{bp} = k_{bp} \cdot m$ = (+2.53 K/molal) • 0.0496 molal = 0.126 °C

Boiling point = 80.10 °C + 0.126 °C = 80.23 °C

14-52. *Step 1:* Calculate the molality of the solution from $\Delta t_{bp} = k_{bp} \cdot m$

80.51 °C - 80.10 °C = (+2.53 deg/molal) • $m$

$m$ = 0.162 molal

*Step 2:* Calculate the number of moles and then grams of $C_{10}H_{10}Fe$

0.162 mol/kg • 0.050 kg = 0.00810 mol $C_{10}H_{10}Fe$

$0.00810 \text{ mol } C_{10}H_{10}Fe \cdot 186.0 \text{ g/mol} = 1.51 \text{ g } C_{10}H_{10}Fe$

14-54.

| **Solution** | **Molality of Particles** |
|---|---|
| 0.20 *m* ethylene glycol | 0.20 |
| 0.12 *m* $Na_2SO_4$ | 0.36 |
| 0.10 *m* $CaCl_2$ | 0.30 |
| 0.12 *m* KBr | 0.24 |

0.20 m glycol < 0.12 *m* KBr < 0.10 *m* $CaCl_2$ < 0.12 *m* $Na_2SO_4$

—increasing effect of solute on solution→

—increasing lowering of vapor pressure→

—increasing boiling point→

14-56. *Step 1:* Calculate the molality of the particles in solution from $\Delta t_{bp} = k_{bp} \cdot i \cdot m$ where $i$ = 2 in this case.

$63.20\ ^\circ\text{C} - 61.70\ ^\circ\text{C} = 1.50\ ^\circ\text{C} = (+3.63 \text{ deg/molal}) \cdot 2 \cdot m$

$m = 0.207 \text{ mol/kg}$

*Step 2:* Calculate the number of moles and then grams of salt actually present.

$0.207 \text{ mol/kg} \cdot 0.025 \text{ kg} \cdot 341.9 \text{ g/mol} = 1.77 \text{ g } (C_4H_9)_4N^+ClO_4^-$

14-58. *Step 1:* Calculate the molality of the solution from $\Delta t_{bp} = k_{bp} \cdot i \cdot m$ (where $i$ = 1, since the compound is not ionic).

$80.39\ ^\circ\text{C} - 80.10\ ^\circ\text{C} = (+2.53 \text{ deg/molal}) \cdot m$

$m = 0.12 \text{ mol/kg}$

*Step 2:* Calculate the moles of caffeine dissolved.

$0.12 \text{ mol/kg} \cdot 0.0250 \text{ kg} = 0.0029 \text{ mol}$

*Step 3:* Calculate the molar mass of caffeine

$0.563 \text{ g} / 0.0029 \text{ mol} = 2.0 \times 10^2 \text{ g/mol}$

(The actual molar mass of caffeine, $C_8H_{10}N_4O_2$, is 194.19 g/mol)

14-60. *Step 1:* Calculate the molality of the solution from $\Delta t_{bp} = k_{bp} \cdot i \cdot m$ (where $i = 1$, since the compound is not ionic).

$80.34\ ^\circ C - 80.10\ ^\circ C = (+2.53\ deg/molal) \cdot m$

$m = 0.095\ mol/kg$

*Step 2:* Calculate the moles of caffeine dissolved.

$0.095\ mol/kg \cdot 0.0300\ kg = 0.0028\ mol$

*Step 3:* Calculate the molar mass of caffeine

$0.500\ g / 0.0028\ mol = 180\ g/mol$

(The actual molar mass of anthracene, $C_{14}H_{10}$, is 178.2 g/mol)

## FREEZING POINT DEPRESSION

14-62. *Step 1:* Calculate the molality of the solution from $\Delta t_{fp} = k_{fp} \cdot i \cdot m$ (where $i = 1$, since the compound is not ionic).

$-15\ ^\circ C = (-1.86\ deg/molal) \cdot m$

$m = 8.1\ mol/kg$

*Step 2:* Calculate the moles and then grams of ethylene glycol

$8.1\ mol/kg \cdot 5.0\ kg = 40.\ mol$

$40.\ mol \cdot 62.1\ g/mol = 2.5 \times 10^3\ g$

Step 3: Calculate the boiling point of the solution

$\Delta t_{bp} = k_{bp} \cdot m = (+0.512\ deg/molal)(8.1\ molal) = 4.1\ ^\circ C$

Boiling point = 104.1 °C

14-64. *Step 1:* Calculate moles of $CaCl_2$ used.

$35.0\ g\ CaCl_2 \cdot 1\ mol / 111.0\ g = 0.315\ mol$

*Step 2:* Calculate the molality of the solution

$0.315\ mol / 0.150\ kg = 2.10\ molal$

*Step 3:* Calculate the freezing point depression from $\Delta t_{fp} = k_{fp} \cdot i \cdot m$, where $i = 3$ in this case.

$\Delta t_{fp} = k_{fp} \cdot i \cdot m = (-1.86\ deg/molal)(3)(2.10\ molal) = -11.7\ ^\circ C$

Therefore, the freezing point of the solution is -3.91 °C.

*Step 4:* Find the number of grams of NaCl required to lower the freezing point of 150. g of water to -3.91 °C.

$\Delta t_{fp} = k_{fp} \cdot i \cdot m$

-11.7 deg = (-1.86 deg/molal)(2) • $m$

$m$ = 3.15 mol/kg

3.15 mol/kg • 0.150 kg • 58.44 g/mol = 27.6 g NaCl required

14-66. *Step 1:* Calculate the molality of the solution from $\Delta t_{fp} = k_{fp} \cdot i \cdot m$ (where $i = 1$, since the compound is not ionic).

-0.0920 °C = (-1.86 deg/molal) • $m$

$m$ = 0.0495 mol/kg

*Step 2:* Calculate the moles of vanillin present

0.0495 mol/kg • 0.100 kg = 0.00495 mol vanillin

*Step 3:* Calculate the molar mass of vanillin

0.750 g / 0.00495 mol = 152 g/mol

(Vanillin is $C_8H_8O_3$, molar mass = 152.14 g/mol)

14-68. *Step 1:* Calculate the molality of the solution from $\Delta t_{fp} = k_{fp} \cdot i \cdot m$ (where $i = 1$, since the compound is not ionic).

5.41 °C - 5.50 °C = (-4.90 deg/molal) • m

$m$ = 0.018 mol/kg

*Step 2:* Calculate the moles of unknown present

0.018 mol/kg • 0.01565 kg = $2.9 \times 10^{-4}$ mol

*Step 3:* Calculate the molar mass of the unknown

0.125 g / $2.9 \times 10^{-4}$ mol = 430 g/mol

*Step 4:* Determine the actual formula of the compound.

The molar mass of the unit $(C_2H_5)_2AlF$ is 104.1 g/mol. The molar mass of the actual species in solution is 4 times as large, so the correct molecular formula is $[(C_2H_5)_2AlF]_4$.

| 14-70. **Solution** | **Molality of Particles** |
|---|---|
| 0.20 *m* ethylene glycol | 0.20 |
| 0.12 *m* $K_2SO_4$ | 0.36 |
| 0.10 *m* $MgCl_2$ | 0.30 |
| 0.12 *m* KBr | 0.24 |

0.20 *m* ethylene glycol > 0.12 *m* KBr > 0.10 *m* $MgCl_2$ > 0.12 *m* $K_2SO_4$

—increasing effect of solute on solution→

—increasing freezing point lowering (increasing $\Delta T$)→

—*de*creasing freezing point→

## OSMOTIC PRESSURE

14-72. If seawater, with a high concentration of salts, is ingested, body fluids would be more concentrated in solute than the fluid inside cells. Water would flow out of the cells in an attempt to bring the two concentrations into balance.

14-74. Osmotic pressure = $\Pi$ = MRT

To obtain M, we assume (although it is not a good assumption for more concentrated solutions) that it is the same as molality, where

$\Delta t_{fp} = k_{fp} \cdot i \cdot m$ (where $i = 1$, since the compound is not ionic)

$-2.30\ ^\circ C = (-1.86\ ^\circ C) \cdot m$

$m = 1.24$ mol/kg

Therefore, assuming that $m = M = 1.24$ mol/L

$\Pi = (1.24\ \text{mol/L}) \cdot (0.082057\ \text{L} \cdot \text{atm/K} \cdot \text{mol}) \cdot (298\ \text{K})$

$\Pi = 30.2$ atm

## GENERAL QUESTIONS

14-76. One liter of water contains 997 g of $H_2O$ or 55.4 mol. Therefore, the molality is

$m = 55.4\ \text{mol}/0.997\ \text{kg} = 55.6$ molal

and the molarity is

$M = 55.4\ \text{mol}/1.00\ \text{L} = 55.4$ molar

14-78. $I_2$, a nonpolar solid, dissolves preferentially in the nonpolar liquid $CCl_4$, rather than in highly polar water. The solubility of $I_2$ in water is only 0.030 g/100 mL water at 25 °C, whereas it is 2.91 g/100 mL of $CCl_4$ at 25 °C. (Remember the saying that "like dissolves like.")

14-80. *Step 1:* Calculate the molality of the solution

$9.41 \text{ g} \cdot 1 \text{ mol}/104.1 \text{ g} = 0.0904 \text{ mol}$

$m = 0.0904 \text{ mol} / 1.00 \text{ kg} = 0.0904 \text{ mol/kg}$

*Step 2:* Calculate the value of *i* in $\Delta t_{fp} = k_{fp} \cdot i \cdot m$

$-0.33 \text{ °C} = (-1.86 \text{ °C}) \cdot i \cdot (0.0904 \text{ mol/kg})$

$i = 2.0$

*Step 3:* Decide which reaction is correct.

Since $i = 2$, this must mean that the compound dissolves to give $Na^+$ and $HSO_3^-$ ions.

14-82. According to Figure 14.11 approximately *125 g* of $NaNO_3$ will dissolve in 100 g of water at 60 °C. Since the solubility of $NaNO_3$ at 20 °C is about 85 g/100 g water, about *40 g* of $NaNO_3$ will precipitate on cooling.

14-86.

| **Solution** | **Molality of Particles** |
|---|---|
| 0.35 $m$ $C_2H_4(OH)_2$ | 0.35 |
| 0.50 $m$ sugar | 0.50 |
| 0.20 $m$ KBr | 0.40 |
| 0.20 $m$ $Na_2SO_4$ | 0.60 |

0.35 $m$ $C_2H_4(OH)_2$ < 0.12 $m$ KBr < 0.50 $m$ sugar < 0.12 $m$ $Na_2SO_4$

—increasing boiling point→

14-89. 150. g KBr added - 46.0 g KBr not dissolved = 104.0 g KBr dissolved

14-91. *Step 1:* Calculate the moles of each component of the solution

$1.00 \times 10^3$ g $C_2H_4(OH)_2$ • 1 mol / 62.07 g = 16.1 mol $C_2H_4(OH)_2$

$2.00 \times 10^3$ g water • 1 mol / 18.02 g = 111 mol water

*Step 2:* Calculate the mole fraction water

$$X_{water} = \frac{111 \text{ mol}}{16.1 \text{ mol} + 111 \text{ mol}} = 0.873$$

*Step 3:* Calculate the vapor pressure of water at 90 °C from Raoult's law

$P_{water} = X_{water} \cdot P°_{water} = 0.873 \cdot 525.8$ mmHg = 459 mmHg

14-93. *Step 1:* Calculate the molality of the solution

$\Delta t_{bp} = k_{bp} \cdot m$

80.78 °C - 80.10 °C = (+2.53 deg/mol) • $m$

$m$ = 0.27 mol/kg

*Step 2:* Calculate the number of moles of unknown present in solution

0.27 mol/kg • 0.01523 kg = 0.0041 mol

*Step 3:* Determine the molar mass of the unknown

0.500 g / 0.0041 mol = 122 g/mol

The compound $B_{10}H_{14}$ has a molar mass of 122.2 g/mol, and so must be the unknown compound.

14-95. Here the solution is assumed to be ideal, so the total vapor pressure over the solution is the sum of the vapor pressures of the individual components.

$P_{total} = X_{ben}P°_{ben} + X_{tol}P°_{tol}$

266 mmHg = $X_{ben}$ (388 mmHg) + (1 - $X_{tol}$) (137 mmHg)

$X_{ben}$ = 0.514

14-97. Here the solution is assumed to be ideal, so the total vapor pressure over the solution is the sum of the vapor pressures of the individual components. We simply have to find the mole fraction of each component and then calculate the vapor pressure of each component.

*Step 1:* Calculate the moles of each component

$$5.00 \text{ mL } CH_2Cl_2 \cdot 1.34 \text{ g/mL} \cdot 1 \text{ mol} / 84.93 \text{ g} = 0.0789 \text{ mol } CH_2Cl_2$$

$$10.0 \text{ mL } CHCl_3 \cdot 1.49 \text{ g/mL} \cdot 1 \text{ mol} / 119.4 \text{ g} = 0.125 \text{ mol } CHCl_3$$

*Step 2:* Calculate the mole fraction of each component

$$X(CHCl_3) = \frac{0.125 \text{ mol}}{0.125 \text{ mol} + 0.0789 \text{ mol}} = 0.614$$

$$X(CH_2Cl_2) = 1 - X(CHCl_3) = 0.386$$

*Step 3:* Calculate the total pressure

$$P_{total} = X(CHCl_3) \cdot P^{\circ}(CHCl_3) + X(CH_2Cl_2) \cdot P^{\circ}(CH_2Cl_2)$$

$$= (0.614)(435 \text{ mmHg}) + (0.386)(195 \text{ mmHg})$$

$$= 288 \text{ mmHg}$$

## SUMMARY QUESTIONS

14-100. PART (a) — Determine the empirical formula

$$0.0151 \text{ g unknown} \rightarrow 0.0111 \text{ g } H_2O + 0.0449 \text{ g } CO_2$$

*Step 1:* Calculate moles and then grams of C

$$0.0449 \text{ g } CO_2 \cdot 1 \text{ mol } CO_2 / 44.01 \text{ g} \cdot 1 \text{ mol C}/1 \text{ mol } CO_2 = 0.00102 \text{ mol C}$$

$$0.00102 \text{ mol C} \cdot 12.01 \text{ g C}/1 \text{ mol C} = 0.0123 \text{ g C}$$

*Step 2:* Calculate moles and then grams of H

$$0.0111 \text{ g } H_2O \cdot 1 \text{ mol } H_2O / 18.02 \text{ g} \cdot 2 \text{ mol H}/1 \text{ mol } H_2O = 0.00123 \text{ mol H}$$

$$0.00123 \text{ mol H} \cdot 1.008 \text{ g H/mol H} = 0.00124 \text{ g H}$$

*Step 3:* Calculate the mass and then moles of O in the unknown:

Total mass = 0.0151 g = mass of O + 0.0123 g C + 0.00124 g H

Mass of O = 0.00161 g O

Moles of O = 0.00161 g O • 1 mol/16.00 g = 0.000100 mol O

*Step 4:* This leads us finally to the combining ratios of the elements:

0.001243 mol H / 0.000100 mol O = 12.3 mol H / 1.00 mol O

0.00102 mol C / 0.000100 mol O = 10.2 mol C / 1.00 mol O

*Step 5:* Write the empirical formula

The combining ratios lead to $C_{10.2}H_{12.3}O$, which simplifies to $C_{10}H_{12}O$.

PART (b) — Calculate the molar mass of the unknown

*Step 1:* Calculate the molality of the solution

$\Delta t_{fp} = k_{fp} \cdot m$

177.05 °C - 179.75 °C = -2.70 °C = (-40.0 deg/molal) • $m$

$m$ = 0.0675 mol/kg

*Step 2:* Calculate the moles of unknown present

0.0675 mol/kg • 0.100 kg = 0.00675 mol unknown

*Step 3:* Calculate the molar mass of the unknown

1.00 g / 0.00675 mol unknown = 148 g/mol

PART (c) — Determine the molecular formula

The mass of the empirical formula ($C_{10}H_{12}O$) is 148.2 g/mol, so the molecular formula must be the same as the empirical formula.

# PART IV

# THE CONTROL OF CHEMICAL REACTIONS

# CHAPTER 15

# CHEMICAL KINETICS: RATES AND MECHANISMS OF CHEMICAL REACTIONS

## INSTRUCTOR'S NOTES

This chapter has been very thoroughly revised, with input from several faculty colleagues. One major aspect of the revision was to move the discussion of mechanisms that have an equilibrium as a first step to the next chapter on chemical equilibria.

The discussion of kinetics and mechanisms has been receiving more attention lately in introductory chemistry, although there is still some argument about its value at this level. Nonetheless, we spend 3-4 lecture hours on the material. Our emphasis is on the contrast in speeds of chemical reactions and the factors that go into determining those speeds. We also discuss the idea of a rate law and the important point that it must be experimentally determined. After discussing the first order rate law in particular, we spend about one half a lecture or so on catalysis.

In talking about kinetics our preference might have been to first describe common mechanisms and then see how they are evidenced by rate laws. It would seem that students would come to a better appreciation of kinetics by taking such an approach. However, this does not seem to work in practice. The idea of a mechanism, as a hypothetical pathway for the reaction, is not an easy concept for them to grasp.

Another area of difficulty for students when studying kinetics is the contrast between the fact that the rate law is related to the mechanism but not necessarily related to the stoichiometric equation. When discussing mechanisms, we always stress this point.

## LIST OF TRANSPARENCIES FOR THIS CHAPTER

| No. | Figure | Page | Subject |
|---|---|---|---|
| 83 | 15.5 | 622 | Illustration of half-life |
| 84 | 15.7 | 625 | Concentrations vs. time curves for $NO_2$ decomposition |
| 85 | 15.11 | 629 | Reaction energy diagram for an exothermic process |

## SUGGESTED DEMONSTRATIONS

The burning of lycopodium powder in Figure 13.2 is a good demonstration, although it is certainly more dramatic to explode lycopodium powder as outlined in Volume 1 of "Chemical Demonstrations" by Shakhashiri (page 103).

Some of the examples in this chapter are based on hydrogen peroxide decomposition because we commonly do this reaction as a demonstration The details were outlined in Chapter 4 of this *Manual.*

In Volume 2 of "Chemical Demonstrations" by Shakhashiri the last chapter outlines many oscillating chemical reactions, which could be used as demonstrations in lectures on kinetics.

The following is a list of possible demonstrations on kinetics; the list was obtained from the Project SERAPHIM database (see the introduction to this *Manual*).

- Brice, L.K. "Rossini, William Tell, and the Iodine Clock Reaction" *J. Chem. Educ.* **1980**, *57*, 152 .
- Alyea, H.N. "The Old Nassau Reaction" *ibid.* **1977**, *54*, 167 .
- Moss, A. "The Landolt, Old Nassau, and Variant Reactions" *ibid.* **1978**, *55*, 244.
- Mishra, H.C.; Singh, C.M. "Belousov's Oscillating Reaction in Acidic Medium other than Sulfuric Acid" *ibid.* **1977**, *54*, 377 .
- Shigematsu, E. "A Few Chemical Tricks Based on the Clock Reaction" *ibid.* **1979**, *56*, 184.
- Field, R.J.; Winfree, A.T. "Traveling Waves of Chemical Activity in the Zaikin-Zhabotinskii-Winfree Reagent" *ibid.* **1979**, *56*, 754.
- McNaught, I.J. "Thermodynamic versus Kinetic Control: A Lecture Demonstration" *ibid.* **1978**, *55*, 722.
- Littlejohn, D.G.; Fanning, J.C.; Jostansek, E.C. "The Reaction of $CN^-$ Ions with $Ni(EDTA)_2$ Ions" *ibid.* **1972**, *49*, A26.
- Briggs, T.S.; Rauscher, W.C. "An Oscillating Iodine Clock" *ibid.* **1973**, *50*, 486.
- Field, R J. "A Reaction Periodic in Time and Space" *ibid.* **1972**, *49*, 308.
- Siwon, J. "Effect on Temperature on Reaction Rate" *ibid.* **1972**, *49*, A85.

# SOLUTIONS TO SELECTED STUDY QUESTIONS

## REACTION RATES

15-21. (a) $-\Delta[HC\equiv CH]/\Delta t = -\Delta[HCN]/\Delta t = +\Delta[H_2C\text{=}CHCN]/\Delta t$

(b) $-\Delta[CH_4]/\Delta t = -\frac{1}{2}\Delta[O_2]/\Delta t = +\Delta[CO_2]/\Delta t = +\frac{1}{2}\Delta[H_2O]/\Delta t$

(c) $-\frac{1}{2}\Delta[C_2H_6]/\Delta t = -{}^1\!/_7\,\Delta[O_2]/\Delta t = +{}^1\!/_4\,\Delta[CO_2]/\Delta t = +{}^1\!/_6\,\Delta[H_2O]/\Delta t$

15-23. Plot of the concentration of phenyl acetate versus time.

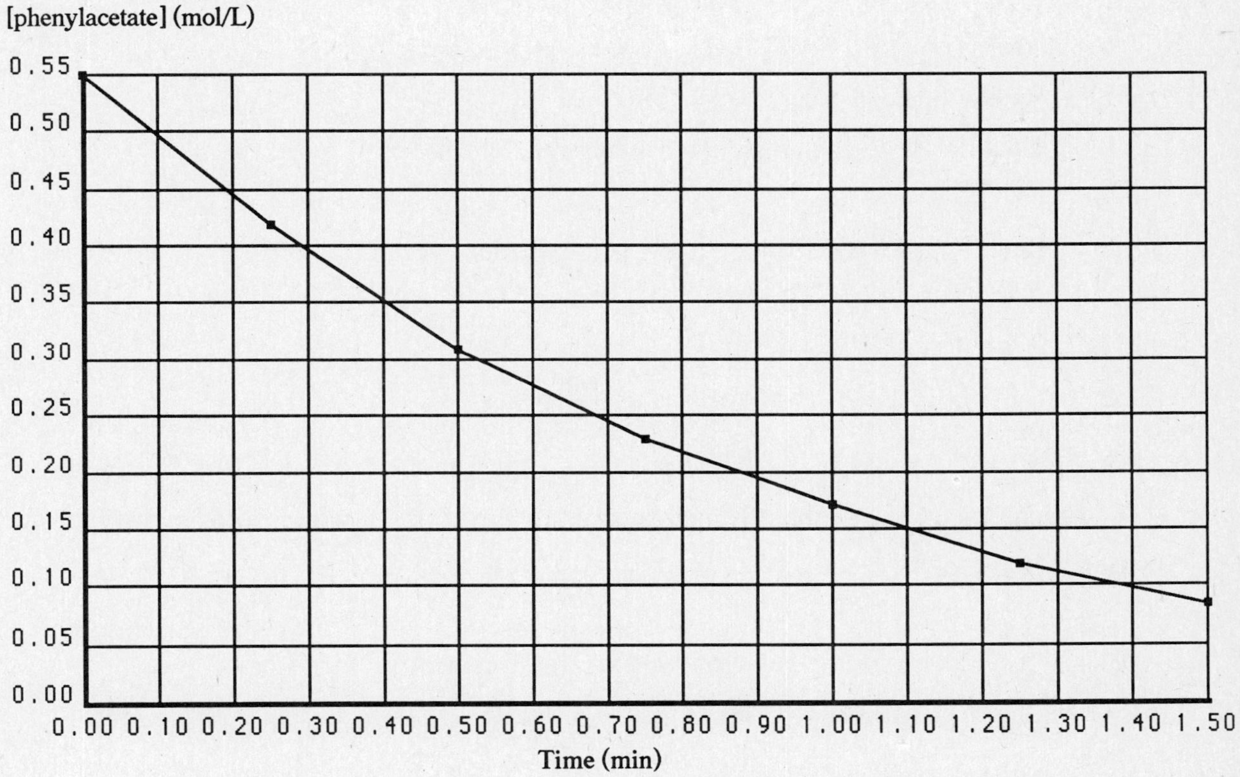


(a) The curve shape is a decreasing exponential, as for $[Pt(NH_3)_2Cl_2]$ in Figure 15.4 on page 610.

(b) For the time period from 0.20 minutes to 0.40 minutes, the rate of change of [phenyl acetate] is

$\Delta$[phenyl acetate] / $\Delta t$ = - (0.10 mol/L) / 0.20 min = - 0.50 mol/L • min

For the time period from 1.2 minutes to 1.4 minutes, the rate of change of [phenyl acetate] is

$\Delta$[phenyl acetate] / $\Delta$t = - (0.030 mol/L) / 0.20 min = - 0.15 mol/L • min

The rate of change for the later time period is lower because [phenyl acetate] is lower at that time.

(c) For the time period 1.00 to 1.25 minutes:

Rate of change for [phenol] = - rate of change of [phenyl acetate]

= - [- (0.05 mol/L) / 0.25 min]

= + 0.20 mol/L • min

## CONCENTRATION AND RATE EXPRESSIONS

15-25. (a) Rate = $k[NO]^2[Br_2]$

(b) Rate will also triple.

(c) Doubling [NO] will raise the reaction rate by a factor of four.

15-27. (c) Rate = $k[A][B]^2$. This rate law is third order overall.

15-29. Rate = $k[H_2][NO]^2$

15-31. Rate = (0.137 L/mol • sec) • (0.025 mol/L) • ( 0.025 mol/L) = $8.6 \times 10^{-5}$ mol/L • sec

The stoichiometry shows that the methanol appears in solution as rapidly as methyl acetate disappears.

15-33. (a) To deduce the rate law, we shall examine experiments 1 and 2.

$$\frac{\text{Rate of 1}}{\text{Rate of 2}} = \left(\frac{[A]\text{ in 1}}{[A]\text{ in 2}}\right)^x \qquad \frac{6.0 \times 10^{-3}}{1.2 \times 10^{-2}} = \left(\frac{0.10}{0.20}\right)^x$$

The value of x derived from the expression above is 1. Therefore, the reaction is first order in A. Experiments 2 and 3 yield the same result. Experiments 2 and 4 have a concentration value of [A], while [B] changes by a factor of 3.

$$\frac{\text{Rate of 4}}{\text{Rate of 2}} = \left(\frac{[B]\text{ in 4}}{[B]\text{ in 2}}\right)^x \qquad \frac{1.1 \times 10^{-1}}{1.2 \times 10^{-2}} = \left(\frac{0.15}{0.05}\right)^x$$

The value of x derived from this expression is 2, so the reaction is second order with respect to B. This means the rate law is

$$\text{Rate} = k[A][B]^2$$

This means the reaction is *third order* overall.

(b) To calculate k, use the data from Experiment 1.

Initial rate = $6.0 \times 10^{-3}$ mol/L • s = k [0.10 mol/L] [0.05 mol/L]$^2$

k = 20 L$^2$/mol$^2$ • s

15-35. (a) $-\Delta[O_2]/\Delta t = +\frac{1}{2}\Delta[CH_2O]/\Delta t$

(b) On going from Experiment 1 to Experiment 2, the concentration of $C_2H_4$ is held constant, while $[O_3]$ triples. At the same time the initial rate of change of $[CH_2O]$ triples. Therefore, we know that the reaction is *first order* with respect to $[O_3]$. In the first and third data sets, both reactants change in concentration. Based on the previous conclusion, we know that doubling $[O_3]$ should lead to a doubling of the initial rate. Since the rate increases by a factor of 4 on going from Experiment 1 to Experiment 3, this must mean, therefore, that there is a *first order* dependence on $[C_2H_4]$ as well. Thus, the rate law, which is *second order overall,* is

$$\text{Initial rate} = k[O_3][C_2H_4]$$

(c) To calculate the rate constant, substitute values into the rate law:

$1.0 \times 10^{-12}$ mol/L • s = k ($0.50 \times 10^{-7}$ mol/L) • ($1.0 \times 10^{-8}$ mol/L)

k = $2.0 \times 10^{3}$ L/mol • s

(d) Reaction rate = ($2.0 \times 10^{3}$ L/mol • s) • ($2.0 \times 10^{-7}$ mol/L) • ($2.0 \times 10^{-7}$ mol/L)

= $8.0 \times 10^{-11}$ mol/L • s

15-37. (a) On going from Experiment 1 to 2, [NO] is constant, but $[O_2]$ triples. At the same time the initial rate triples. Therefore, the reaction is *first order* in $[O_2]$. Comparison of Experiments 1 and 3 gives the same result. We can deduce the order with respect to [NO] from Experiments 3 and 5. Here $[O_2]$ is constant, while [NO] increases by a factor of 3. At the same time the initial rate increases by a factor of

9. Therefore, there must be a *second order* dependence on [NO]. Based on the conclusions above, we conclude that the reaction rate law is

Initial rate $= k\,[O_2]\,[NO]^2$

(b) The overall order is 3.

(c) Initial rate $= k\,[O_2]\,[NO]^2$

(d) Rate of appearance of $NO_2$ when $[NO] = [O_2] = 0.005$ mol/L.

In order to perform this calculation, we first need a value for k, the rate constant. Therefore, using the data from Experiment 2, we have

$14 \times 10^{-6}$ mol/L • s $= k \cdot (0.002$ mol/L$) \cdot (0.001$ mol/L$)^2$

$k = 7000$ $L^2/mol^2 \cdot s$

With the value k determined, we can find the initial under the specified conditions.

Initial rate $= (7000\ L^2/mol^2 \cdot s) \cdot (0.005\ mol/L) \cdot (0.005\ mol/L)^2$

$= 9 \times 10^{-4}$ mol/L • s

## CONCENTRATION/TIME EQUATIONS AND REACTION HALF-LIFE

15-39. $\ln \{[SO_2Cl_2] / [SO_2Cl_2]_0\} = -k \cdot t$

$\ln (0.31 \times 10^{-3}) / (1.25 \times 10^{-3}) = -(0.17/hr) \cdot t$

$t = 8.2$ hours

Alternatively, we see that the concentration drops to 0.25 of its original value, and so requires 2 half-lives. From $t_{\frac{1}{2}} = 0.693/k$, we know that the half-life is 4.1 hours, so the time required is 8.2 hours.

15-41. (a) $t_{\frac{1}{2}} = 0.693/k = 0.693 / (5.0 \times 10^{-4}/s) = 1.4 \times 10^3$ seconds

(b) $\ln (0.10/1.0) = -(5.0 \times 10^{-4}/s)t$

$t = 4.6 \times 10^3$ seconds

15-43. $k = 0.693 / t_{\frac{1}{2}} = 0.693 / 4.1$ hours $= 0.17$/hour

$\ln (2.00 \times 10^{-4}$ moles $/ 1.6 \times 10^{-3}$ moles$) = -(0.17/hr)t$

$t = 12$ hours

In this case, the ratio $[A]/[A]_0 = 0.125$. That is, the concentration has dropped to 1/8 of its original value. This means that three half-lives have elapsed. Since the half-life is 4.1 hours, 12.3 hours must have elapsed.

15-45. (a) $k = 0.693/t_{\frac{1}{2}} = 0.693/1.07 \text{ hours} = 0.648/\text{hr}$

(b) $\ln\{[A]/[A]_0\} = \ln(\text{fraction remaining}) = -k \cdot t = -(0.648/\text{hr}) \cdot (3.21 \text{ hours})$

$\ln(\text{fraction remaining}) = -2.08$

fraction remaining = 0.125

We would have obtained this same result by recognizing 3.21 hours represents 3 half-lives. After 3 half-lives the concentration should drop to 1/8 (0.125) of the original value.

(c) Rate = k[*trans*] = (0.648/hr) • (0.125 mol/L) = 0.0810 mol/L • hr

(d) $\ln(0.010/1.00) = -(0.648/\text{hr}) \cdot t$

t = 7.11 hours

15-47. (a) Find the partial pressure of each reactant and product after 1 half-life (245 minutes).

| | $P(SO_2Cl_2)$ | $P(SO_2)$ | $P(Cl_2)$ |
|---|---|---|---|
| Initial P | 25 | 0 | 0 |
| Change in P | -12.5 | +12.5 | +12.5 |
| Final P | 12.5 | 12.5 | 12.5 |

The total pressure after one half-life is 37.5 mmHg.

(b) Find the partial pressure of each reactant after 12 hours (720 minutes).

$k = 0.693/245 \text{ min} = 2.83 \times 10^{-3}/\text{min}$

$\ln(\text{fraction remaining}) = -k \cdot t = -(2.83 \times 10^{-3}/\text{min}) \cdot (720 \text{ minutes})$

fraction remaining = 0.13

Pressure of $SO_2Cl_2$ after 720 minutes = 25 mmHg (0.13) = 3.3 mmHg

Therefore, the pressure of each of the products is 21.7 mmHg for a total pressure at this point of 46.7 mmHg.

15-49.

$$\frac{1}{0.0095 \text{ M}} - \frac{1}{0.010} = (0.137 \text{ L/mol} \cdot \text{sec}) \cdot t \text{ and so } t = 38 \text{ seconds}$$

## GRAPHICAL ANALYSIS

15-51.

| Time (min) | [Phenyl acetate] | ln [Phenyl acetate] | 1 / [Phenyl acetate] |
|---|---|---|---|
| 0 | 0.55 | -0.59 | 1.9 |
| 0.25 | 0.42 | -0.86 | 2.4 |
| 0.50 | 0.31 | -1.17 | 3.2 |
| 0.75 | 0.23 | -1.47 | 4.3 |
| 1.0 | 0.17 | -1.77 | 5.9 |
| 1.25 | 0.12 | -2.12 | 8.3 |
| 1.50 | 0.085 | -2.47 | 11.8 |

**Plot of ln[phenyl acetate] versus time**

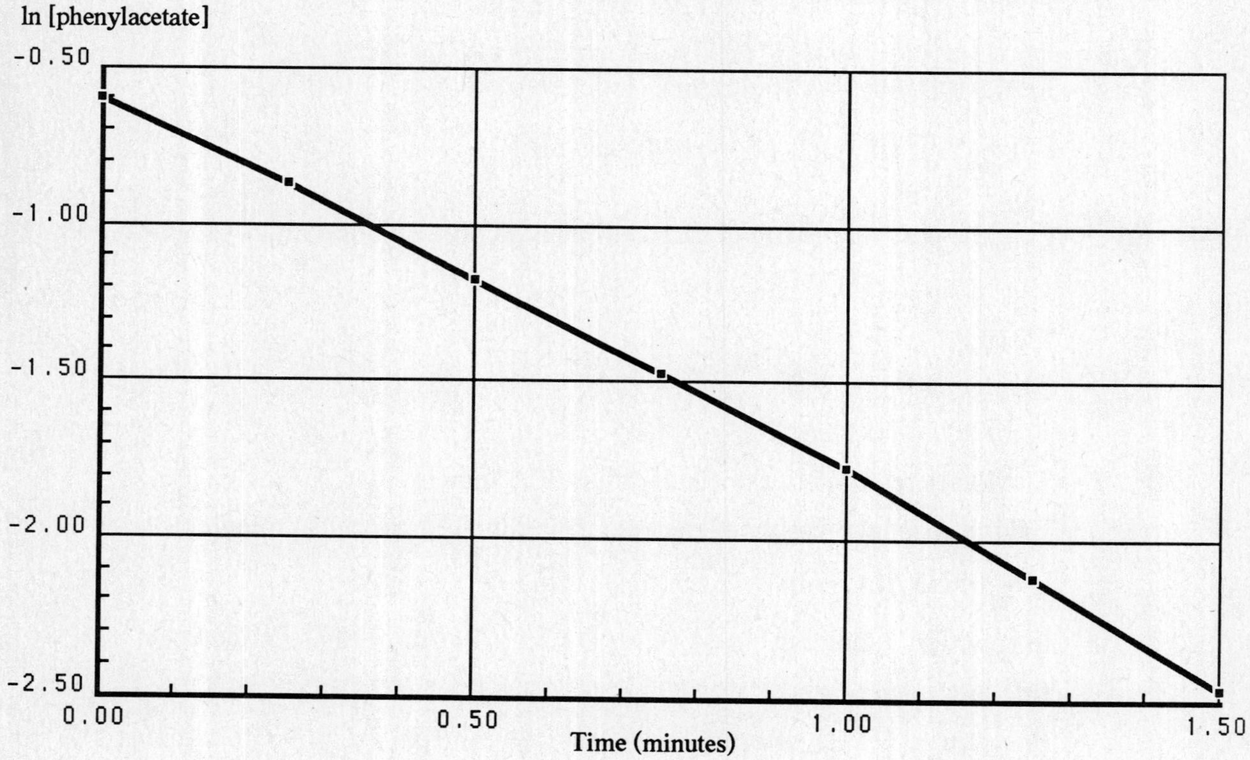

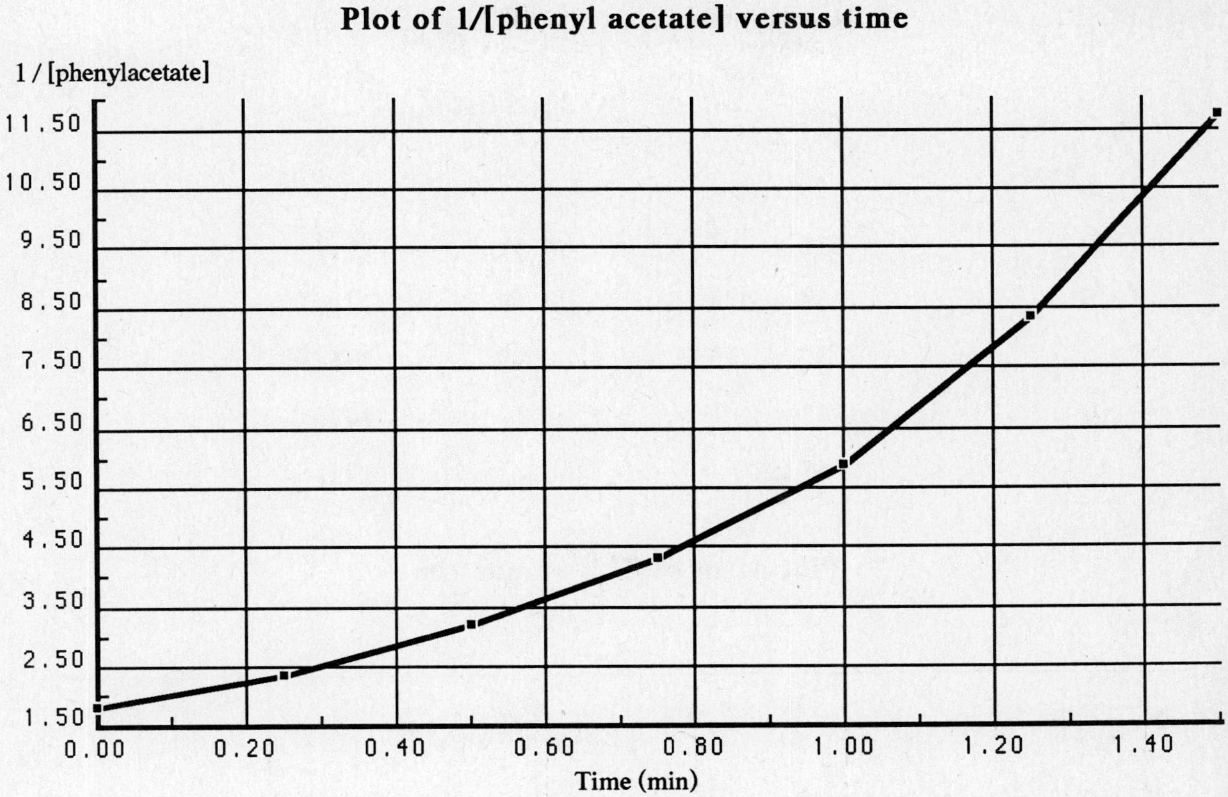


Since the plot of ln[PA] versus time is a straight line, this indicates that the reaction is first order in phenyl acetate.

(b) Slope of the plot = [(-2.47) - (-0.60) / (1.50 - 0) = $-1.25\ min^{-1}$

$k$ = - slope = $1.25\ min^{-1}$

$t_{\frac{1}{2}}$ = 0.693/k = 0.555 min

15-53.

| TIME (sec) | $[C_4H_6]$(mol/L) | ln $[C_4H_6]$ | 1 / $[C_4H_6]$ |
|---|---|---|---|
| 195 | $1.62 \times 10^{-2}$ | -4.12 | 61.7 |
| 604 | $1.47 \times 10^{-2}$ | -4.22 | 68.0 |
| 1246 | $1.29 \times 10^{-2}$ | -4.35 | 77.5 |
| 2180 | $1.10 \times 10^{-2}$ | -4.51 | 90.9 |
| 4140 | $0.84 \times 10^{-2}$ | -4.78 | 119.0 |
| 4655 | $0.80 \times 10^{-2}$ | -4.83 | 125 |
| 6210 | $0.68 \times 10^{-2}$ | -4.99 | 147 |
| 8135 | $0.57 \times 10^{-2}$ | -5.17 | 175 |

**Plot of ln $[C_4H_6]$ versus time**

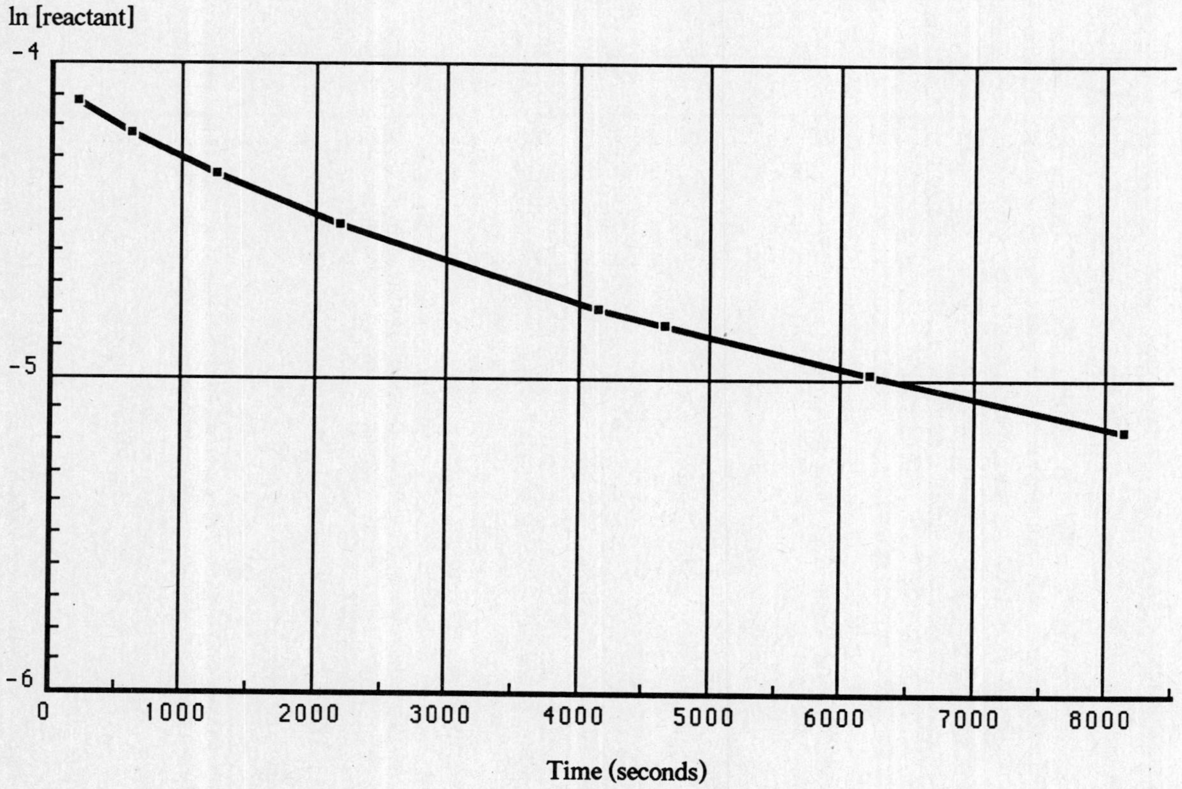

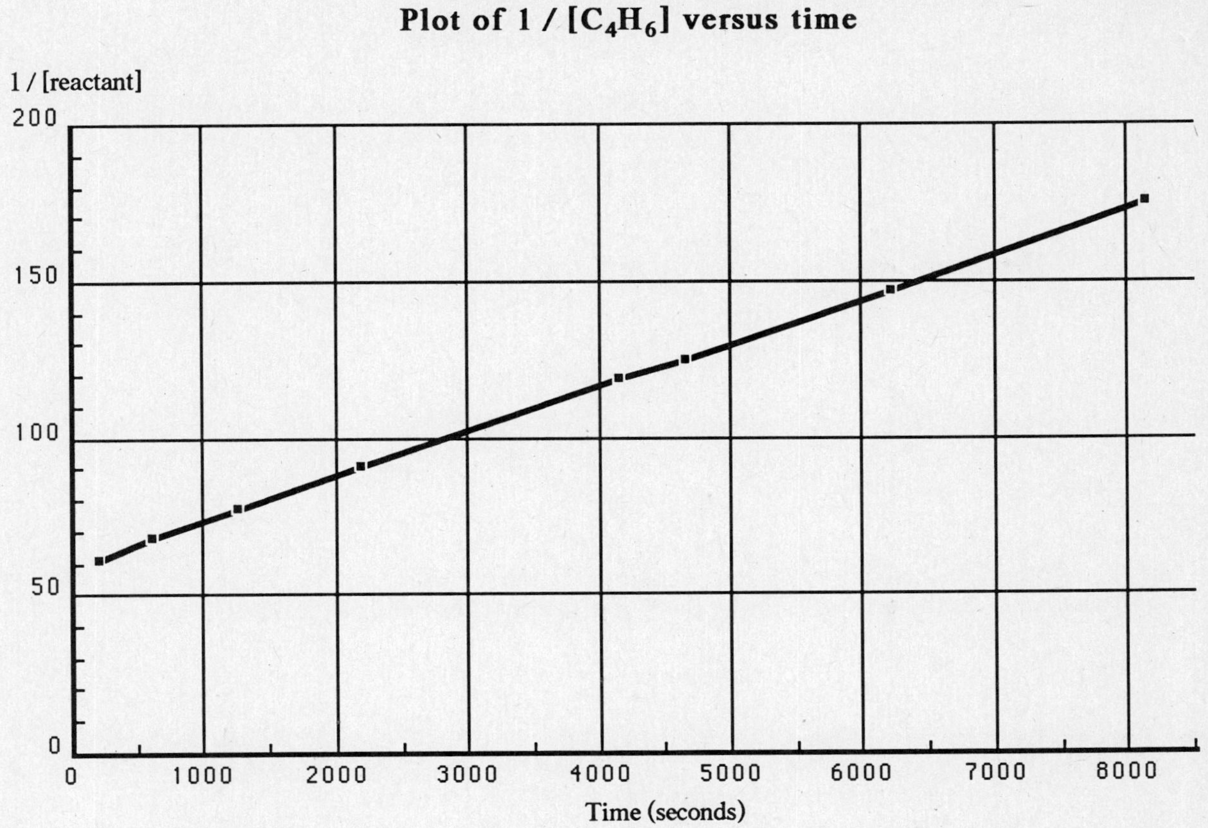


The plot of 1/ $[C_4H_6]$ versus time is linear, so the reaction is second order in $C_4H_6$.

The slope of the plot = k = (147 - 61.7) / (6210 - 195) = $1.42 \times 10^{-2}$ $M^{-1}$ $s^{-1}$

## KINETICS AND ENERGY

15-55. (a) Endothermic

(b) $\Delta E_f$ = 25 kJ/mol

(c) $E^*_f$ = 65 kJ/mol; $E^*_r$ = 40 kJ/mol

(d) $E^*_r$ is lowered by 10 kJ/mol

15-57. E* for a reaction if k triples when the temperature rises from 300. K to 310. K.

$\ln (k_2/k_1) = (-E^*/R)\ [(1/T_2) - (1/T_1)]$

$\ln (3/1) = (-E^* / 8.3145 \times 10^{-3}\ \text{kJ/mol} \cdot \text{K})\ [(1/310) - (1/300)]$

$1.10 = (-E^* / 8.3145 \times 10^{-3}\ \text{kJ/mol} \cdot \text{K})\ [-1.08 \times 10^{-4} /\text{K}]$

E* = 85 kJ/mol

## MECHANISMS

15-59. (a) Step 2 is rate determining

(b) Rate = k $[CHCl_3]$ [ Cl]

(c) Step 1 is unimolecular, while steps 2 and 3 are bimolecular.

15-61. (a) The "alternate step 2" is rate-determining.

(b) Rate = k $[NH_2Cl]$ $[OH^-]$

15-63. Mechanism 1 has the rate law: Rate = k $]NO_2]$ [CO]

Mechanism 2 has the rate law: Rate = k $[NO_2]^2$

Mechanism 3 has the rate law: Rate = k $[NO_2]$

*Mechanism 2* is the only one of the three that is consistent with experiment.

## CATALYSIS

15-65. A finely divided heterogeneous catalyst has more surface area on which the molecules can react than does a block of the catalyst of the same mass.

15-67. $H^+$ is apparently a catalyst in the reaction.

## GENERAL QUESTIONS

15-70. **Reaction energy diagram that proceeds through two steps, with the second step rate-determining.**

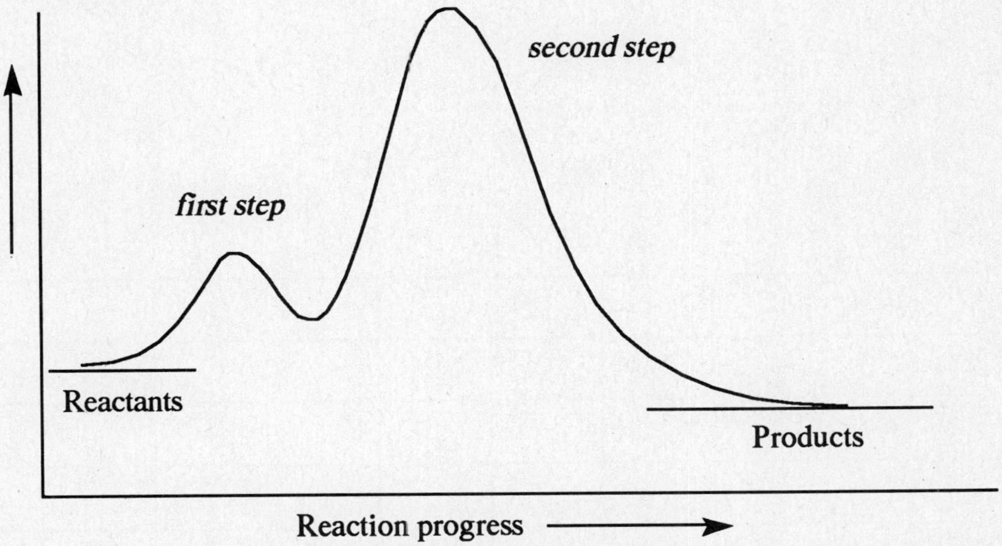


The second step is rate determining here. If the first step were rate determining, the energy of the first step — the first "hump" in the curve — would be higher than the second.

15-73. *Step 1:* Calculate the rate constant.

$k = 0.693 / t_{\frac{1}{2}} = 0.693 / 10.2 \text{ years} = 0.0680 \text{ /yr}$

*Step 2:* Find the fraction of pesticide remaining after 4.00 years.

$\ln(\text{fraction remaining}) = -k \cdot t = -(0.0680/\text{yr}) \cdot (4.00 \text{ yr})$

fraction remaining = 0.762

*Step 3:* Find the concentration 4.00 years ago (= $[A]_0$).

fraction remaining = $[A]/[A]_0$

$0.762 = (3.1 \times 10^{-5} \text{ g/mL}) / [A]_0$

$[A]_0 = 4.07 \times 10^{-5} \text{ g/mL}$

15-75. (a) True (c) False "... the rate decreases."

(b) True (d) True

| 15-77. T(K) | 1/T | k(per second) | ln k |
|---|---|---|---|
| 338 | $2.96 \times 10^{-3}$ | $4.87 \times 10^{-3}$ | -5.32 |
| 328 | $3.05 \times 10^{-3}$ | $1.50 \times 10^{-3}$ | -6.50 |
| 318 | $3.14 \times 10^{-3}$ | $4.98 \times 10^{-4}$ | -7.60 |
| 308 | $3.25 \times 10^{-3}$ | $1.35 \times 10^{-4}$ | -8.91 |
| 298 | $3.36 \times 10^{-3}$ | $3.46 \times 10^{-5}$ | -10.3 |
| 273 | $3.66 \times 10^{-3}$ | $7.87 \times 10^{-7}$ | -14.1 |

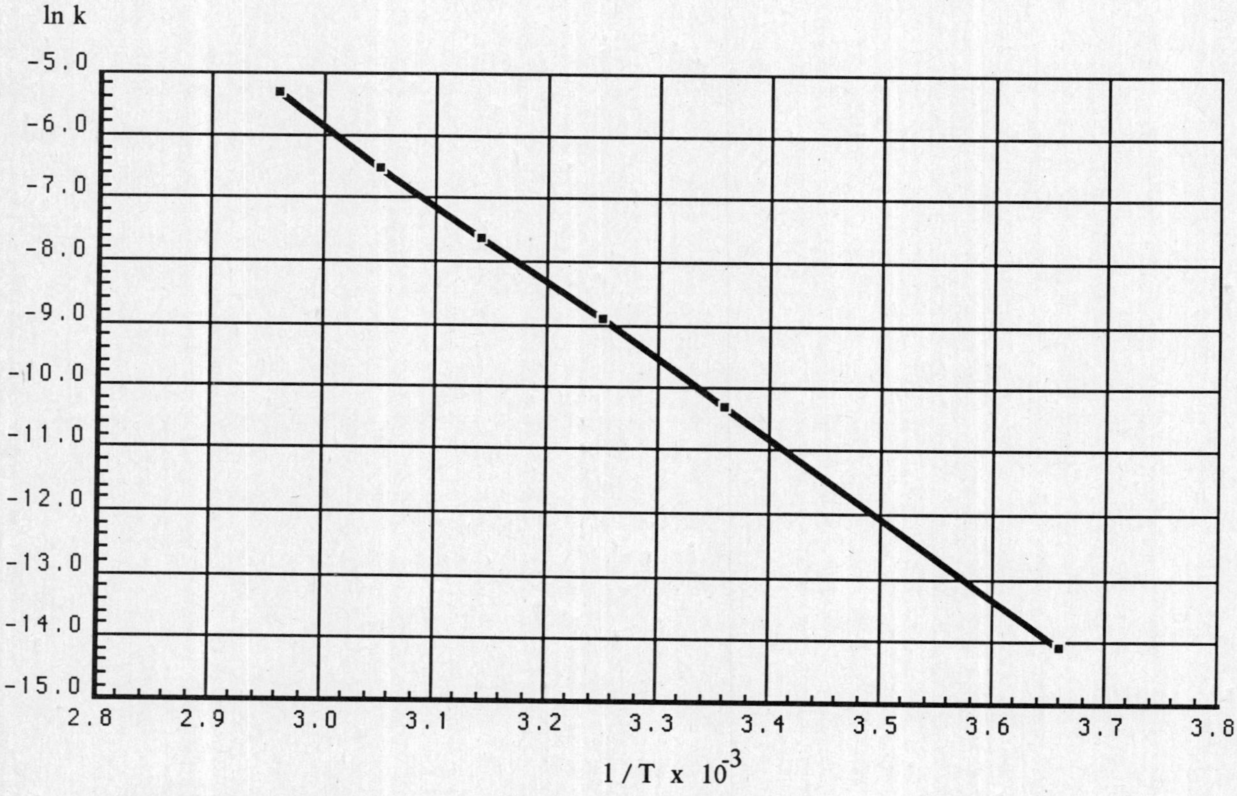


We shall calculate the slope using the data for $T_1$ = 338 K and $T_2$ = 298 K.

$$\text{Slope} = \frac{\ln k_2 - \ln k_1}{(1/T_2) - (1/T_1)} = \frac{(-10.27) - (-5.32)}{(0.00336 - 0.00296)} = -1.24 \times 10^4 \text{ K}$$

Slope = $-1.24 \times 10^4$ K = $-E^*/R$

Since R = $8.3145 \times 10^{-3}$ kJ/K • mol

$E^*$ = 103 kJ/mol

# CHAPTER 16

# CHEMICAL EQUILIBRIA: GENERAL CONCEPTS

## INSTRUCTOR'S NOTES

Introductory courses have always been the point at which students began their study of chemical equilibria, so Chapters 16-19 are quite important. Chapter 16 covers the most basic aspects of equilibria, and we usually spend about 3 lecture hours on the material.

One somewhat unusual aspect of this book is the graphical presentation of equilibria. We have found this a useful approach, since the students tie this in with previous discussions of phase equilibria and because they can readily see that there are many combinations of concentrations that can define an equilibrium system. Further, it is easy to see when a system is or is not at equilibrium. Hence, the concept of the reaction quotient was introduced after describing Figure 16.6. Clearly, however, it is not necessary to take this approach, since the reaction quotient can be described without reference to an equilibrium plot.

You will notice that we have not restricted ourselves to gas phase equilibrium systems here but have introduced a few simple systems in aqueous solution. This was done deliberately so that students see that the principles apply to all types of chemical systems.

## LIST OF TRANSPARENCIES FOR THIS CHAPTER

There are no overhead transparencies designated for this chapter.

## SUGGESTED DEMONSTRATIONS

We have not found as many demonstrations of chemical equilibria for the introductory lectures on the topic as we would like. We have demonstrated the $CO_2/Ca(OH)_2$ system (see pages 806-807) in class with reasonable success. (This is taken from Volume 1 of "Chemical Demonstrations" by Shakhashiri, page 329.) The use of the $Fe^{3+}/SCN^-$ system on page 661 was done to tie the book to the classical laboratory experiment on this equilibrium system.

The illustration of the $NO_2/N_2O_4$ equilibrium system in Figure 16.6 (page 681) is simple to perform. Nitrogen dioxide, generated in the copper/nitric acid reaction, can be sealed into flasks.

To see a distinct color difference it is necessary to have as large a temperature difference as possible.

The so-called "silver one-pot" reaction illustrated on page 809 is also a useful demonstration of chemical equilibria, as is the reaction

$$2\ CrO_4^{2-}(aq)\ +\ 2\ H_3O^+(aq) \rightleftarrows Cr_2O_7^{2-}(aq)\ +\ 3\ H_2O(\ell)$$

The following are possible demonstrations of chemical equilibria taken from the Project SERAPHIM database (see the introduction to this *Manual*).

- Ophardt, C.E "Cobalt Complexes in Equilibrium" *J. Chem. Educ.* **1980**, *57*, 453.
- Burke, B.L "Chemical Equilibrium" *ibid.* **1977**, *54*, 29.

# SOLUTIONS TO SELECTED STUDY QUESTIONS

## WRITING EQUILIBRIUM CONSTANT EXPRESSIONS

16-8. (a) $$K = \frac{[O_3]^2}{[O_2]^3}$$

(b) $$K = \frac{[H_2O]^2}{[SiH_4]\,[O_2]^2}$$

(c) $$K = \frac{1}{[SO_2]\,[O_2]^{\frac{1}{2}}}$$

(d) $$K = \frac{[SO_2]^2}{[O_2]^3}$$

16-10. (a) $$K = \frac{P^3(HF) \cdot P^{\frac{1}{2}}(N_2) \cdot P^{\frac{1}{2}}(Cl_2)}{P(NH_3) \cdot P(ClF_3)}$$

(b) $$K = \frac{[ClO_3^-]^2\,[Cl^-]}{[ClO_2^-]^3}$$

(c) $$K = \frac{[IO_6^{5-}]\,[Cl^-]^2}{[IO_3^-]\,[OH^-]^6\,P(Cl_2)}$$

16-12. (d) $(K_c)^3 = K'_c$

16-14. $K = (1.7 \times 10^{-4})^{\frac{1}{2}} = 1.3 \times 10^{-2}$

16-16. $K_{net} = K_1 \cdot K_2 \cdot K_3$

16-18. $K'_c = (1 / K_c)^{\frac{1}{2}} = (1 / 1.7 \times 10^{-5})^{\frac{1}{2}} = 240$

## CALCULATING AND USING EQUILIBRIUM CONSTANTS

16-20. (a) Calculate K

$$K = \frac{[FeSCN^{2+}]}{[Fe^{3+}]\,[SCN^-]} = \frac{(2.5 \times 10^{-5})}{(9.8 \times 10^{-4})\,(1.8 \times 10^{-4})} = 140$$

(b) Product is favored over reactants.

16-22. $K = [CO]^2 / [CO_2] = (0.10\ mol/2.0\ L)^2 / (0.20\ mol/2.0\ L) = 0.025$

16-24.

| | [HI] | $[H_2]$ | $[I_2]$ |
|---|---|---|---|
| Initial [...] | 2.0 | 0.0 | 0.0 |
| Δ [...] | -0.20 (2.0) | $+\frac{1}{2}(0.20)(2.0)$ | $+\frac{1}{2}(0.20)(2.0)$ |
| Equilibrium [...] | 1.6 | 0.20 | 0.20 |

$$K_c = \frac{[H_2]\,[I_2]}{[HI]^2} = \frac{(0.20)(0.20)}{(1.6)^2} = 0.016$$

$K_p = K_c(RT)^{\Delta n}$

Here $\Delta n = 0$, so $K_p = K_c$

16-26.

| | $[H_2]$ | $[S_2]$ | $[H_2S]$ |
|---|---|---|---|
| Equilibrium [...] | 0.208 M | $1.13 \times 10^{-6}$ M | 0.725 M |

$$K_c = \frac{[H_2S]^2}{[H_2]^2\,[S_2]} = \frac{(0.725)^2}{(0.208)^2\,(1.13 \times 10^{-6})} = 1.08 \times 10^7$$

$K_p = K_c(RT)^{\Delta n}$

Here $\Delta n = 2 - 3 = -1$, so $K_p = (1.08 \times 10^7)\,[(0.082057)\,(273 + 700)]^{-1} = 1.35 \times 10^5$

16-28. $P_{total} = 705\ mmHg \cdot 1\ atm/760\ mmHg = 0.928\ atm$

$P_{total} = P(NH_3) + P(HI)$

Since $P(NH_3) = P(HI)$, this means that $P(NH_3) = P(HI) = \frac{1}{2} P_{total} = 0.464$ atm

$$K_p = P(NH_3) \cdot P(HI) = (0.464)^2 = 0.215$$

16-30.

$$Q = \text{reaction quotient} = \frac{P(NO) \cdot P^{\frac{1}{2}}(Br_2)}{P(NOBr)} = \frac{(0.80)\,(0.40)^{\frac{1}{2}}}{1.0} = 0.51$$

Here $Q > K_p$, so the reaction is *not* at equilibrium, and a net reaction should occur. Too much product is present relative to reactant, so NO is consumed (along with $Br_2$) as more NOBr is formed.

16-32. (a)

$$Q = \text{reaction quotient} = \frac{[NO_2]^2}{[NO]^2\,[O_2]} = \frac{(3.25)^2}{(2.25)^2\,(1.25)} = 1.67$$

Q is greater than $K_c$, so the reaction is not at equilibrium.

(b) The concentration of $NO_2$ should decrease on proceeding to equilibrium.

16-34.

| | $[C_6H_{12}]$ | $[C_5H_9CH_3]$ |
|---|---|---|
| Initial [..] | 0.016 | 0 |
| Δ [..] | -x | +x |
| Equilibrium [..] | 0.016-x | x |

$$K_c = 0.12 = \frac{[C_5H_9CH_3]}{[C_6H_{12}]} = \frac{x}{0.016-x}$$

$[C_5H_9CH_3] = x = 1.7 \times 10^{-3}$ M

$[C_6H_{12}] = 0.016\text{-}x = 0.014$

16-36.

| | $[COBr_2]$ | $[CO]$ | $[Br_2]$ |
|---|---|---|---|
| Equilibrium [..] | 0.0060 | x | x |

$$K_c = 0.190 = \frac{[CO]\,[Br_2]}{[COBr_2]} = \frac{x \cdot x}{0.0060}$$

$x = [CO] = [Br_2] = 0.034\ M$

16-40. (b) $K = (x)(\frac{1}{2}x)^{\frac{1}{2}}/(0.030 - x)$

16-40. $K_c = 1.2 = 1/[CO_2]$

$[CO_2] = 0.83\ mol/L$

$0.83\ mol/L \cdot 2.0\ L = 1.7\ mol\ CO_2$

16-42.

| | $[CO_2]$ | $[H_2]$ | $[CO]$ | $[H_2O]$ |
|---|---|---|---|---|
| Initial [..] | 0.10 | 0.10 | 0 | 0 |
| Δ [..] | -x | -x | +x | +x |
| Equilibrium [..] | 0.10-x | 0.10-x | x | x |

$$K_c = 1.6 = \frac{[CO][H_2O]}{[CO_2][H_2]} = \frac{x^2}{(0.10 - x)^2}$$

$x = [CO] = [Br_2] = 0.056\ M$

$0.10\text{-}x = [CO_2] = [H_2] = 0.044\ M$

16-44. (a)

| | $[NH_3]$ | $[N_2]$ | $[H_2]$ |
|---|---|---|---|
| Initial [..] | 1.80 | 0 | 0 |
| Δ [..] | -2x | +x | +3x |
| Equilibrium [..] | 1.80-2x | x | 3x |

$$K_c = 6.56 \times 10^{-3} = \frac{[N_2][H_2]^3}{[NH_3]^2} = \frac{(x)(3x)^3}{(1.80\text{-}2x)^2} = \frac{27x^4}{(1.80\text{-}2x)^2}$$

Take the square root of both sides of the equation —

$$0.0810 = 5.20\ x^2 / (1.80\text{-}2x)$$

which simplifies to

$$5.20\ x^2 + 0.162\ x - 0.146 = 0$$

To solve this equation we use the quadratic equation, and find that

$x = [N_2] = 0.153$ M

and so $[NH_3] = 1.80\text{-}2x = 1.49$ M and $[H_2] = 3x = 0.459$ M

(b) Total concentration of gases = 2.14 mol/L

Total moles of gas = 2.14 mol/L • 2.00 L = 4.28 mol

$$P = nRT/V = \frac{(4.28\text{ mol}) \cdot (0.082057\text{ L} \cdot \text{atm/K} \cdot \text{mol}) \cdot (633 + 273)\text{K}}{2.00\text{ L}}$$

P = 159 atm

16-46. (a)

| | $[COBr_2]$ | [CO] | $[Br_2]$ |
|---|---|---|---|
| Initial [..] | x | 0 | 0 |
| Δ [..] | -y | +y | +y |

(Since the equilibrium concentration of CO and $Br_2$ is 0.0500 M, this is the value of y.)

| | | | |
|---|---|---|---|
| Equilibrium [..] | x - 0.500 | 0.0500 | 0.0500 |

$$K_c = 0.190 = \frac{[CO]\,[Br_2]}{[COBr_2]} = \frac{(0.0500)^2}{x - 0.0500}$$

$x = [COBr_2]_{initial} = 0.0632$ M

0.0632 mol/L • 5.0 L • 187.8 g/mol = 59 g $COBr_2$

16-48. $K_c = 1.1 \times 10^{-10} = [Ba^{2+}]\,[SO_4^{2-}] = [Ba^{2+}]\,(5.2 \times 10^{-3})$

$[Ba^{2+}] = 2.1 \times 10^{-8}$ M

## LE CHATELIER'S PRINCIPLE

16-50.

| **Change** | **Direction of Shift in Equilibrium** |
|---|---|
| Addition of more $Br_2(g)$ | left (to consume "excess" $Br_2$) |
| Removal of some NOBr(g) | left (to produce more NOBr) |
| Increase in the container volume | right (favors the side with the larger total moles of gas) |
| Decrease in temperature | right (to produce more heat) |

16-52.

| **Change** | **Direction of Shift in Equilibrium** |
| --- | --- |
| Add more $O_2(g)$ | right (to consume the added $O_2$) |
| Add more $NO_2(g)$ | left (to consume the added $NO_2$) |
| Increase V of the reaction flask | left (Increasing the volume of the flask lowers the pressure. Therefore, the shift is left to raise the pressure in the reacting system.) |
| Lower the temperature | right (The reaction is exothermic, so the reaction shifts right to raise the temperature.) |

16-54. $NH_4HS(s) \rightleftarrows NH_3(g) + H_2S(g)$

(a) $K_p = 0.11 = P(NH_3) \cdot P(H_2S)$

Since $P(NH_3) = P(H_2S)$, this means that

$K_p = P^2(NH_3) = P^2(H_2S)$

Therefore, $P(NH_3) = P(H_2S) = 0.33$ atm

(b) $K_p = 0.11 = P(NH_3) \cdot P(H_2S) = [0.50 + P(H_2S)] \cdot P(H_2S)$

$P^2(H_2S) + 0.50 \cdot P(H_2S) - 0.11 = 0$

$P(H_2S) = 0.17$

Therefore, $P(NH_3) = 0.67$

16-56. (b) $K = (0.35 - x) / [(x)(0.15 + \frac{1}{2}x)^{\frac{1}{2}}]$

16-58. $COBr_2(g) \rightleftarrows CO(g) + Br_2(g)$

| (a) | $[COBr_2]$ | $[CO]$ | $[Br_2]$ |
| --- | --- | --- | --- |
| Initial [..] | 0.750 | 0 | 0 |
| Δ [..] | -x | +x | +x |
| Equilibrium [..] | 0.750-x | x | x |

$$K_c = 0.190 = \frac{[CO]\,[Br_2]}{[COBr_2]} = \frac{x \cdot x}{0.750-x}$$

$0.143 - 0.190x = x^2$

$x^2 + 0.190x - 0.143 = 0$

$x = [CO] = [Br_2] = 0.295$ M and so $[COBr_2] = 0.750 - 0.295 = 0.455$ M

(b) Add CO to the system at equilibrium. Its initial concentration is 0.500 M. In solving this problem, we assume that the CO is added to the original system.

| | $[COBr_2]$ | $[CO]$ | $[Br_2]$ |
|---|---|---|---|
| Initial [..] | 0.750 | 0.500 | 0 |
| Δ [..] | -x | +x | +x |
| Equilibrium [..] | 0.750-x | 0.500+x | x |

$$K_c = 0.190 = \frac{[CO][Br_2]}{[COBr_2]} = \frac{(0.500+x) \cdot x}{0.750-x}$$

$x^2 + 0.690x - 0.143 = 0$

$x = [Br_2] = 0.167$ M and so $[COBr_2] = 0.750 - 0.167 = 0.583$ M and $[CO] = 0.500 + 0.167 = 0.667$ M.

## EQUILIBRIUM AND KINETICS

16-60. (a) Step 2 is the rate determining step. Its rate law is: Rate $= k \cdot [O_3] \cdot [O]$

(b) The equilibrium constant for the first step is

$K = [O_2][O]/[O_3]$

and so $[O] = K[O_3]/[O_2]$

Substituting this into the rate law we have

Rate $= k \cdot [O_3] \cdot \{K[O_3]/[O_2]\} = (k \cdot K)[O_3]^2/[O_2]$

This is equivalent to the experimentally derived rate law.

16-62. Mechanism 2 agrees with the experimental rate law.

Rate law for rate determining step: Rate $= k \cdot [N_2O_2] \cdot [H_2]$

Equilibrium constant for step 1: $K = [N_2O_2]/[NO]^2$

Rearranging the equilibrium expression we have $[N_2O_2] = K \cdot [NO]^2$

and substituting it into the rate law we find: Rate = k • K • $[NO]^2$ • $[H_2]$

This is equivalent to the experimentally observed rate law.

## GENERAL PROBLEMS

16-64. $PCl_3(g) + Cl_2(g) \rightleftarrows PCl_5(g)$

The question gives the value of $K_p$, but we need $K_c$ in order to solve the problem. Therefore, we first find $K_c$ from

$$K_p = K_c(RT)^{\Delta n}$$

where $\Delta n = 1 - 2 = -1$ in this case.

$0.087 = K_c [(0.082057)(300 + 273)]^{-1}$

$K_c = 4.1$

| | $[PCl_3]$ | $[Cl_2]$ | $[PCl_5]$ |
|---|---|---|---|
| Initial [..] | 0.20 | 0.20 | 0 |
| Δ [..] | -x | -x | +x |
| Equilibrium [..] | 0.20 - x | 0.20 - x | x |

$$K_c = \frac{[PCl_5]}{[PCl_3][Cl_2]} = \frac{x}{(0.20 - x)^2} = 4.1$$

$x = [PCl_5] = 0.070$

16-66. *Step 1:* Calculate Q, the reaction quotient, and compare with K.

$Q = [n]/[iso] = 1.40$

*Q is less than K*, so the system is *not* at equilibrium. It proceeds to equilibrium by consuming *n*-butane and producing *iso*-butane.

*Step 2:* Calculate the equilibrium concentrations of *n*- and *iso*-butane.

| | $[n]$ | $[iso]$ |
|---|---|---|
| Initial [..] | 1.75 | 1.25 |
| Δ [..] | -x | +x |
| Equilibrium [..] | 1.75 - x | 1.25 + x |

$K_c = 2.5 = (1.25 + x)/(1.75 - x)$ and so $x = 0.89$

Therefore, [*n*] = 0.86 and [*iso*] = 2.14

16-68. $N_2(g) + O_2(g) \rightleftarrows 2\ NO(g)$

(a) Calculate $K_c$ knowing the value of $K_p$. The equation for doing this is $K_p = K_c(RT)^{\Delta n}$ where $\Delta n = 0$ in this case. Therefore, $K_p = K_c$.

(b)
$$Q = \frac{[NO]^2}{[N_2]\,[O_2]} = \frac{(0.0042)^2}{(0.25)\,(0.25)} = 2.8 \times 10^{-4}$$

Here *Q is less than K*, so the system is not at equilibrium.

(c) Since Q < K, the reaction shifts to the right to produce more product and consume reactants.

(d)

| | $[N_2]$ | $[O_2]$ | [NO] |
|---|---|---|---|
| Initial [..] | 0.25 | 0.025 | 0.0042 |
| Δ [..] | -x | -x | +2x |
| Equilibrium [..] | 0.25 - x | 0.25 - x | 0.0042 + 2x |

$$K_c = 4.0 \times 10^{-4} = \frac{[NO]^2}{[N_2]\,[O_2]} = \frac{(0.0042 + 2x)^2}{(0.25 - x)^2}$$

To solve for x, first take the square root of both sides. This gives $x = 4.0 \times 10^{-4}$

Therefore, $[N_2] = [O_2] \approx 0.25$ M and [NO] = $5.0 \times 10^{-3}$ M

16-70. $PCl_5(g) \rightleftarrows PCl_3(g) + Cl_2(g)$

| | $[PCl_5]$ | $[PCl_3]$ | $[Cl_2]$ |
|---|---|---|---|
| Initial [..] | 3.0 | 0.30 | 0.30 |
| Δ [..] | -x | +x | +x |
| Equilibrium [..] | 3.0 - x | 0.30 + x | 0.30 + x |

Using $K_c = 4.0 \times 10^{-2}$ from Example 16.1,* we have

$$K_c = \frac{[PCl_3][Cl_2]}{[PCl_5]} = \frac{(0.30 + x)^2}{3.0 - x} = 4.0 \times 10^{-2}$$

$x^2 + 0.64\,x - 0.030 = 0$

$x = 0.044$

$[PCl_3] = [Cl_2] = 0.30 + 0.044 = 0.34$ M

$[PCl_5] = 3.0 - x = 3.0$ M

16-72. $C(s) + CO_2(g) \rightleftarrows 2\,CO(g)$

$K_p = P^2(CO) / P(CO_2) = 1.72$

(a) $P_{total} = P(CO) + P(CO_2) = 4.70$ atm

$P(CO_2) = 4.70 - P(CO)$

$$K_p = 1.72 = \frac{P^2(CO)}{P(CO_2)} = \frac{P^2(CO)}{4.70 - P(CO)}$$

$P^2(CO) + 1.72 \cdot P(CO) - 8.08 = 0$

$P(CO) = 2.11$ atm

$P(CO_2) = 2.59$ atm

(b)

| | $P(CO_2)$ | $P(CO)$ |
|---|---|---|
| Initial P | 1.11 | 0 |
| $\Delta$ P | -x | +2x |
| Equilibrium P | 1.11 - x | 2x |

$$K_p = 1.72 = \frac{P^2(CO)}{P(CO_2)} = \frac{(2x)^2}{1.11 - x}$$

---

* **NOTE:** The value of $K_c$ in Example 16.1 should be 0.245 in order to be consistent with the values of $K_p$ quoted in other places in the chapter. Using this value, we find that $x^2 + 0.845\,x - 0.645 = 0$. This gives $x = 0.485$. In second and subsequent printings of the book, the value 0.245 will be used in Example 16.1 and thus in this problem.

$4x^2 + 1.72x - 1.91 = 0$

$x = 0.51$

Therefore, $P(CO_2) = 1.11 - 0.51 = 0.60$ atm and $P(CO) = 1.02$ atm

16-74. $N_2O_4(g) \rightleftarrows 2\,NO_2(g)$

(a) Calculate fraction dissociated when $P_{total} = 1.50$ atm

$K_p = 0.113 = P^2(NO) / P(N_2O_4)$

$P_{total} = 1.50 \text{ atm} = P(NO) + P(N_2O_4)$

$P(N_2O_4) = 1.50 - P(NO)$

$K_p = 0.113 = P^2(NO) / [1.50 - P(NO)]$

$P^2(NO) + 0.113 \cdot P(NO) - 0.170 = 0$

$P(NO) = 0.360$ atm

Therefore, $P(N_2O_4) = 1.14$ atm at equilibrium

Initially, $P(N_2O_4)$ must have been 1.14 atm + $\frac{1}{2} \cdot 0.360$ atm = 1.32 atm

fraction $N_2O_4$ dissociated = (1.32 atm - 1.14 atm) / 1.32 atm = 0.136

(b) Calculate fraction dissociated when $P_{total} = 1.00$ atm

$K_p = 0.113 = P^2(NO) / [1.00 - P(NO)]$

$P(NO) = 0.284$ atm

Therefore, $P(N_2O_4) = 0.716$ atm at equilibrium

Initially, $P(N_2O_4)$ must have been 0.716 atm + $\frac{1}{2} \cdot 0.284$ atm = 0.858 atm

fraction $N_2O_4$ dissociated = (0.858 atm - 0.716 atm) / 0.858 atm = 0.166

Note that when the total pressure dropped, the fraction dissociated *increased.*

16-76. $L{-}B(CH_3)_3(g) \rightleftarrows L(g) + B(CH_3)_3(g)$

(a) The complex providing the largest partial pressure of $B(CH_3)_3$ would be that with the largest value of $K_p$. In this case, that would be $H_3N{-}B(CH_3)_3$.

(b) *Step 1:* Calculate the pressure of $H_3N—B(CH_3)_3$ before dissociation.**

$0.73\ g \cdot 1\ mol / 72.9\ g = 0.010\ mol$

$$P = nRT/V = \frac{(0.010\ mol)(0.082057\ L \cdot atm/K \cdot mol)(373\ K)}{0.010\ L} = 3.1\ atm$$

*Step 2:* Calculate the partial pressure of $H_3N—B(CH_3)_3$ when equilibrium has been established.

| | $P(H_3N—B(CH_3)_3)$ | $P(H_3N)$ | $P(B(CH_3)_3)$ |
|---|---|---|---|
| Initial P | 3.1 | 0 | 0 |
| Δ P | -x | +x | +x |
| Equilibrium P | 3.1 - x | x | x |

$$K_p = \frac{P(H_3N) \cdot P(B(CH_3)_3)}{P(H_3N—B(CH_3)_3)} = \frac{x^2}{3.1 - x} = 4.62$$

$x^2 + 4.62\,x - 14.1 = 0$

$x = 2.1$ atm

The equilibrium partial pressures are:

$P(H_3N—B(CH_3)_3) = 3.1 - x = 1.0$ atm

$P(H_3N) = P(B(CH_3)_3) = 2.1$ atm

$P_{total} = 1.0\ atm + 2 \cdot 2.1\ atm = 5.2\ atm$

The % dissociation = $(2.1 / 3.1) \cdot 100 = 68\%$

16-78. $PCl_5(g) \rightleftarrows PCl_3(g) + Cl_2(g)$

| (a) | $P(PCl_5)$ | $P(PCl_3)$ | $P(Cl_2)$ |
|---|---|---|---|
| Initial [..] | 1.50 | 0 | 0 |
| Δ [..] | -x | +x | +x |
| Equilibrium [..] | 1.50 - x | x | x |

** [NOTE: In the first printing of the book, the volume was given as 100 L. It should be 100 *mL*. If the problem is solved using 100 L, the % dissociation is almost 100%.]

$$K_c = \frac{P(PCl_3)\,(Cl_2)}{P(PCl_5)} = \frac{(x)^2}{1.50 - x} = 11.5$$

$x^2 + 11.5\,x - 17.3 = 0$

$x = 1.35 \text{ atm} = P(PCl_3) = (Cl_2)$

$P(PCl_5) = 1.50 - x = 0.15 \text{ atm}$

% dissociation $= (1.35 / 1.50) \cdot 100 = 90.\%$

(b) Now compare the fraction of $PCl_5$ dissociated in part (a) with the fraction that dissociates if the initial pressure of $PCl_5$ is doubled to 3.00 atm.

$$K_c = \frac{P(PCl_3)\,(Cl_2)}{P(PCl_5)} = \frac{(x)^2}{3.0 - x} = 11.5$$

$x^2 + 11.5\,x - 34.5 = 0$

$x = 2.5 \text{ atm} = P(PCl_3) = (Cl_2)$

$P(PCl_5) = 3.0 - x = 0.5 \text{ atm}$

% dissociation $= (2.5 / 3.0) \cdot 100 = 83\%$

# CHAPTER 17

# THE CHEMISTRY OF ACIDS AND BASES

## INSTRUCTOR'S NOTES

The chapters on acid-base chemistry (Chapters 17 and 18) precede that on the solubility of salts (Chapter 19), largely because that seems to be the most commonly used order. However, we often actually cover the material in the reverse order. We believe solubility equilibria are conceptually easier for our students and that solubility equilibria are a good introduction to aqueous equilibria. Therefore, the three chapters on aqueous equilibria were written with the idea that they could be used in either order.

We cover aqueous equilibria in the second semester of our course. Approximately 9 lectures are devoted to the entire subject of equilibrium (Chapters 16-19). We do not give much coverage to polyprotic acids or to equilibria involving complex ions. Lewis acid/base concepts are mentioned briefly.

One aspect of the lectures on Chapter 17 that we especially emphasize is the concept of conjugate acids and bases and relative acid and base strengths. In lecture we perform some acid-base reactions to build a portion of Table 17.2, the chart of relative strengths of acids and bases.

In revising the first edition of this text we thought long and hard about better ways to help the students solve problems involving the ionization of weak acids and bases. In the first edition we introduced our so-called "rule of thumb" about when they could or could not neglect $x$ in the term $[C]_0 - x$ in the common expression $K = x^2 / [C]_0 - x$ (see page 573 of the first edition). However, we felt that simply applying a rule did not leave the students with a sense of the actual chemistry involved or for the reasons that the approximate expression could be used. Therefore, although we have retained our "rule" in this edition, we have given much more atention to the reasons for using the appproximation. Further, we are advocating the use of the ***method of successive approximations*** for solving for $x$ in $K = x^2 / [C]_0 - x$. In-class experience with the method has been good. However, as outlined in Appendix A, there are times when this method breaks down. Experience shows that the mathematical conditions are such that the method works with the usual problems invlving weak acids and bases. However, it can lead to a nonconverging solution for problems of the type encountered more commonly in Chapter 16.

As we are all aware students have some difficulty with the general topic of aqueous equilibria. The successful student is generally one who takes the time to work through as many examples, exercises, and study questions as possible. Therefore, it ia useful to assign as many homework problems as possible and point out extra work that may be done. The *Study Guide* that accompanies this text is especially useful for the students studying these chapters. The author, our colleague Harry Pence, has given special attention to this topic. Finally, the *Problems Book* by R. Ragsdale gives students many additional problems.

## LIST OF TRANSPARENCIES FOR THIS CHAPTER

| No. | Figure | Page | Subject |
|---|---|---|---|
| 86 | 17.3 | 713 | pH of some common solutions |
| 87 | Table 17.3 | 716 | Acid-base indicators |
| 88 | Table 17.4 | 718 | Ionization constants of weak acids and bases |
| 89 | Table 17.5 | 731 | Acid-base properties of some common ions |

## SUGGESTED DEMONSTRATIONS

Acid-base chemistry can be introduced by performing a reaction such aa $HCO_3^-$ or $CO_3^{2-}$ + acid. (An Alka-Seltzer tablet in a petri dish on an overhead projector ia an effective way to do this. See Figure 18.1 on page 754.)

The concept of strong and weak acids and bases ia a topic introduced early in these lectures. Although the demonstration was done for Chapter 3, we like to show again how conductivity differs for aqueous solutions of strong and weak acids and bases.

To demonstrate pH (Section 17.3) in a large classroom we use a variety of solutions in petri dishes on the overhead projector. Universal indicator is added to a variety of solutions. To show acidic pH's we have used the following: dilute HCl; a metal ion that hydrolyzes in water such aa $Cu^{2+}$, $Zn^{2+}$, $Cd^{2+}$, or $Al^{3+}$; and $NH_4Cl$ (Figure 17.7, page 721). We also breathe into a flask containing universal indicator or add a piece of dry ice. Finally, we add some indicator to a flask or dish of a soft drink such as 7-Up or other consumer product (Figure 17.4, page 714). For pH's on the basic aide we use: sodium acetate, sodium carbonate, NH3, and dilute NaOH (Figure 17.8, page 727).

The amphoteric nature of certain hydroxides [$Al(OH)_3$ or $Zn(OH)_2$ (Figure 17.9, page 737)] ia introduced in the section on Lewis acids and bases. We find the reaction depicted in Figure 17.10 to be particularly useful to demonstrate. In addition, many other metal ion/Lewis base complexes can be illustrated in lecture. One of the simplest is the formation of $Cu(NH_3)_4^{2+}$, but others would include Prussian blue [$Fe(CN)_6^{4-}$ + $Fe^{3+}$], the deep red iron(III)-thiocyanate complex, or $Ni^{2+}$ and dimethylglyoxime. Conditions for forming these and other complexes may be found in any qualitative analysis book.

Finally, Volume 3 of the "Chemical Demonstrations" series by Professor Shakhashiri contains a useful discussion of acid-base chemistry along with 32 demonstrations for this topic. In addition, the videotape, which was prepared by Professor Shakhashiri and which accompanies this text, shows some useful demonstrations.

# SOLUTIONS TO SELECTED STUDY QUESTIONS

## GENERAL ACID-BASE THEORY

17-15.

| Conjugate Acid | Name |
|---|---|
| $NH_4^+$ | ammonium ion |
| $H_2CO_3$ | carbonic acid |
| $H_2S$ | hydrogen sulfide |
| HBr | hydrobromic acid |
| $H_2SO_4$ | sulfuric acid |

17-17.

| Species | Conjugate | Name |
|---|---|---|
| HCOOH | $HCOO^-$ | formate ion |
| $H_2CO_3$ | $HCO_3^-$ | hydrogen carbonate ion |
| $ClO^-$ | HClO | hypochlorous acid |
| $SO_3^{2-}$ | $HSO_3^-$ | hydrogen sulfite ion |

17-19. $NH_4^+(aq) + H_2O(\ell) \rightleftarrows NH_3(aq) + H_3O^+(aq)$

The $Br^-$ ion does not affect the acidity of the solution because the ion is the conjugate base of a strong acid.

17-21. Brønsted acid: $HC_2O_4^-(aq) + H_2O(\ell) \rightleftarrows C_2O_4^{2-}(aq) + H_3O^+(aq)$

Brønsted base: $HC_2O_4^-(aq) + H_2O(\ell) \rightleftarrows H_2C_2O_4(aq) + OH^-(aq)$

17-23. (a) $CH_3COOH(aq) + C_5H_5N(aq) \rightleftarrows CH_3COO^-(aq) + C_5H_5NH^+(aq)$

acid A — base B — conjugate base of A — conjugate acid of B

(b) $N_2H_4(aq) + HSO_4^-(aq) \rightleftarrows N_2H_5^+(aq) + SO_4^{2-}(aq)$

base B — acid A — conjugate acid of B — conjugate acid of A

(c) $[Al(H_2O)_6]^{3+}(aq) + OH^-(aq) \rightleftarrows [Al(H_2O)_5OH]^{2+}(aq) + H_2O(\ell)$

acid A — base B — conjugate base of A — conjugate acid of B

17-25. $HC_2H_3O_2(aq) + HCO_3^-(aq) \rightleftarrows C_2H_3O_2^-(aq) + H_2CO_3(aq)$

This reaction occurs only to a small extent, since both acetic acid and $H_2CO_3$ are both weak acids. But note that the decomposition of $H_2CO_3$ will occur ($H_2CO_3 \rightleftarrows H_2O + CO_2$) and, if the $CO_2$ is free to escape, the reaction will proceed extensively.

17-27. $HSO_4^-(aq) + NH_3(aq) \rightleftarrows SO_4^{2-}(aq) + NH_4^+(aq)$

The $HSO_4^-$ ion is a stronger acid than $NH_4^+$, so the equilibrium will lie to the right.

17-29. (a) $NH_4^+(aq) + Br^-(aq) \rightleftarrows NH_3(aq) + HBr(aq)$

HBr is a stronger acid than $NH_4^+$, so the equilibrium lies predominantly to the left.

(b) $C_6H_5OH(aq) + CH_3COO^-(aq) \rightleftarrows C_6H_5O^-(aq) + CH_3COOH(aq)$

$C_6H_5OH$ is a weaker acid than $CH_3COOH$, so the equilibrium lies predominantly to the left.

(c) $NH_2^-(aq) + H_2O(\ell) \rightleftarrows NH_3(aq) + OH^-(aq)$

$NH_2^-$ is a stronger base than $NH_3$, so the equilibrium lies predominantly to the right.

17-31. (a) HCOOH is the strongest acid and $C_6H_5OH$ is the weakest.

(b) HCOOH

(c) $C_6H_5OH$

17-33. (a) $CN^-$ is the strongest base, and $NO_2^-$ is the weakest.

(b) $H_2S$

(c) $NO_2^-$ has the strongest conjugate acid, while $HS^-$ has the weakest conjugate acid.

17-35. $CN^-$ is the strongest of these three bases [(a) $CN^-$, (b) $NH_3$, (c) $SO_4^{2-}$], and $SO_4^{2-}$ is the weakest (Table 17.2). Therefore, $SO_4^{2-}$ has the strongest conjugate acid ($HSO_4^-$).

## pH CALCULATIONS

17-37. pH = 10.5

$[H_3O^+] = 10^{-pH} = 10^{-10.5} = 3 \times 10^{-11}$ M

$[OH^-] = K_w / [H_3O^+] = 3 \times 10^{-4}$ M, and so the solution is basic.

17-39. $[OH^-] = 1.2 \times 10^{-4}$

The pOH = 3.92, and so pH = 10.08

$H_3O^+] = 10^{-pH} = 10^{-10.08} = 8.3 \times 10^{-11}$ M

17-41. pH = 10.66

$[H_3O^+] = 10^{-pH} = 10^{-10.66} = 2.2 \times 10^{-11}$ M

$[OH^-] = K_w / [H_3O^+] = 4.6 \times 10^{-4}$ M

17-43.

| | pH | $[H_3O^+]$ M | $[OH^-]$ M | Acid/Base |
|---|---|---|---|---|
| (a) | 7.17 | $6.7 \times 10^{-8}$ | $1.5 \times 10^{-7}$ | basic |
| (b) | 7.34 | $4.5 \times 10^{-8}$ | $2.2 \times 10^{-7}$ | basic |
| (c) | 5.25 | $5.6 \times 10^{-6}$ | $1.8 \times 10^{-9}$ | acidic |
| (d) | 1.60 | $2.5 \times 10^{-2}$ | $4.0 \times 10^{-13}$ | acidic |
| (e) | 11.75 | $1.8 \times 10^{-12}$ | $5.6 \times 10^{-3}$ | basic |

17-45. (a) $[OH^-] = 3.2 \times 10^{-3}$ M

The pOH = 2.49, and so pH = 11.51

(b) The solution is basic, so phenolphthalein is *pink* while alizarin is *violet* (as it is changing color from red to purple).

## GENERAL ACID-BASE EQUILIBRIA

17-47. (a) pH = 10.09

$[H_3O^+] = 10^{-pH} = 10^{-10.09} = 8.1 \times 10^{-11}$ M

$[OH^-] = 1.2 \times 10^{-4}$ M

(b) Assuming the base is weak, we can write the expression

$K_b \approx [OH^-]^2/0.015 = 1 \times 10^{-6}$

**This shows that the base is moderately weak.**

17-49. pH = 1.95

$[H_3O^+] = 10^{-pH} = 10^{-1.95} = 1.1 \times 10^{-2}$ M

$ClCH_2COOH(aq) + H_2O(\ell) \rightleftarrows ClCH_2COO^-(aq) + H_3O^+(aq)$

| | $[ClCH_2COOH]$ | $[ClCH_2COO^-]$ | $[H_3O^+]$ |
|---|---|---|---|
| Initial [..] | 0.10 | 0 | 0 |
| Δ [..] | -0.011 | +0.011 | +0.011 |
| Equilibrium [..] | 0.09 | 0.011 | 0.011 |

$$K_a = \frac{[ClCH_2COO^-][H_3O^+]}{[ClCH_2COOH]} = \frac{(0.011)(0.011)}{0.09} = 1 \times 10^{-3}$$

17-51. pH = 3.44

$[H_3O^+] = 10^{-pH} = 10^{-3.44} = 3.6 \times 10^{-4}$ M

$Acid(aq) + H_2O(\ell) \rightleftarrows Conjugate\ Base(aq) + H_3O^+(aq)$

| | [Acid] | [Conj. Base] | $[H_3O^+]$ |
|---|---|---|---|
| Initial [..] | 0.010 | 0 | 0 |
| Δ [..] | -0.00036 | +0.00036 | +0.00036 |
| Equilibrium [..] | 0.010 | 0.00036 | 0.00036 |

$$K_a = \frac{[Conj.\ Base][H_3O^+]}{[Acid]} \approx \frac{(0.00036)^2}{0.010} = 1.3 \times 10^{-5}$$

17-53. $CH_3NH_2(aq) + H_2O(\ell) \rightleftarrows CH_3NH_3^+(aq) + OH^-(aq)$

pH = 11.76, which gives a pOH of 2.24. Using the pOH we find $[OH^-] = 5.8 \times 10^{-3}$ M.

| | $[CH_3NH_2]$ | $[CH_3NH_3^+]$ | $[OH^-]$ |
|---|---|---|---|
| Initial [..] | 0.065 | 0 | 0 |
| Δ [..] | -0.0058 | +0.0058 | +0.0058 |
| Equilibrium [..] | 0.059 | 0.0058 | 0.0058 |

$$K_b = \frac{[CH_3NH_3^+][OH^-]}{[CH_3NH_2]} = \frac{(0.0058)^2}{0.059} = 5.7 \times 10^{-4}$$

17-55. $CH_3COOH(aq) + H_2O(\ell) \rightleftarrows CH_3COO^-(aq) + H_3O^+(aq)$

| | $[CH_3COOH]$ | $[CH_3COO^-]$ | $[H_3O^+]$ |
|---|---|---|---|
| Initial [..] | 0.20 | 0 | 0 |
| Δ [..] | -x | +x | +x |
| Equilibrium [..] | 0.20 - x | x | x |

$$K_a = \frac{[CH_3COO^-][H_3O^+]}{[CH_3COOH]} = \frac{x \cdot x}{0.20 - x} = 1.8 \times 10^{-5}$$

Assume that x is much smaller than 0.20, so $1.8 \times 10^{-5} = x^2/0.20$

This gives $x = [CH_3COO^-] = [H_3O^+] = 1.9 \times 10^{-3}$ M

Therefore, $[CH_3COOH] \approx 0.20$ M

17-57. $C_6H_5OH(aq) + H_2O(\ell) \rightleftarrows C_6H_5O^-(aq) + H_3O^+(aq)$

0.780 g • 1 mol/ 94.11 g = 0.00829 mol
0.00829 mol / 0.500 L = 0.0166 M

| | $[C_6H_5OH]$ | $[C_6H_5O^-]$ | $[H_3O^+]$ |
|---|---|---|---|
| Initial [..] | 0.0166 | 0 | 0 |
| Δ [..] | -x | +x | +x |
| Equilibrium [..] | 0.0166 - x | x | x |

$$K_a = \frac{[C_6H_5O^-][H_3O^+]}{[C_6H_5OH]} = \frac{x \cdot x}{0.0166 - x} = 1.3 \times 10^{-10}$$

Assume that x is much smaller than 0.0166, so $1.3 \times 10^{-10} = x^2/0.0166$

This gives $x = [C_6H_5O^-] = [H_3O^+] = 1.5 \times 10^{-6}$ M ⇒ pH = 5.83

Therefore, $[C_6H_5OH] \approx 0.0166$ M

17-59. $C_8H_5O_4^-(aq) + H_2O(\ell) \rightleftarrows C_8H_4O_4^{2-}(aq) + H_3O^+(aq)$

2.55 g $KC_8H_5O_4$ • 1 mol / 204.2 g = 0.0125 mol

0.0125 mol / 0.250 L = 0.0500 M

| | $[C_8H_5O_4^-]$ | $[C_8H_4O_4^{2-}]$ | $[H_3O^+]$ |
|---|---|---|---|
| Initial [..] | 0.0500 | 0 | 0 |
| Δ [..] | -x | +x | +x |
| Equilibrium [..] | 0.0500 - x | x | x |

$$K_a = 2.0 \times 10^{-7} = \frac{[C_8H_4O_4^{2-}][H_3O^+]}{[C_8H_5O_4^-]} \approx \frac{x^2}{0.0500 - x}$$

Assume that x is small relative to 0.0500, so $2.0 \times 10^{-7} \approx x^2/0.0500$

$x = 1.0 \times 10^{-4} = [C_8H_4O_4^{2-}] = [H_3O^+]$

pH = 4.00

17-61. **NOTE:** There was an error in the first printing of the book for the value of $K_a$ for HOBr. This value should be $2.5 \times 10^{-9}$ (Appendix E). Therefore, answers are given below for both the first printing and for subsequent printings.

***First Printing***

Order of increasing strength: Valeric acid < Glutaric acid < Hypobromous acid

Order of increasing pH: Hypobromous acid < Glutaric acid < Valeric acid

***Second and Subsequent Printings***

Order of increasing strength: Hypobromous acid < Valeric acid < Glutaric acid

Order of increasing pH: Glutaric acid < Valeric acid < Hypobromous acid

17-63. $MOH(aq) \rightleftarrows M^+(aq) + OH^-(aq)$

| | [MOH] | $[M^+]$ | $[OH^-]$ |
|---|---|---|---|
| Initial [..] | 0.15 | 0 | 0 |
| Δ [..] | -x | +x | +x |
| Equilibrium [..] | 0.15 - x | x | x |

$$K_b = 5.0 \times 10^{-4} = \frac{[M^+][OH^-]}{[MOH]} = \frac{x^2}{0.15 - x}$$

Solving by the *method of successive approximations* (Appendix A), we find x = $8.4 \times 10^{-3}$ and so [MOH] = 0.14 M

17-65. $CH_3NH_2(aq) + H_2O(\ell) \rightleftarrows CH_3NH_3^+(aq) + OH^-(aq)$

| | $[CH_3NH_2]$ | $[CH_3NH_3^+]$ | $[OH^-]$ |
|---|---|---|---|
| Initial [..] | 0.25 | 0 | 0 |
| Δ [..] | -x | +x | +x |
| Equilibrium [..] | 0.25 - x | x | x |

$$K_b = 5.0 \times 10^{-4} = \frac{[CH_3NH_3^+][OH^-]}{[CH_3NH_2]} = \frac{x^2}{0.25 - x}$$

Solving by the *method of successive approximations* (Appendix A), we find x = $[CH_3NH_3^+]$ = $[OH^-]$ = 0.011 M and so $[CH_3NH_2]$ = 0.25 M. This calculation leads to a pOH = 2.0 and a pH = 12.0.

17-67. $NH_2OH(aq) + H_2O(\ell) \rightleftarrows NH_3OH^+(aq) + OH^-(aq)$

| | $[NH_2OH]$ | $[NH_3OH^+]$ | $[OH^-]$ |
|---|---|---|---|
| Initial [..] | 0.051 | 0 | 0 |
| Δ [..] | -x | +x | +x |
| Equilibrium [..] | 0.051 - x | x | x |

$$K_b = 6.6 \times 10^{-9} = \frac{[NH_3OH^+][OH^-]}{[NH_2OH]} = \frac{x^2}{0.051 - x}$$

Making the very reasonable assumption that x is very small relative to 0.051, we find x = $1.8 \times 10^{-5}$ M = $[NH_3OH^+]$ = $[OH^-]$. Therefore, the hydroxide ion concentration gives pOH = 4.74 and pH = 9.26.

17-69. Acid + $H_2O$ ⇄ Conjugate Base + $H_3O^+$

0.125 g • 1 mol / 139.1 g = $8.99 \times 10^{-4}$ mol

$8.99 \times 10^{-4}$ mol / 0.250 L = 0.00359 M

| | [Acid] | [Base] | $[H_3O^+]$ |
|---|---|---|---|
| Initial [..] | 0.00359 | 0 | 0 |
| Δ [..] | -x | +x | +x |
| Equilibrium [..] | 0.00359 - x | x | x |

$$K_a = 7.0 \times 10^{-8} = \frac{[Base][H_3O^+]}{[Acid]} = \frac{x^2}{0.00359 - x}$$

Solving by the *method of successive approximations* (Appendix A), we find x = [Base] = $[H_3O^+]$ = $1.6 \times 10^{-5}$ M. This leads to a pH of 4.80

17-71. $HF(aq) + H_2O(\ell) \rightleftarrows F^-(aq) + H_3O^+(aq)$

| | [HF] | $[F^-]$ | $[H_3O^+]$ |
|---|---|---|---|
| Initial [..] | 0.0010 | 0 | 0 |
| Δ [..] | -x | +x | +x |
| Equilibrium [..] | 0.0010 - x | x | x |

$$K_a = 7.2 \times 10^{-4} = \frac{[F^-][H_3O^+]}{[HF]} = \frac{x^2}{0.0010 - x}$$

Solving by the *method of successive approximations* (Appendix A), we find x = $[F^-]$ = $[H_3O^+]$ = $5.6 \times 10^{-4}$ M. This leads to a pH of 3.25

17-73. (a) $NaNO_3$ — Neutral. Neither $Na^+$ nor $NO_3^-$ affects the pH of the solution.

(b) $NaC_7H_5O_2$ — Basic. $Na^+$ has no effect on the pH, but $C_7H_5O_2^-$, which is the conjugate base of a weak acid, does make the solution basic.

(c) $Na_2HPO_4$ — Basic. $Na^+$ has no effect on the pH, but $HPO_4^{2-}$, which is the conjugate base of the weak acid $H_2PO_4^-$, does make the solution basic.

17-75. (a) $AlCl_3$ — Acidic, owing to the $Al^{3+}$ ion.

(b) $Na_2S$. — Basic, due to the hydrolysis of the $S^{2-}$ ion, the conjugate base of the weak acid $HS^-$.

(c) $KClO_4$. — Neutral. Neither $K^+$ nor $ClO_4^-$ will affect the pH.

(b) $NaH_2PO_4$. — The $H_2PO_4^-$ ion is a weak acid, so the pH will less than 7.

17-77. (a) KCN — Basic due to the hydrolysis of $CN^-$, the conjugate base of the weak acid HCN.

(b) $FeCl_3$ — Acidic due to hydrolysis of the $Fe^{3+}$ ion.

(c) $MgCl_2$ — Neutral. Neither the $Mg^{2+}$ nor the $Cl^-$ affects the pH of the solution.

(d) $(NH_4)_2S$ — Basic due to the hydrolysis of $S^{2-}$, the conjugate base of the weak acid $HS^-$.

17-79. $CH_3COO^-(aq) + H_2O(\ell) \rightleftarrows CH_3COOH(aq) + OH^-(aq)$

| | $[CH_3COO^-]$ | $[CH_3COOH]$ | $[OH^-]$ |
|---|---|---|---|
| Initial [..] | 0.015 | 0 | 0 |
| Δ [..] | -x | +x | +x |
| Equilibrium [..] | 0.015 - x | x | x |

$$K_b = 5.6 \times 10^{-10} = \frac{[CH_3COOH]\,[OH^-]}{[CH_3COO^-]} = \frac{x^2}{0.015 - x}$$

$[CH_3COOH] = [OH^-] = x = 2.9 \times 10^{-6}$ M

The pOH = 5.54, and so pH = 8.46

This leads to $[H_3O^+] = 3.5 \times 10^{-9}$ M

17-81. The sodium ion has no effect on the pH of the solution, but the $C_3H_5O_2^-$ ion is the conjugate base of a weak acid. Therefore, the solution will be weakly basic.

$C_3H_5O_2^-(aq) + H_2O(\ell) \rightleftarrows HC_3H_5O_3(aq) + OH^-(aq)$

| | $[C_3H_5O_2^-]$ | $[HC_3H_5O_2]$ | $[OH^-]$ |
|---|---|---|---|
| Initial [..] | 0.10 | 0 | 0 |
| Δ [..] | -x | +x | +x |
| Equilibrium [..] | 0.10 - x | x | x |

$$K_b = \frac{[HC_3H_5O_3]\,[OH^-]}{[C_3H_5O_2^-]} = \frac{x^2}{0.10 - x} = 7.1 \times 10^{-10}$$

$x = [HC_3H_5O_3] = [OH^-] = 8.4 \times 10^{-6}$ M. This gives a pOH of 5.07 and a pH of 8.93.

17-83. $C_3H_6O_3(aq) + H_2O(\ell) \rightleftarrows C_3H_5O_3^-(aq) + H_3O^+(aq)$

$0.056\text{ g} \cdot 1\text{ mol}/90.1\text{ g} = 6.2 \times 10^{-4}\text{ mol}$

$6.2 \times 10^{-4}\text{ mol}/0.250\text{ L} = 0.0025\text{ M}$

| | $[C_3H_6O_3]$ | $[C_3H_5O_3^-]$ | $[H_3O^+]$ |
|---|---|---|---|
| Initial [..] | 0.0025 | 0 | 0 |
| Δ [..] | -x | +x | +x |
| Equilibrium [..] | 0.0025 - x | x | x |

$$K_a = 1.4 \times 10^{-4} = \frac{[C_3H_5O_3^-][H_3O^+]}{[C_3H_6O_3]} = \frac{x^2}{0.0025 - x}$$

Solving by the *method of successive approximations* (Appendix A), we find x = $[C_3H_5O_3^-] = [H_3O^+] = 5.3 \times 10^{-4}$ M. This leads to a pH of 3.28

## POLYPROTIC ACIDS

17-85. The pH of the solution is determined by the first ionization of the acid.

$0.0050\text{ g} \cdot 1\text{ mol}/176\text{ g} = 2.8 \times 10^{-5}\text{ mol}$

$2.8 \times 10^{-5}\text{ mol}/0.0010\text{ L} = 0.028\text{ M}$

$C_6H_8O_6(aq) + H_2O(\ell) \rightleftarrows C_6H_8O_6^-(aq) + H_3O^+(aq)$

| | $[C_6H_8O_6]$ | $[C_6H_8O_6^-]$ | $[H_3O^+]$ |
|---|---|---|---|
| Initial [..] | 0.028 | 0 | 0 |
| Δ [..] | -x | +x | +x |
| Equilibrium [..] | 0.028 - x | x | x |

$$K_{a1} = \frac{[C_6H_8O_6^-][H_3O^+]}{[C_6H_8O_6]} = \frac{x^2}{0.028 - x} = 7.9 \times 10^{-5}$$

$x = [C_6H_8O_6^-] = [H_3O^+] = 1.5 \times 10^{-5}$ M, and this leads to a pH = 2.84.

17-87. The pH of the solution is controlled by the first ionization step.

$H_2NC_2H_4NH_2(aq) + H_2O(\ell) \rightleftarrows H_2NC_2H_4NH_3^+(aq) + OH^-(aq)$

| | $[H_2NC_2H_4NH_2]$ | $[H_2NC_2H_4NH_3^+]$ | $[OH^-]$ |
|---|---|---|---|
| Initial [..] | 0.15 | 0 | 0 |
| Δ [..] | -x | +x | +x |
| Equilibrium [..] | 0.15 - x | x | x |

$$K_{b1} = \frac{[H_2NC_2H_4NH_3^+][OH^-]}{[H_2NC_2H_4NH_2]} = \frac{x^2}{0.15 - x} = 8.5 \times 10^{-5}$$

$x = [H_2NC_2H_4NH_3^+] = [OH^-] = 3.6 \times 10^{-5}$

This gives a pOH of 2.45 and a pH of 11.55. The concentration of the dipotonated ion, $H_3NC_2H_4NH_3^{2+}$, is given by the second ionization. That is, $[H_3NC_2H_4NH_3^{2+}] = K_{b2} = 2.7 \times 10^{-8}$ M.

## LEWIS ACIDS AND BASES

17-89. (a) $BCl_3$ is a Lewis acid.

(b) $H_2N—NH_2$, hydrazine, is a Lewis base.

(c) $CN^-$ in the reaction $Au^+(aq) + 2\ CN^-(aq) \rightleftarrows [Au(CN)_2]^-(aq)$ in a Lewis base.

17-91. :C≡O: is a Lewis base in its complexes with Fe, Ni, and other transition metals.

17-93. $Fe(CO)_5$ is a Lewis base in this reaction, since it acts as an electron pair donor in forming an Fe-H bond.

## FORMATION CONSTANTS

17-95. The formation constant for $[AlF_6]^{3-}$ is 5.0 x $10^{23}$, while that for $[Al(OH)_4]^-$ is 7.7 x $10^{33}$. The more stable complex is that with the larger value of $K_{formation}$. Therefore, to most completely remove $Al^{3+}$ from solution use $OH^-$.

17-97. $Cd^{2+}(aq) + 4\ Cl^-(aq) \rightleftarrows [CdCl_4]^{2-}(aq)$

| | $[Cd^{2+}]$ | $[Cl^-]$ | $\{[CdCl_4]^{2-}\}$ |
|---|---|---|---|
| Initial [..] | 0.010 | 1.0 | 0 |
| Now allow $[CdCl_4]^{2-}$ to form stochiometrically. | | | |
| [..] after initial rxn. | 0 | 1.0 - 4(0.010) | 0.010 |
| Now allow $[CdCl_4]^{2-}$ to dissociate back to reactants. | | | |
| Δ [..] | +x | +4x | -x |
| Equilibrium [..] | x | 0.96 + 4x | 0.010 - x |

$$K_{form} = \frac{\{[CdCl_4]^{2-}\}}{[Cd^{2+}]\,[Cl^-]^4} = \frac{0.010 - x}{(x)(0.96 + 4x)^4} = 1.0 \times 10^4$$

To solve for x, assume that x is very small relative to 0.96 and 0.010. This gives x = 1.2 x $10^{-6}$. Therefore, at equilibrium $[Cd^{2+}]$ = 1.2 x $10^{-6}$ M; $[Cl^-]$ = 0.96 M; and $\{[CdCl_4]^{2-}\}$ = 0.010 M.

17-99. $[CuCl_2]^-(aq) \rightleftarrows Cu^+(aq) + 2\ Cl^-(aq)$

| | $\{[CuCl_2]^-\}$ | $[Cu^+]$ | $[Cl^-]$ |
|---|---|---|---|
| Initial [..] | 0.015 | 0 | 0 |
| Δ [..] | -x | +x | +2x |
| Equilibrium [..] | 0.015 - x | x | 2x |

$$K = 1/K_{form} = 1/1.0 \times 10^5 = \frac{[Cu^+][Cl^-]}{\{[CuCl_2]^-\}} = \frac{(x)(2x)^2}{0.015 - x}$$

Solving the equation by the *method of successive approximations*, leads to x = $[Cu^+]$ = 3.1 x $10^{-3}$ M, and so $[Cl^-]$ = 6.2 x $10^{-3}$ M and $\{[CuCl_2]^-\}$ = 0.012 M

## GENERAL PROBLEMS

17-101. (a) Balanced equation for the auto-ionization process of liquid ammonia.

$$2\ NH_3(\ell) \rightleftarrows NH_4^+ + NH_2^-$$

(b) The conjugate acid of $NH_3$ is $NH_4^+$ and its conjugate base is $NH_2^-$.

(c) If $NaNH_2$ is dissolved in liquid ammonia, it is a base.

17-103.

| **Ion** | **Acid or Base** | **Conjugate** |
|---|---|---|
| $PO_4^{3-}$ | base | $HPO_4^{2-}$ |
| $H^-$ | base | $H_2$ |
| $NH_4^+$ | acid | $NH_3$ |
| $CO_3^{2-}$ | base | $HCO_3^-$ |
| OH- | base | $H_2O$ |

17-105. 0.50 g $Ca(OH)_2$ • 1 mol / 74.1 g = 0.0067 mol

0.0067 mol / 1.0 L = 0.0067 M

$[OH^-]$ = 2 • 0.0067 M = 0.014 M

pOH = 1.87 and so pH = 12.13

17-107. (a) The pH of an *ethylamine* solution is higher than that of the ethanolamine solution.

(a) *Ethylamine* is a stronger base than ethanolamine.

Ethylamine + $H_2O$ ⇄ EthylamineH$^+$ + $OH^-$

$$K_b = 5.6 \times 10^{-4} = \frac{[\text{EthylamineH}^+][OH^-]}{[\text{Ethylamine}]} = \frac{x^2}{0.10 - x}$$

x = [ EthylamineH$^+$] = $[OH^-]$ = 0.0075 M. This gives pOH = 2.13 and so pH = 11.87.

17-109. $[A]_1$ = initial concentration of weak acid in original solution

$[A]_2$ = initial concentration of weak acid in diluted solution

$[H_3O^+]_1$ = equiibrium hydronium ion concentration in original solution

$[H_3O^+]_2$ = equiibrium hydronium ion concentration in diluted solution

If the fraction ionized doubles, then $[H_3O^+]_2 = 2 \cdot [H_3O^+]_1$

Since $K \approx [H_3O^+]^2 / [C]$, we can say that

$$[H_3O^+]_1^2 / [A]_1 = [H_3O^+]_2^2 / [A]_2$$

$$[H_3O^+]_1^2 / [A]_1 = \{2 \cdot [H_3O^+]_1\}^2 / [A]_2$$

Therefore, $[A]_2 / [A]_1 = 4$

Thus, in order to double the percent ionization of the acid, you would have to dilute the 100 mL of solution to 400 mL of solution.

17-111. $HCl < NH_4Cl < NaCl\ (pH = 7) < Na(CH_3COO) < KOH$

*—increasing pH→*

17-113.

| | |
|---|---|
| NaOH | Arrhenius base |
| $Na_2CO_3$ | Brønsted base |
| $Ca(OH)_2$ | Arrhenius base |
| KCN | Brønsted base |
| $NaC_2H_5O$ | Brønsted base |

17-115.

$$H_2C_2O_4(aq) + H_2O(\ell) \rightleftarrows HC_2O_4^-(aq) + H_3O^+(aq) \qquad K_{a1} = 5.9 \times 10^{-2}$$

$$HC_2O_4^-(aq) + H_2O(\ell) \rightleftarrows C_2O_4^{2-}(aq) + H_3O^+(aq) \qquad K_{a2} = 6.4 \times 10^{-5}$$

---

$$H_2C_2O_4(aq) + 2\,H_2O(\ell) \rightleftarrows C_2O_4^{2-}(aq) + 2\,H_3O^+(aq) \qquad K_{net} = 3.8 \times 10^{-6}$$

17-117. $9.0\ g \cdot 1\ mol / 144\ g = 0.063\ mol$

$0.063\ mol / 0.500\ L = 0.13\ M$

$C_6H_5COO^-(aq) + H_2O(\ell) \rightleftarrows C_6H_5COOH(aq) + OH^-(aq)$

$$K_b = \frac{[C_6H_5COOH][OH^-]}{[C_6H_5COO^-]} = \frac{x^2}{0.13 - x} = 1.6 \times 10^{-10}$$

$x = [C_6H_5COOH] = [OH^-] = 4.5 \times 10\text{-}6$ M

pOH = 5.35 and so pH = 8.65

17-119. $AgCN(s) + CN^-(aq) \rightleftarrows [Ag(CN)_2]^-(aq)$

The reaction above shows that AgCN can react as a Lewis acid.

$AgCN(s) + H_3O^+(aq) \rightleftarrows Ag^+(aq) + HCN(aq) + H_2O(\ell)$

In the reaction above AgCN behaves as a Brønsted base.

AgCN is amphoteric as it can react with both an acid ($H^+$) and with a base ($CN^-$).

17-121. (a) $NH_4^+(aq) + H_2O(\ell) \rightleftarrows NH_3(aq) + H_3O^+(aq)$ $K_w/K_b$

$CN^-(aq) + H_2O(\ell) \rightleftarrows HCN(aq) + OH^-(aq)$ $K_w/K_a$

$H_3O^+(aq) + OH^-(aq) \rightleftarrows 2\,H_2O(\ell)$ $1/K_w$

---

$NH_4^+(aq) + CN^-(aq) \rightleftarrows NH_3(aq) + HCN(aq)$ $K_{net} = K_w / K_aK_b$

(b) From the ionization of the ammonium ion we can write the equation:

$$[H_3O^+] = \frac{K_w / K_b \cdot [NH_4^+]}{[NH_3]}$$

From the ionization of HCN we can write the equation:

$$[H_3O^+] = \frac{K_a \cdot [HCN]}{[CN^-]}$$

Now we combine these equations into an expression for $[H_3O^+]^2$:

$$[H_3O^+]^2 = \frac{K_w / K_b \cdot [NH_4^+]}{[NH_3]} \cdot \frac{K_a \cdot [HCN]}{[CN^-]}$$

In any solution of $NH_4CN$, we know that $[NH_4^+] = [CN^-]$ and that $[HCN] = [NH_3]$. Therefore, the equation above can be simplified to

$[H_3O^+]^2 = [K_w \cdot K_a / K_b]$ and so $[H_3O^+] = [K_w \cdot K_a / K_b]^{\frac{1}{2}}$

(c) $[H_3O^+] = [K_w \cdot K_a / K_b]^{\frac{1}{2}} = [(1.0 \times 10^{-14})(4.0 \times 10^{-10}) / (1.8 \times 10^{-5})]^{\frac{1}{2}}$

$[H_3O^+] = 4.7 \times 10^{-10}$

pH = 9.33

## SUMMARY QUESTIONS

17-123. (a) 150.0 g = 0.85 • x

x = 176 g sulfanilic acid if there is a 100% yield

176 g • 1 mol / 173.2 g = 1.02 mol sulfanilic acid

1.02 mol of the acid requires 1.02 mol aniline

1.02 mol aniline • 93.13 g/mol = 94.9 g

94.9 g aniline • 1 mL / 1.02 g = 93.0 mL aniline

(b)

| **Angle** | **Value** |
|---|---|
| C—N—H | 109° |
| H—N—H | 109° |
| O—S—C | 109° |
| O—S—O | 109° |

(c) 1.25 g • 1 mol / 195.2 g = 0.00640 mol

0.00640 mol / 0.100 L = 0.0640 M

$H_2NC_6H_4SO_3^-(aq) + H_2O(\ell) \rightleftarrows H_2NC_6H_4SO_3H(aq) + OH^-(aq)$

$$K_b = K_w / K_a = \frac{1.0 \times 10^{-14}}{5.9 \times 10^{-4}} = \frac{[H_2NC_6H_4SO_3H]\,[OH^-]}{[H_2NC_6H_4SO_3^-]} = \frac{x^2}{0.0640 - x}$$

$x = [H_2NC_6H_4SO_3H] = [OH^-] = 1.0 \times 10^{-6}$ M

pOH = 6.0 and pH = 8.00

# CHAPTER 18

# REACTIONS BETWEEN ACIDS AND BASES

## INSTRUCTOR'S NOTES

This chapter is an extension of the concepts introduced in Chapter 17. As noted there we usually plan on about 9 lectures for all of the material on aqueous equilibria, with about three of those assigned to this chapter.

## LIST OF TRANSPARENCIES FOR THIS CHAPTER

| No. | Figure | Page | Subject |
|---|---|---|---|
| 90 | 18.5 | 770 | pH change in strong acid-strong base titration |
| 91 | 18.7 | 772 | pH change in titration of weak acid with strong base |

## SUGGESTED DEMONSTRATIONS

This chapter further describes the nature of acid-base reactions, so a lead-off demonstration of such a reaction would be helpful. If not used before, the Alka-Seltzer reaction in water can be done effectively in a flask or in a petri dish on an overhead projector (see Figure 18.21 on page 754).

It was not easy to find a demonstration of the common ion effect that could be photographed. However, the photo on page 761 shows one possible reaction.

The photograph on page 763 shows a demonstration of buffering that we have used in lecture; the color changes can be seen effectively in a large lecture room.

The following demonstrations on acids and bases were obtained from the Project SERAPHIM database (see the introduction to this *Manual*):

- Corso, C. R.; DeOliveira, J.E. "Acid-Base Reactions in the Gaseous State" *J. Chem. Educ.* **1978**, *55*, 244.
- Lerman, C.L.; Doyle, R.R. "An Effective Demonstration of the Behavior of Indicators and Biochemicals in Buffers" *ibid.* **1976**, *53*, 634.
- Chang, J.C. "A Buffer Solution and its Action" *ibid.* **1976**, *53*, 228.

- Ophardt, C.E. "Weak and Strong Acids and Buffers" *ibid.* **1976**, *53*, 229.
- Schultz, C.W.; Spannuth, S.L. "Comparison of Strong Acid and Weak Acid Titration Curves" *ibid.* **1979**, *56*, 194.

## COMPUTER SOFTWARE

Computers are an excellent way to represent some of the concepts of this chapter. We have used some simple programs to plot acid-base titration curves. The image of the computer screen is projected onto a screen in the front of the lecture room, and the students can then see how the value of $K_a$ of a weak acid, for example, affects the shape of the titration curve.

"The Acid-Base Package" is a program for MS-DOS/PC compatibles available from *JCE: Software.* The program is a powerful method of displaying acid-base titration curves. Other, less elegant programs for displaying titrations are also available from Project SERAPHIM.

For those interested in computer applications, see *Dynamic Models in Chemistry: A Workbook of Computer Simulations Using Electronic Worksheets* from N. Simonson and Company. (The authors are D.E. Atkinson, D.C. Brower, R.W. McClard, and D.S. Barkley.) This book, which uses Microsoft Excel, shows the student how to use spreadsheets to build models in "Chemical Equilibrium" and "Acids, Bases, and Ions," among other topics.

# SOLUTIONS TO SELECTED STUDY QUESTIONS

## ACID-BASE REACTIONS AND TITRATIONS

18-9. This is the reaction of a strong acid and a weak base.

$C_5H_5N(aq) + H_2O(\ell) \rightleftarrows C_5H_5NH^+(aq) + OH^-(aq)$ $K_b = 1.5 \times 10^{-9}$

$H_3O^+(aq) + OH^-(aq) \rightleftarrows 2\,H_2O(\ell)$ $K = 1/K_w$

------------------------------------------------------------ --------------------

$C_5H_5N(aq) + H_3O^+(aq) \quad C_5H_5NH^+(aq) + H_2O(\ell)$ $K_{net} = 1.5 \times 10^5$

18-11. $NH_3 + HCl \rightleftarrows NH_4^+ + Cl^-$

$0.050 \text{ L} \cdot 0.40 \text{ M} = 0.020 \text{ mol } NH_3$

This requires 0.020 mol HCl, which has a volume of 50.0 mL. 0.020 mol of $NH_4^+$ is produced, and this is dissolved in 100.0 mL of solution. Therefore,

$[NH_4^+] = 0.20 \text{ mol}/0.100 \text{ L} = 0.20 \text{ M}$

$NH_4^+(aq) + H_2O(\ell) \rightleftarrows NH_3(aq) + H_3O^+(aq)$

$[H_3O^+] = \{K_a \cdot [NH_4^+]\}^{1/2} = [5.6 \times 10^{-10} \cdot 0.20]^{1/2} = 1.1 \times 10^{-5} \text{ M}$

pH = 4.98

18-13. (a) 0.45 M $H_2SO_4$ with 0.90 M NaOH (both $H^+$ ions of $H_2SO_4$ are titrated)

All of the acid and base are consumed to leave a solution of $Na_2SO_4$. Since $SO_4^{2-}$ is a very weak base, the solution will be neutral to *very* slightly basic. pH ≥ 7

(b) 0.050 M formic acid, HCOOH, is titrated with 0.034 M NaOH

The titration produces a solution of $Na^+$ and $HCOO^-$. Since the latter is the conjugate base of a weak acid, the solution will be basic. pH > 7

(c) 0.052 M $H_2C_2O_4$ (oxalic acid) is titrated with 0.065 M NaOH (both $H^+$ ions of oxalic acid are titrated)

The titration produces a solution of $Na^+$ and $C_2O_4^{2-}$. The latter is the conjugate base of a weak acid, so the solution will be basic. pH > 7

18-15. (a) The pH of the solution when 25.0 mL of 0.015 M NaOH is mixed with 25.0 mL of 0.015 M formic acid, HCOOH.*

Mixing these reagents will lead to 50.0 mL of 0.0075 M NaHCOO (sodium formate). The formate ion, $HCOO^-$, is the conjugate base of a weak acid, so the solution will be basic.

$HCOO^-(aq) + H_2O(\ell) \rightleftarrows HCOOH(aq) + OH^-(aq)$

$[OH^-] = \{K_b \cdot [HCOO^-]\}^{1/2} = [(5.6 \times 10^{-11}) \cdot (0.0075)]^{1/2} = 6.5 \times 10^{-7}$ M

pOH = 6.19 and so pH = 7.81

(b) The pH of the solution when 25.0 mL of 0.015 M NaOH is mixed with 0.046 g of benzoic acid, $C_6H_5COOH$.

0.046 g $C_6H_5COOH$ • 1 mol / 122 g = $3.8 \times 10^{-4}$ mol

$3.8 \times 10^{-4}$ mol / 0.025 L = 0.015 M = $[C_6H_5COOH]$

Mixing these reagents will lead to 25.0 mL of 0.015 M $NaC_6H_5COO$ (sodium benzoate). The benzoate ion, $C_6H_5COO^-$, is the conjugate base of a weak acid, so the solution will be basic.

$C_6H_5COO^-(aq) + H_2O(\ell) \rightleftarrows C_6H_5COOH(aq) + OH^-(aq)$

$[OH^-] = \{K_b \cdot HCOO^-]\}^{1/2} = [(1.6 \times 10^{-10}) \cdot (0.015)]^{1/2} = 1.6 \times 10^{-6}$ M

pOH = 5.81 and so pH = 8.19

18-17. *Step 1:* Calculate the moles of benzoic acid

0.235 g • 1 mol / 122.1 g = 0.00192 mol

*Step 2:* Calculate moles and then volume of NaOH solution needed.

0.00192 mol benzoic acid require 0.00192 mol NaOH

0.00192 mol NaOH • 1 L / 0.108 mol = 0.0178 L

*Step 3:* Calculate the moles and then the "original" concentrations of the ions at the equivalence point.

The total volume at the equivalence point = 0.100 L + 0.0178 L = 0.118 L

---

* **NOTE:** In the first printing of the book the concentration of formic acid was omitted. It should be the same as that of NaOH, that is, 0.015 M.

| **"Original" moles at the equivalence point** | | **[..] at the equivalence point** |
|---|---|---|
| $Na^+$ | 0.00192 mol | 0.0163 M |
| $C_6H_5COO^-$ | 0.00192 mol | 0.0163 M |

*Step 4:* Calculate the concentrations of all species at the equivalence point.

$C_6H_5COO^-(aq) + H_2O(\ell) \rightleftarrows C_6H_5COOH(aq) + OH^-(aq)$

$[OH^-] = \{K_b \cdot [C_6H_5COO^-]\}^{1/2} = [(1.6 \times 10^{-10}) \cdot (0.0163)]^{1/2} = 1.6 \times 10^{-6}$ M

$[OH^-] = [C_6H_5COOH] = 1.6 \times 10^{-6}$ M and $[C_6H_5COO^-] = 0.0163$ M

pOH = 5.79 and so pH = 8.21

18-19. *Step 1:* Calculate the moles of NaOH used, the moles of formic acid titrated, and the moles of sodium formate produced.

0.275 M • 0.02243 L = 0.00617 mol NaOH used

Therefore, 0.00167 mol of HCOOH are titrated and 0.00167 mol of NaHCOO are produced. This means the solution will be basic because the solution after the titration contains the formate ion ($HCOO^-$), the conjugate base of a weak acid.

*Step 2:* Calculate the concentration of the formate ion at the equivalence point.

$[HCOO^-] = 0.00617$ mol/(0.025 L + 0.02243 L) = 0.13 M

*Step 3:* Calculate the concentrations of the ions at the equivalence point.

$HCOO^-(aq) + H_2O(\ell) \rightleftarrows HCOOH(aq) + OH^-(aq)$

$[OH^-] = \{K_b \cdot [HCOO^-]\}^{1/2} = [(5.6 \times 10^{-11}) \cdot (0.13)]^{1/2} = 2.7 \times 10^{-6}$ M

$[OH^-] = [HCOO^-] = 2.7 \times 10^{-6}$ M

pOH = 5.57 and so pH = 8.43

*Step 4:* Calculate the concentration of formic acid in the original solution.

0.00617 mol / 0.025 L = 0.247 M

## THE COMMON ION EFFECT

18-21. (a) $H_2C_2O_4(aq) + 2\,H_2O(\ell) \rightleftarrows C_2O_4^{2-}(aq) + 2\,H_3O^+(aq)$

On adding the oxalate salt to the solution should lower the hydronium ion concentration and raise the pH of the solution.

(b) $NH_4^+(aq) + H_2O(\ell) \rightleftarrows NH_3(aq) + H_3O^+(aq)$

On adding ammonium chloride to HCl, there may be a *slight* increase in pH.

(c) No change in pH

18-23. See page 760 for the lactic acid equilibrium and the value of $K_a$.

*Step 1:* Calculate the concentration of the lactate ion.

$2.75 \text{ g } NaC_3H_5O_3 \cdot 1 \text{ mol} / 112.1 \text{ g} = 0.0245 \text{ mol}$

$0.0245 \text{ mol} / 0.500 \text{ L} = 0.0491 \text{ M}$

*Step 2:* Calculate the pH of the solution

$C_3H_6O_3(aq) + H_2O(\ell) \rightleftarrows C_3H_5O_3^-(aq) + H_3O^+(aq)$

$$[H_3O^+] = \frac{[C_3H_6O_3]}{[C_3H_5O_3^-]} K_a = \frac{0.100}{0.0491} \cdot (1.4 \times 10^{-4}) = 2.85 \times 10^{-4} \text{ M}$$

pH = 3.55

*Step 3:* Compare the pH of the sodium lactate/lactic acid solution with the solution of just 0.100 M lactic acid.

$[H_3O^+] = \{K_a \cdot [C_3H_6O_3]\}^{1/2} = 3.74 \times 10^{-3} \text{ M}$

pH = 2.43

18-25. (a) Calculate the pH of a 0.050 M solution of HF

$HF(aq) + H_2O(\ell) \rightleftarrows F^-(aq) + H_3O^+(aq)$

$$K_a = \frac{[F^-][H_3O^+]}{[HF]} = \frac{x^2}{0.050 - x} = 7.2 \times 10^{-4}$$

Solving this equation for x using the method of successive approximations, we find

$x = [F^-] = [H_3O^+] = 5.7 \times 10^{-3} \text{ M}$

pH = 2.25

(b) Calculate the pH after adding 1.58 g of NaF

$1.58 \text{ g NaF} \cdot 1 \text{ mol} / 41.99 \text{ g} = 0.0376 \text{ mol}$

$0.0376 \text{ mol} / 0.250 \text{ L} = 0.150 \text{ M}$

$$K_a = \frac{[F^-][H_3O^+]}{[HF]} = \frac{(0.150) \cdot x}{0.050 - x} = 7.2 \times 10^{-4}$$

Solving this equation for x using the method of successive approximations, we find

$x = [F^-] = [H_3O^+] = 2.4 \times 10^{-4}$ M

pH = 3.62

As expected, the pH increases when a common ion ($F^-$) is added to HF.

18-27. (a) After reaction, the sodium hydroxide will have been consumed and the solution will contain oxalic acid and its salt. The pH of the solution will have increased (owing to a decrease in hydronium ion concentration).

(b) After reaction, the HCl will have been consumed, and the solution will contain pyridine, a weak base, and its salt. The pH will decrease owing to a decrease in the $OH^-$ concentration.

## BUFFER SOLUTIONS

18-29. *Step 1:* Calculate the hydronium ion concentration desired.

$[H_3O^+] = 10^{-pH} = 10^{-4.5} = 3 \times 10^{-5}$ M

*Step 2:* Calculate the concentration of $Na(CH_3COO)$ required.

$$[H_3O^+] = 3 \times 10^{-5}\text{ M} = \frac{[CH_3COOH]}{[CH_3COO^-]} \cdot K_a = \frac{0.15}{[CH_3COO^-]} \cdot (1.8 \times 10^{-5})$$

$[CH_3COO^-] = 0.09$ M

*Step 3:* Calculate the grams of sodium acetate required.

0.09 mol/L · 1.00 L · 82 g/mol = 7 g

18-31. *Step 1:* Calculate the hydronium ion concentration desired.

$[H_3O^+] = 10^{-pH} = 10^{-3.88} = 1.3 \times 10^{-4}$ M

Step 2: Calculate the moles of sodium benzoate in the solution.

14.4 g · 1 mol / 144.1 g = 0.0999 mol

*Step 2:* Calculate the moles of $C_6H_5COOH$ required.

$$[H_3O^+] = 1.3 \times 10^{-4}\,M = \frac{C_6H_5COOH}{C_6H_5COO^-} \cdot K_a = \frac{C_6H_5COOH}{0.0999\ mol} \cdot (6.3 \times 10^{-5})$$

$C_6H_5COOH = 0.209$ mol

*Step 3:* Calculate the grams of benzoic acid required.

0.209 mol • 122.1 g/mol = 25.5 g

18-33. Calculate the pH of the original solution.

$NH_3(aq) + H_2O(\ell) \rightleftarrows NH_4^+(aq) + OH^-(aq)$

*Step 1:* Calculate the number of moles of each component.

5.15 g • 1 mol / 80.04 g = 0.0643 mol $NH_4^+$

0.10 L • 0.15 mol/L = 0.015 mol $NH_3$

*Step 2:* Calculate the hydroxide ion concentration and then the pH.

$$[OH^-] = \frac{[NH_3]}{[NH_4^+]} \cdot K_b = \frac{0.015\ mol}{0.0643\ mol} \cdot (1.8 \times 10^{-5})$$

$[OH^-] = 4.2 \times 10^{-6}$ M pOH = 5.38 and pH = 8.62

Calculate the pH after diluting the solution to 500 mL.

Adding water to the solution only serves to the change the concentrations of $NH_4^+$ and $NH_3$ and not to change the *ratio* of those concentrations. Therefore, the value of $[OH^-]$ will not change.

18-35. Assume that [conjugate base] / [acid] = 1, so $[H_3O^+] = K_a$.

(a) $H_3PO_4 / NaH_2PO_4$ $[H_3O^+] = 7.5 \times 10^{-3}$ pH = 2.12

(b) $NaH_2PO_4 / Na_2HPO_4$ $[H_3O^+] = 6.2 \times 10^{-8}$ pH = 7.21

(c) $Na_2HPO_4 / Na_3PO_4$ $[H_3O^+] = 3.6 \times 10^{-13}$ pH = 12.44

The best choice here is (b), the $NaH_2PO_4/Na_2HPO_4$ buffer.

18-37. Part (a) — Calculate the pH of the buffer solution.

$CH_3COOH(aq) + H_2O(\ell) \rightleftarrows CH_3COO^-(aq) + H_3O^+(aq)$

*Step 1:* Calculate moles of each component

4.95 g $Na(CH_3COO)$ • 1 mol / 82.03 g = 0.0603 mol $CH_3COONa$

0.250 L • 0.150 mol/L = 0.0375 mol $CH_3COOH$

*Step 2:* Calculate the pH of the solution.

$$[H_3O^+] = \frac{CH_3COOH}{CH_3COO^-} \cdot K_a = \frac{0.0375 \text{ mol}}{0.0603 \text{ mol}} \cdot (1.8 \times 10^{-5})$$

$[H_3O^+] = 1.1 \times 10^{-5}$ M

pH = 4.95

Part (b) — Now add 80. mg of NaOH to 100. mL of the solution.

*Step 1:* If we take 100. mL of the buffer solution, we are taking

0.0603 mol $CH_3COONa$ • 100. mL / 250. mL = 0.0241 mol $CH_3COONa$

0.0375 mol $CH_3COOH$ • 100. mL / 250. mL = 0.0150 mol $CH_3COOH$

*Step 2:* Calculate the moles of acid and conjugate base after adding NaOH

0.080 g NaOH • 1 mol / 40.0 g = 0.0020 mol NaOH

| | mol $CH_3COOH$ | mol $CH_3COO^-$ |
|---|---|---|
| mol before reaction | 0.0150 | 0.0241 |
| Δ mol | -0.0020 | +0.0020 |
| mol after reaction | 0.0130 | 0.0261 |

*Step 3:* Calculate the pH of the solution.

$$[H_3O^+] = \frac{CH_3COOH}{CH_3COO^-} \cdot K_a = \frac{0.0130 \text{ mol}}{0.0261 \text{ mol}} \cdot (1.8 \times 10^{-5})$$

$[H_3O^+] = 9.0 \times 10^{-6}$ M

pH = 5.05

18-39. *Step 1:* Calculate the moles of each component of the solution.

$0.150 \text{ mol/L} \cdot 1.00 \text{ L} = 0.150 \text{ mol } NH_3$

$10.0 \text{ g } NH_4Cl \cdot 1 \text{ mol} / 53.49 \text{ g} = 0.187 \text{ mol } NH_4Cl$

*Step 2:* Calculate the pH of the buffer

$NH_4^+(aq) + H_2O(\ell) \rightleftarrows NH_3(aq) + H_3O^+(aq)$

$$[H_3O^+] = \frac{[NH_4^+]}{[NH_3]} \cdot K_b = \frac{0.187 \text{ mol}}{0.150 \text{ mol}} \cdot 5.6 \times 10^{-10}$$

$[H_3O^+] = 7.0 \times 10^{-10}$ M  pH = 9.16

*Step 3:* Calculate the moles of $NH_4^+$ and $NH_3$ that are required to have a pH of 8.16.

The final pH should be one unit lower than 9.16, that is, pH = 8.16. This means that $[H_3O^+] = 7.0 \times 10^{-9}$ M. To achieve this, we add HCl to consume $NH_3$ and produce $NH_4^+$. The moles of $NH_3$ consumed = x, and the moles of $NH_4^+$ produced = x.

$$[H_3O^+] = 7.0 \times 10^{-9} = \frac{0.187 \text{ mol} + x}{0.150 \text{ mol} - x} \cdot 5.6 \times 10^{-10}$$

x = 0.125 mol $NH_3$ consumed and 0.125 mol $NH_4^+$ produced.

Therefore, we add 0.125 mole of HCl to consume the $NH_3$ and produce $NH_4^+$.

## TITRATION CURVES AND INDICATORS

18-41. Titration of $Na_2CO_3$ with HCl. The curve begins at a basic pH because the carbonate ion is the conjugate base of a weak acid. The equivalence point is less than 7 because the solution at this point contains a weak acid, $HCO_3^-$. Given below is a *very* rough curve for this titration.

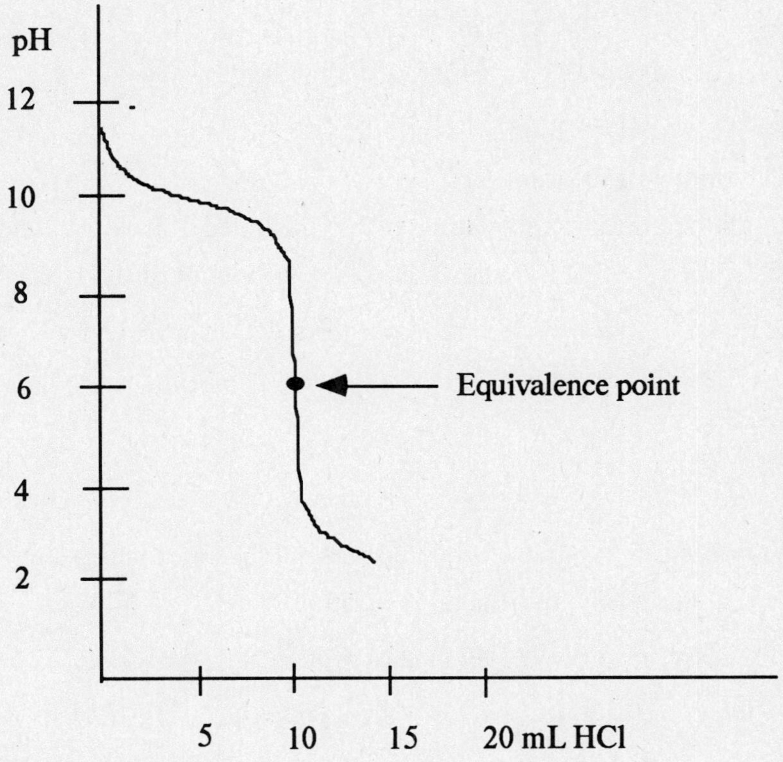


18-43. Part (a) — pH before titration

$$C_6H_5NH_3^+(aq) + H_2O(\ell) \rightleftarrows C_6H_5NH_2(aq) + H_3O^+(aq)$$

$$K_a = K_w / K_b = 1.0 \times 10^{-14} / 4.2 \times 10^{-10} = 2.4 \times 10^{-5}$$

$$[H_3O^+] = \{K_a \cdot [C_6H_5NH_3^+]\}^{1/2} = 1.5 \times 10^{-3}\ M$$

$$pH = 2.81$$

Part (b) — pH at the equivalence point

*Step 1:* Calculate moles of $[C_6H_5NH_3]Cl$ before titration.

$$0.0250\ L \cdot 0.100\ mol/L = 0.00250\ mol\ [C_6H_5NH_3]Cl$$

*Step 2:* Calculate volume of NaOH required to reach the equivalence point.

This requires 0.00250 mol NaOH for titration to the equivalence point.

0.00250 mol NaOH • 1 L / 0.115 mol = $2.17 \times 10^{-2}$ L or 21.7 mL of NaOH

*Step 3:* Calculate the concentration of $C_6H_5NH_2$ at the equivalence point.

0.00250 mol $C_6H_5NH_2$ / (0.0250 L + 0.0217 L) = 0.0535 M $C_6H_5NH_2$

*Step 4:* Calculate the pH at the equivalence point.

$[OH^-] = \{K_b \cdot [C_6H_5NH_2]\}^{1/2} = \{4.2 \times 10^{-10} \cdot 0.0535\}^{1/2} = 4.7 \times 10^{-6}$ M

pOH = 5.32 pH = 8.68

Part (c) — pH at the midpoint of the titration

Half way through the titration, the volume of NaOH added is 10.9 mL, and the solution consists of $[C_6H_5NH_3]Cl$ and an equivalent amount of $C_6H_5NH_2$. Therefore, the hydronium ion concentration of the solution is given by $K_a$ of $[C_6H_5NH_3]Cl$.

$[H_3O^+] = K_a = 2.4 \times 10^{-5}$ pH = 4.61

Part (d) — Indicator.

Methyl orange or bromcresol green would be reasonable choices for an indicator.

Part (e) — Calculate pH at various points along the titration with 0.115 M NaOH.

The reaction begins with 0.00250 mol $[C_6H_5NH_3]Cl$.

| **mL NaOH Added** | **mol NaOH Used** | **mol $C_6H_5NH_2$ Produced** | **mol $C_6H_5NH_3^+$ Left** | **$[H_3O^+]$** | **pH** |
|---|---|---|---|---|---|
| 5.00 | $5.75 \times 10^{-4}$ | $5.75 \times 10^{-4}$ | 0.00193 | $8.06 \times 10^{-5}$ | 4.09 |
| 10.0 | $1.15 \times 10^{-3}$ | $1.15 \times 10^{-3}$ | 0.00135 | $2.82 \times 10^{-5}$ | 4.55 |
| 15.0 | $1.73 \times 10^{-3}$ | $1.73 \times 10^{-3}$ | 0.000775 | $1.08 \times 10^{-5}$ | 4.97 |
| 20.0 | $2.30 \times 10^{-3}$ | $2.30 \times 10^{-3}$ | 0.000200 | $2.09 \times 10^{-6}$ | 5.68 |

After 24.0 mL of NaOH has been added, the solution consists only of $C_6H_5NH_2$ and *excess* NaOH, and the pH depends only on the excess NaOH. When 24.0 mL of NaOH has been added, there are 24.0 - 21.7 = 2.3 mL of 0.115 M NaOH in excess. This gives a hydroxide concentration of $5.4 \times 10^{-3}$ M and a pH of 11.73. After 30.0 mL of NaOH, there are 30.0 - 21.7 = 8.3 mL of NaOH in excess. This gives [OH-] = $1.7 \times 10^{-2}$ M and a pH of 12.24.

The plot below was done using only those points calculated above. The curve would be more smooth, and the rise in pH near the equivalence point more steep, if more points were included.

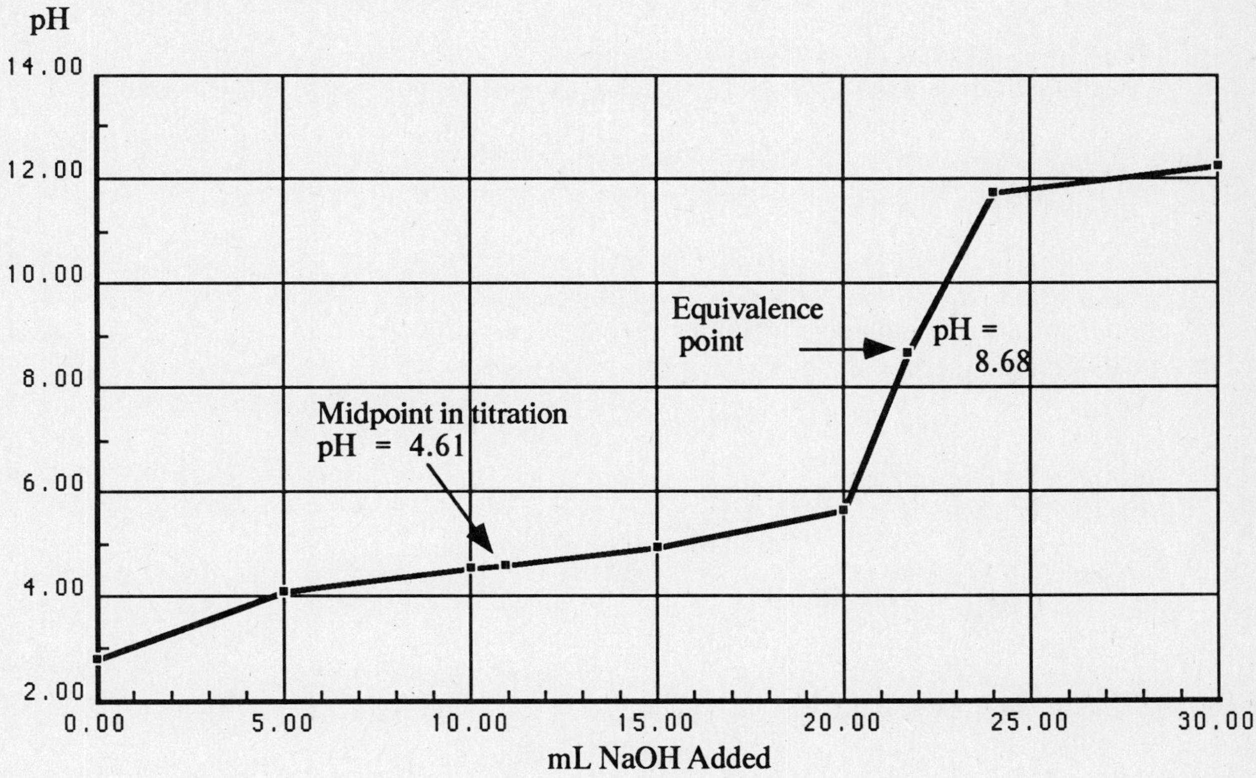


18-45.

| Titration | pH at Equiv. Point | Possible Indicator |
|---|---|---|
| $CO_3^{2-}$ titrated with HCl | >7 (about 8-9) | phenol red |
| HOCl with NaOH | >7 (about 8-10) | phenolphthalein |
| $(CH_3)_3N$ with HCl | <7 (about 5-6) | bromcresol green |

## GENERAL QUESTIONS

18-47. $NH_3(aq) + H_2O(\ell) \rightleftarrows NH_4^+(aq) + OH^-(aq)$

| | $[NH_3]$ | $[NH_4^+]$ | $[OH^-]$ |
|---|---|---|---|
| Initial [..] | 0.20 | 0 | 0.040 (from NaOH) |
| Δ [..] | -x | +x | +x |
| Equilibrium [..] | 0.20 - x | +x | 0.040 + x |

$$K_b = 1.8 \times 10^{-5} = \frac{[NH_4^+] \cdot [OH^-]}{[NH_3]} = \frac{x \cdot (0.040 + x)}{0.20 - x}$$

The ionization of ammonia produces a negligible amount of OH- in solution, so we assume x is very small relative to 0.040 and 0.20. Therefore,

$$x = [NH_4^+] = 9.0 \times 10^{-5}\ M$$

In summary, the ion concentration at equilibrium are:

$$[Na^+] = [OH^-] = 0.040\ M$$

$$[NH_3] = 0.20\ M \text{ and } [NH_4^+] = 9.0 \times 10^{-5}\ M$$

18-49. *Step 1:* Calculate the moles of pyridine

$0.453\ g \cdot 1\ mol / 79.10\ g = 0.00573\ mol$

*Step 2:* Calculate the quantity of HCl required in the titration

0.00573 mol pyridine require 0.00573 mol HCl

$0.00573\ mol\ HCl \cdot 1.00\ L / 0.205\ mol = 0.0279\ L$ of HCl required

*Step 3:* Calculate the concentration of $C_5H_5NH^+$ at the equivalence point

$0.00573\ mol / (0.100\ L + 0.0279\ L) = 0.0448\ M$

*Step 4:* The ion $C_5H_5NH^+$ is the conjugate acid of a weak base.

$$C_5H_5NH^+(aq) + H_2O(aq) \rightleftarrows C_5H_5N(aq) + H_3O^+(aq)$$

$$K_a = K_w / K_b = 1.0 \times 10^{-14} / 1.5 \times 10^{-9} = 6.7 \times 10^{-6}$$

$$[H_3O^+] = \{K_a \cdot [C_5H_5NH^+]\}^{1/2} = 5.47 \times 10^{-4}\ M$$

pH at the equivalence point = 3.26

| *Summary:* | **Ion** | **Concentration** |
|---|---|---|
| | $C_5H_5NH^+$ | 0.0448 M |
| | $Cl^-$ | 0.0448 M |
| | $H_3O^+$ | $5.47 \times 10^{-4}$ M |
| | $OH^-$ | $1.83 \times 10^{-11}$ M |

18-51. When pH = 7.4, $[H_3O^+] = 3.98 \times 10^{-8}$ M

$$[H_3O^+] = 3.98 \times 10^{-8}\text{ M} = \frac{[CH_3CH(OH)COOH]}{[CH_3CH(OH)COO^-]} \cdot K_a$$

Since $K_a = 1.4 \times 10^{-4}$, the only way the above can be an equality is if $[CH_3CH(OH)COO^-] \gg [CH_3CH(OH)COOH]$. That is, the lactate ion predominates.

18-53. Calculate the pH of a buffer solution having equimolar amounts of acetic acid and sodium acetate.

$$[H_3O^+] = \frac{[CH_3COOH]}{[CH_3COO^-]} \cdot K_a$$

$[H_3O^+] = K_a = 1.8 \times 10^{-5}$ M when the acid and base concentrations are equal, and so pH = 4.74. If the pH increases by 1 unit, the pH becomes 5.74, and $[H_3O^+]$ would decrease to $1.82 \times 10^{-6}$ M. This would be caused by adding a strong base such as NaOH, which has the effect of increasing $[CH_3COO^-]$ by an amount x and decreasing $[CH_3COOH]$ by an equal amount x. Therefore,

$$[H_3O^+] = 1.82 \times 10^{-6} = \frac{0.10 - x}{0.10 + x} \cdot (1.8 \times 10\text{-}5)$$

Solving for x, we find x = 0.082. Therefore, if we add 0.082 mol of NaOH to a buffer that has 0.10 mol of $CH_3COOH$ and 0.10 mol of $NaCH_3COO$, the pH will increase by 1.0 unit.

18-55. *Step 1:* Calculate moles of HF present and then moles and volume of NaOH required.

0.050 L HF • 0.0500 mol/L = $2.50 \times 10^{-3}$ mol

$2.50 \times 10^{-3}$ mol NaOH required

$2.50 \times 10^{-3}$ mol NaOH • 1.00 L / 0.0955 mol = 0.0262 L NaOH

*Step 2:* Calculate the concentration of NaF at the equivalence point.

$2.50 \times 10^{-3}$ mol NaF / (0.0500 L + 0.0262 L) = 0.0328 M

*Step 3:* Calculate the pH at the equivalence point

$F^-(aq) + H_2O(\ell) \rightleftarrows HF(aq) + OH^-(aq)$ $K_b = 1.4 \times 10^{-11}$

$[OH^-] \approx \{K_b \cdot [F^-]\}^{1/2} = 6.8 \times 10^{-7}$ M

pOH = 6.17. and so pH = 7.83

18-57. When pH = 2.40, then $[H_3O^+] = 10^{-pH} = 3.98 \times 10^{-3}$ M

For a monoprotic acid HA, we know that $K_a = [H_3O^+]^2 / \{[HA]_0 - [H_3O^+]\}$, where $[HA]_0$ is the initial concentration of the acid. Using $[H_3O^+] = 3.98 \times 10^{-3}$ M and $[HA]_0$ = 0.40 M, we find $K_a = 4.0 \times 10^{-5}$. Now we can find the pH of a buffer solution containing equimolar amounts of HA and NaA.

$[H_3O^+] = \{[HA] / [A^-]\} \cdot K_a$

Since $[HA] = [A^-]$, $[H_3O^+] = K_a$. This means the pH = 4.40.

18-59. $C_6H_4(COOH)COO^-(aq) + OH^-(aq) \rightarrow C_6H_5(COO)_2^{2-}(aq) + H_2O(\ell)$

$HP^-$ $P^{2-}$

*Step 1:* Calculate moles of each component

0.043 L NaOH • 0.10 mol/L = 0.0043 mol NaOH

0.050 L • 0.10 mol/L = 0.0050 mol KHP

*Step 2:* Calculate composition of solution after reaction

0.0050 mol KHP + 0.0043 mol NaOH

→ 0.0043 mol $P^{2-}$ + 0.0007 mol KHP left

*Step 3:* Calculate the solution pH

$$HP^-(aq) + H_2O(\ell) \rightleftarrows P^{2-}(aq) + H_3O^+(aq)$$

$$[H_3O^+] = \frac{\text{moles } HP^-}{\text{moles } P^{2-}} \cdot K_a = \frac{0.0007}{0.0043} \cdot (3.9 \times 10^{-6})$$

$[H_3O^+] = 6 \times 10^{-7}$ M  pH = 6.2

18-61. $ProcH^+(aq) + H_2O(\ell) \rightleftarrows Proc(aq) + H_3O^+(aq)$

$[H_3O^+] = 10^{-pH} = 1.0 \times 10^{-6}$ M

$$K_a = \frac{[Proc][H_3O^+]}{[ProcH^+]} = \frac{(1.0 \times 10^{-6})^2}{0.10} = 1.0 \times 10^{-11}$$

18-63. Dissolve 0.221 g of trimethylamine, $(CH_3)_3N$, in *50.0 mL of water* and titrate it with 0.100 M HCl.**

(a) What is the pH of the original amine solution?

$$(CH_3)_3N(aq) + H_2O(\ell) \rightleftarrows (CH_3)_3NH^+(aq) + OH^-(aq)$$

0.221 g • 1 mol / 59.11 g = 0.00374 mol

0.00374 mol / 0.050 L = 0.0748 M

$$K_b = 7.4 \times 10^{-5} = \frac{[(CH_3)_3NH^+][OH^-]}{[(CH_3)_3N]} = \frac{[OH^-]^2}{0.0748 - [OH^-]}$$

$[OH^-] = 2.32 \times 10^{-3}$ M

pOH = 2.64, and so pH = 11.36

(b) What is the pH of the solution at the mid-point of the titration?

At the midpoint of the titration, $[OH^-] = K_b$. Therefore, pOH = 4.13, and so pH = 9.87.

(c) What is the pH at the equivalence point?

0.00374 mol $(CH_3)_3N$ requires 0.00374 mol HCl and produces 0.00374 mol

** NOTE: In the first printing of the book the volume of the amine solution (50.0 mL) was missing.

$(CH_3)_3NH^+$.

0.00374 mol HCl • 1.00 L / 0.100 mol = 0.0374 L HCl

Total solution volume at the equivalence point = 87.4 mL

$[(CH_3)_3NH^+]$ = 0.00374 mol $(CH_3)_3NH^+$ / 0.0874 L = 0.0428 M

At the equivalence point, the ammonium ion is a weak acid —

$$(CH_3)_3NH^+(aq) + H_2O(\ell) \rightleftarrows (CH_3)_3N(aq) + H_3O^+(aq)$$

$$K_a = K_w / K_b = 1.0 \times 10^{-14} / 7.4 \times 10^{-5} = 1.4 \times 10^{-10}$$

$$[H_3O^+] = \{K_a \cdot [(CH_3)_3NH^+]\}^{1/2} = 2.4 \times 10^{-6}\ M$$

pH = 5.62

(d) The titration curve is similar to that in Study Question 18-41.

18-65. (a) What is the pH of the $[NH_3OH]Cl$ solution before the titration begins?

$$NH_3OH^+(aq) + H_2O(\ell) \rightleftarrows NH_2OH(aq) + H_3O^+(aq)$$

$$K_a = K_w / K_b = 1.0 \times 10^{-14} / 6.6 \times 10^{-9} = 1.5 \times 10^{-6}$$

$$K_a = \frac{[NH_2OH]\,[H_3O^+]}{[NH_3OH^+]} = \frac{[H_3O^+]^2}{0.155 - [H_3O^+]}$$

$[H_3O^+]$ = 4.8 x $10^{-4}$ M pH = 3.32

(b) What is the pH at the equivalence point?

0.025 L • 0.155 mol / L = 0.0039 mol $NH_3OH^+$

0.0039 mol $NH_3OH^+$ requires 0.0039 mol NaOH and produces 0.0039 mol $NH_2OH$.

0.0039 mol NaOH • 1.00 L / 0.108 mol = 0.036 L

Total volume at the equivalence point = 61 mL

$[NH_2OH]$ = 0.0039 mol / 0.061 L = 0.064 M

$$K_b = \frac{[NH_3OH^+]\,[OH^-]}{[NH_2OH]} = \frac{[OH^-]^2}{0.064 - [OH^-]} = 6.6 \times 10^{-9}$$

$[OH^-] = 2.1 \times 10^{-5}$ M $\qquad$ pOH = 4.69 and pH = 9.31

(c) What is the pH at the mid-point of the titration?

At the midpoint of the titration, $[NH_2OH] = [NH_3OH^+]$. Therefore, $[H_3O^+] =$ $K_a$ for $NH_3OH^+$.

$[H_3O^+] = K_a = 1.5 \times 10^{-6}$ M and so pH = 5.82

18-66. *Step 1:* Calculate the value of $K_b$ for the base.

The pH at the midpoint of the titration can give $K_b$ for the base.

$HOCH_2CH_2NH_2(aq) + H_2O(\ell) \rightleftarrows HOCH_2CH_2NH_3^+(aq) + OH^-(aq)$

pH = 9.50, so pOH = 4.50. This means $[OH^-] = 3.2 \times 10^{-5}$ M, and so $K_b = 3.2 \times 10^{-5}$.

*Step 2:* Calculate the moles of HCl required and $HOCH_2CH_2NH_3^+$ produced at equivalence point.

$HOCH_2CH_2NH_2(aq) + H_3O^+(aq) \rightarrow HOCH_2CH_2NH_3^+(aq) + H_2O(\ell)$

$0.025 \text{ L } HOCH_2CH_2NH_2(aq) \cdot 0.010 \text{ mol/L} = 2.5 \times 10^{-4}$ mol

$2.5 \times 10^{-4}$ mol $HOCH_2CH_2NH_2$ produces $2.5 \times 10^{-4}$ mol $HOCH_2CH_2NH_3^+$ and requires $2.5 \times 10^{-4}$ mol HCl.

*Step 3:* Calculate $[HOCH_2CH_2NH_3^+]$ at the equivalence point.

Volume of HCl required = $2.5 \times 10^{-4}$ mol HCl $\cdot$ 1.0 L/0.0095 mol
= 0.026 L

Total volume at the equivalence point = 51 mL

$[HOCH_2CH_2NH_3^+] = 2.5 \times 10^{-4} \text{ mol}/0.051 \text{ L} = 0.0049$ M

*Step 4:* Calculate the pH at the equivalence point.

$HOCH_2CH_2NH_3^+(aq) + H_2O(\ell) \rightleftarrows HOCH_2CH_2NH_2(aq) + H_3O^+(aq)$

$$K_a = K_w/K_b = 1.0 \times 10^{-14}/3.2 \times 10^{-5} = \frac{[HOCH_2CH_2NH_2][H_3O^+]}{[HOCH_2CH_2NH_3^+]}$$

$[H_3O^+] = 1.2 \times 10^{-6}$ M, and so pH at equivalence point = 5.91

# CHAPTER 19
# PRECIPITATION REACTIONS

## INSTRUCTOR'S NOTES

As pointed out in Chapter 17, we sometimes cover this chapter (except for Sections 19.9 and 19.10) before those on acid-base reactions. We find that the equilibrium expression for insoluble salts is a good place to begin our discussions of aqueous equilibria in general. However, we recognize that the majority of instructors probably cover acid-base chemistry before insoluble salts, so the book was constructed in that manner. However, the book was also written in such a way that insoluble salts can be covered first.

We usually assign about 2-3 lecture hours to this chapter.

## LIST OF TRANSPARENCIES FOR THIS CHAPTER

| No. | Figure | Page | Subject |
|---|---|---|---|
| 92 | --- | 806 | The carbon cycle in the biosphere |

## SUGGESTED DEMONSTRATIONS

There are many demonstrations that can be done for this topic. To begin the lectures, we bring some mineral samples to class to show how many common minerals are really insoluble salts whose behavior in water can be treated by the methods of this chapter.

Volume 1 of the demonstration books by Shakhashiri contains a number of demonstrations involving insoluble salts. (Many also involve the dissolution of precipitates using complex ion formation.) Indeed, five of these demonstrations were photographed for this text and have been used in class:

| Demonstration | Text Page |
|---|---|
| • Reactions between $CO_2$ and limewater | 807 |
| • Precipitates and complexes of silver(I) | 809 |
| • Precipitates and complexes of lead(II) | 964 |
| • Precipitates and complexes of iron(III) | 1062 |
| • Precipitates and complexes of nickel(II) | 1064 |

To illustrate the possible separation of cations by a precipitating anion, we also use the following demonstration: 3 g of $Cd(NO_3)_2$ and 0.3 g of $Ni(NO_3)_2$ are dissolved in 500 mL of water containing 10 mL of concentrated $HNO_3$. The solution is placed in a large flask on the lecture table. As it is stirred with a magnetic stirring bar, $(NH_4)_2S$ is added slowly. Bright yellow CdS precipitates first ($K_{sp} = 3.6 \times 10^{-29}$) followed later by black NiS ($K_{sp} = 3.0 \times 10^{-21}$).

The photograph in Figure 19.4 that illustrates the common ion effect was not difficult to set up. Silver acetate was freshly precipitated. After washing with water, distilled water was added, the mixture was shaken vigorously, and the test tube was allowed to stand for some time. On adding $AgNO_3$, more silver acetate is clearly precipitated.

We have also used the "Old Nassau" reaction to illustrate both precipitation reactions and reaction kinetics. Three solutions are prepared. *Solution A:* 15 g of $KIO_3$ is dissolved in 1.0 L of water. *Solution B:* Dissolve 4 g of soluble starch in 500 mL of boiling water then add 15 g of $NaHSO_3$ dissolved in 500 mL of water. *Solution C:* 3 g of $HgCl_2$ dissolved in 1.0 L of water. To perform the demonstration, mix equal volumes of B + C + A in that order. The solution turns bright orange (from the precipitation of $HgI_2$) and then blue-black. The colors will appear sooner if less of solution C is used. (See the list below for this reaction.)

The following demonstrations on solubility equilibria were obtained from the Project SERAPHIM database (see the introduction to this *Manual*):

- Alyea, H N. "The Old Nassau Reaction" *J. Chem. Educ.* **1977**, *54*, 167.
- Moss, A. "The Landolt, 'Old Nassau,' and Variant Reactions" *ibid.* **1978**, *55*, 244.
- Bauman, J.B. "Simultaneous Equilibria involving Insoluble Salts" *ibid.* **1977**, *54*, 618.

# SOLUTIONS TO SELECTED STUDY QUESTIONS

## SOLUBILITY GUIDELINES

19-10. (a) $CaSO_4$ and $BaSO_4$

(b) AgBr and $PbBr_2$

(c) NiS and $NiCO_3$

(d) $Al(OH)_3$ and $Al_2(CO_3)_3$

19-12.

| | | | | | |
|---|---|---|---|---|---|
| (a) | $Zn(NO_3)_2$ | soluble | (d) | $PbCl_2$ | insoluble |
| (b) | $Al(OH)_3$ | insoluble | (e) | CuS | insoluble |
| (c) | $Na_2CO_3$ | soluble | (f) | $Ag_2CrO_4$ | insoluble |

19-14. $Na_2S(aq) + CuSO_4(aq) \rightarrow Na_2SO_4(aq) + CuS(s)$

19-16. (a) $(NH_4)_2S + Ni(NO_3)_2 \rightarrow$ ppt. of NiS

(b) $NaOH + FeCl_3 \rightarrow$ ppt. of $Fe(OH)_3$

(c) $Na_2S + CdCl_2 \rightarrow$ ppt. of CdS

(d) $Na_2CO_3 + K_2SO_4 \rightarrow$ no ppt.

## WRITING SOLUBILITY PRODUCT CONSTANT EXPRESSIONS

19-18. (a) $Bi_2S_3(s) \rightleftarrows 2\ Bi^{3+}(aq) + 3\ S^{2-}(aq)$

$K_{sp} = 1.6 \times 10^{-72} = [Bi^{3+}]^2\,[S^{2-}]^3$

(b) $Ni(OH)_2(s) \rightleftarrows Ni^{2+}(aq) + 2\ OH^-(aq)$

$K_{sp} = 2.8 \times 10^{-16} = [Ni^{2+}]\,[OH^-]^2$

(c) $BaF_2(s) \rightleftarrows Ba^{2+}(aq) + 2\ F^-(aq)$

$K_{sp} = 1.7 \times 10^{-6} = [Ba^{2+}]\,[F^-]^2$

(d) $Al(OH)_3(s) \rightleftarrows Al^{3+}(aq) + 3\ OH^-(aq)$

$K_{sp} = 1.9 \times 10^{-33} = [Al^{3+}]\ [OH^-]^3$

19-20. $Cd_2[Fe(CN)_6]$

## DETERMINING $K_{sp}$

19-22. $K_{sp} = [Ca^{2+}]\ [F^-]^2 = (2.1 \times 10^{-4})\ (4.2 \times 10^{-4})^2 = 3.7 \times 10^{-11}$

19-24. (0.93 g / L) • (1 mol / 74.1 g) = 0.013 mol / L

$K_{sp} = [Ca^{2+}]\ [OH^-]^2 = (0.013)\ (0.026)^2 = 7.9 \times 10^{-6}$

19-26. 34.9 mg $Ag_2CO_3$ / 1 L • 1.00 g / 1000 mg • 1 mol / 275.7 g = $1.27 \times 10^{-4}$ mol /L

Therefore, $[Ag^+] = 2.54 \times 10^{-4}$ M and $[CO_3^{2-}] = 1.27 \times 10^{-4}$ M

$K_{sp} = [Ag^+]^2\ [CO_3^{2-}] = 8.14 \times 10^{-12}$

19-28. 0.0597 g $PbI_2$ / 0.100 L • 1 mol / 461.0 g = 1.30 x 10-3 mol / L

Therefore, $[Pb^{2+}] = 1.30 \times 10^{-3}$ M and $[I^-] = 2.60 \times 10^{-3}$ M

$K_{sp} = [Pb^{2+}]\ [I^-]^2 = (1.30 \times 10^{-3})\ (1.30 \times 10^{-3})2 = 8.72 \times 10^{-9}$

19-30. $[H_3O^+] = 10^{-pH} = 10^{-8.92} = 1.2 \times 10\text{-}9$ M

Therefore, $[OH^-] = 8.3 \times 10^{-6}$ M

From stoichiometry we know that $[Pb^{2+}] = \frac{1}{2}\ [OH^-] = 4.2 \times 10^{-6}$ M

$K_{sp} = [Pb^{2+}]\ [OH^-]^2 = (4.2 \times 10^{-6})\ (8.3 \times 10^{-6})2 = 2.9 \times 10^{-16}$

## ESTIMATING SALT SOLUBILITY FROM $K_{sp}$

19-32. $AuI(s) \rightleftarrows Au^+(aq) + I^-(aq)$

$K_{sp} = [Au^+]\ [I^-] = 1.6 \times 10^{-23}$, and so $[Au^+] = [I^-] = 4.0 \times 10^{-12}$ M

19-34. $PbBr_2(s) \rightleftarrows Pb^{2+}(aq) + 2\ Br^-(aq)$

$K_{sp} = [Pb^{2+}]\ [Br^-]^2 = 6.3 \times 10^{-6}$

If $S = [Pb^{2+}]$, then $K_{sp} = (S)\ (2S)^2$

$S = 0.012$ mol $PbBr_2$ / L

(0.012 mol $PbBr_2$ / L) • 367 g / mol = 4.3 g/L

19-36. $Hg(CN)_2(s) \rightleftarrows Hg^{2+}(aq) + 2\ CN^-(aq)$

$K_{sp} = [Hg^{2+}]\ [CN^-]^2 = 3.0 \times 10^{-23}$

(a) If $S = [Hg^{2+}]$ then $K_{sp} = (S)\ (2S)^2$

$S = 2.0 \times 10^{-8}$ mol $Hg(CN)_2$ / L

(b) ($2.0 \times 10^{-8}$ mol $Hg(CN)_2$ / L • (253 g / mol) = $5.0 \times 10^{-6}$ g / L (or $5.0 \times 10^{-3}$ mg)

(c) 1.0 g • 1.0 L / $5.0 \times 10^{-6}$ g • 1000 mL / 1.0 L = $2.0 \times 10^{8}$ mL

19-38. (a) $PbCl_2$ is more soluble than $PbBr_2$

(b) $BiI_3$ is more soluble than $Bi(OH)_3$

(c) FeS is more soluble than HgS

(d) PbS is more soluble than $SnS_2$

19-40.

| Compound | Solubility |
|---|---|
| AgI | $1.2 \times 10^{-8}$ |
| HgS | $5.5 \times 10^{-27}$ |
| $PbI_2$ | $1.3 \times 10^{-3}$ |
| $PbSO_4$ | $1.3 \times 10^{-4}$ |
| $NH_4NO_3$ | 118 g / 100 mL at 0 °C |

HgS < AgI < $PbSO_4$ < $PbI_2$ < $NH_4NO_3$

19-42. (a) 5.0 mg of $MgF_2$ in 125 mL of pure water

$(0.0050 \text{ g } MgF_2/0.125 \text{ L}) \cdot 1 \text{ mol}/62.3 \text{ g} = 6.4 \times 10^{-4} \text{ mol/L}$

$Q = [Mg^{2+}][F^-]^2 = (6.4 \times 10^{-4})(12.8 \times 10^{-4})^2 = 1.1 \times 10^{-9}$

$Q < K_{sp}$ $(6.4 \times 10^{-9})$. Therefore, the salt dissolves.

(b) $(0.50 \text{ g } CaF_2/0.100 \text{ L}) \cdot 1 \text{ mol}/78.1 \text{ g} = 6.4 \times 10^{-2} \text{ mol/L}$

$Q = [Ca^{2+}][F^-]^2 = (6.4 \times 10^{-2})(12.8 \times 10^{-2})^2 = 1.1 \times 10^{-3}$

$Q > K_{sp}$ $(3.9 \times 10^{-11})$. Therefore, not all salt dissolves.

## PRECIPITATIONS

19-44. $PbCl_2(s) \rightleftarrows Pb^{2+}(aq) + 2\ Cl^-(aq)$

$Q = [Pb^{2+}][Cl^-]^2 = (0.0012)(0.010)^2 = 1.2 \times 10^{-7}$

$Q < K_{sp}$ $(1.7 \times 10^{-5})$. Therefore, the salt does not precipitate.

19-46. $(1.20 \text{ g NaCl}/0.100 \text{ L}) \cdot (1 \text{ mol}/58.44 \text{ g}) = 0.205 \text{ mol/L}$

$Q = [Pb^{2+}][Cl^-]^2 = (0.0012)(0.205)^2 = 5.1 \times 10^{-5}$

$Q > K_{sp}$ $(1.7 \times 10^{-5})$. Therefore, the salt precipitates.

19-48. $K_{sp}$ for $CaF_2 = 3.9 \times 10^{-11} = [Ca^{2+}][F^-]^2$

If $[Ca^{2+}] = 2.0 \times 10^{-3}$ M, then $[F^-] = 1.4 \times 10^{-4}$ M

19-50. $K_{sp}$ for PbS $= 8.4 \times 10^{-28} = [Pb^{2+}][S^{2-}]$

When $[Pb^{2+}] = 0.0012$ M, then $[S^{2-}] = 7.0 \times 10^{-25}$ M

$7.0 \times 10^{-25} \text{ mol/L} \cdot 1.0 \text{ L} \cdot 78.0 \text{ g/mol} = 5.5 \times 10^{-23} \text{ g } Na_2S$

19-52. Total volume of solution after mixing = 100.0 mL

$[OH^-]$ after mixing $= 25.0 \text{ mL} \cdot 0.010 \text{ M}/100.0 \text{ mL} = 0.00250$ M

$[Mg^{2+}]$ after mixing $= 75.0 \text{ mL} \cdot 0.10 \text{ M}/100.0 \text{ mL} = 0.0750$ M

$Q = [Mg^{2+}][OH^-]^2 = 4.7 \times 10^{-7}$

$Q > K_{sp}$ ($1.5 \times 10^{-11}$) so precipitation of $Mg(OH)_2$ is observed.

19-54. Total volume of solution after mixing = 150. mL

$[SO_4^{2-}]$ after mixing = 100. mL • 0.10 M / 150. mL = 0.067 M

$[Ba^{2+}]$ after mixing = 50.0 mL • 0.00013 M / 150. mL = $4.3 \times 10^{-5}$ M

$Q = [Ba^{2+}][SO_4^{2-}] = 2.9 \times 10^{-6}$

$Q > K_{sp}$ ($1.1 \times 10^{-10}$) so precipitation of $BaSO_4$ is observed.

**ORDER OF PRECIPITATION**

19-56.

| **Ion** | **$[I^-]$ Required for Pptn.** |
|---|---|
| $Pb^{2+}$ | $2.9 \times 10^{-4}$ M |
| $Ag^+$ | $1.5 \times 10^{-15}$ M |
| $Hg_2^{2+}$ | $2.1 \times 10^{-14}$ M |

Order of pptn.: $Ag^+$ first, $Hg_2^{2+}$ second, and then $Pb^{2+}$

19-58.

| **Ion (assume 0.10 M)** | **$[CO_3^{2-}]$ Required for Pptn. as $MCO_3$** |
|---|---|
| $Mg^{2+}$ | $4.0 \times 10^{-4}$ M |
| $Ca^{2+}$ | $3.8 \times 10^{-8}$ M |
| $Sr^{2+}$ | $9.4 \times 10^{-9}$ M |
| $Ba^{2+}$ | $8.1 \times 10^{-8}$ M |

Order of pptn.: $Sr^{2+}$ first, $Ca^{2+}$ second, $Ba^{2+}$ third, and finally $Mg^{2+}$.

19-60.

| **Ion (assume 0.001 M)** | **$[SO_4^{2-}]$ Required for Pptn. as $MSO_4$** |
|---|---|
| $Mg^{2+}$ | $MgSO_4$ is a soluble salt (Epsom salt) |
| $Ca^{2+}$ | 0.0024 M |
| $Ba^{2+}$ | $1.1 \times 10^{-7}$ M |

Order of pptn.: $Ba^{2+}$ first and $Ca^{2+}$ second. $MgSO_4$ is a soluble salt.

19-62. $K_{sp} = [Ag^+]^2 [CO_3^{2-}] = 8.1 \times 10^{-12}$

(a) Solubility of $Ag_2CO_3$ in pure water

If $S = [CO_3^{2-}]$, then $K_{sp} = (2S)^2 (S)$

$S = 1.3 \times 10^{-4}$ mol $Ag_2CO_3$/L

(b) Solubility when 0.15 g $Na_2CO_3$ is added to 225 mL.

(0.15 g $Na_2CO_3$/0.225 L) • (1 mol/106 g) = 0.0063 mol/L

Initial $[CO_3^{2-}] = 0.0063$ M

Equilibrium $[CO_3^{2-}] = 0.0063 + S'$

Equilibrium $[Ag^+] = 2S'$

$K_{sp} = (2S')^2 (0.0063 + S')$

Assuming that S′ is much smaller than 0.0063, we have $S' = 1.8 \times 10^{-5}$ M

$Ag_2CO_3$ dissolves to the extent of only $1.8 \times 10^{-5}$ mol/L when $Na_2CO_3$ has been added.

19-64. $K_{sp} = [Ba^{2+}] [F^-]^2 = 1.7 \times 10^{-6}$

(a) Solubility of $BaF_2$ in pure water

If $S = [Ba^{2+}]$, then $K_{sp} = (S)(2S)^2$

$S = 7.5 \times 10^{-3}$ M

$7.5 \times 10^{-3}$ mol $BaF_2$/L • 1 L/1000 mL • 173 g/mol = 0.0013 g $BaF_2$/mL

(b) Solubility when 5.0 mg KF is added per mL.

(0.0050 g KF/0.0010 L) • (1 mol/58.1 g) = 0.086 mol/L

Initial $[F^-] = 0.086$ M

If $[Ba^{2+}] = S'$ then equilibrium $[F^-] = 0.086 + 2S'$

$K_{sp} = (S')(0.086 + 2S')^2$

$S' = 2.3 \times 10^{-4}$ M

$2.3 \times 10^{-4}$ mol $BaF_2$/L • 1 L/1000 mL • 173 g/mol = $4.0 \times 10^{-5}$ g $BaF_2$/mL

19-66. 2.8 mg $Na_2SO_4$ / 1.0 L • 1 g / 1000 mg • 1 mol / 142 g = $2.0 \times 10^{-5}$ mol / L

$K_{sp} = [Ba^{2+}][SO_4^{2-}] = 1.1 \times 10^{-10}$, so the solubility of $BaSO_4$ before adding $Na_2SO_4$ is $1.1 \times 10^{-5}$ M. After adding $Na_2SO_4$, $K_{sp} = (S')(2.0 \times 10^{-5} + S')$. This gives $S' = [Ba^{2+}] = 4.5 \times 10^{-6}$ M and $[SO_4^{2-}] = 2.5 \times 10^{-5}$ M. Note that the concentration of $Ba^{2+}$ has dropped, as expected, on adding $Na_2SO_4$.

19-68. *Step 1:* Calculate the concentrations of $Cl^-$ and $Pb^{2+}$ after mixing.

Total volume after mixing = 150.0 mL

$[Pb^{2+}]$ = (50.0 mL • 0.053 M) / 150.0 mL = 0.018 M

$[Cl^-]$ = (100.0 mL • 0.012 M) / 150.0 mL = 0.0080 M

*Step 2:* Decide if precipitation occurs.

$Q = [Pb^{2+}][Cl^-]^2 = 1.1 \times 10^{-6}$

$K_{sp}$ $(1.7 \times 10^{-5})$ > Q. Therefore, pptn. does NOT occur.

*Step 3:* Summary of concentrations

$[Pb^{2+}]$ = 0.018 M and so $[NO_3^-]$ = 0.036 M

$[Cl^-]$ = 0.0080 M = $[H_3O+]$

## SEPARATIONS

19-70. (a) Concentration of sulfide ion that will precipitate the maximum amount of $Co^{2+}$ ion without precipitating MnS.

*Step 1:* Calculate the $[S^{2-}]$ necessary to begin the pptn. of MnS

$K_{sp} = [Mn^{2+}][S^{2-}] = 5.1 \times 10^{-15}$

If $[Mn^{2+}]$ = 0.10 M, then $[S^{2-}]$ required is $5.1 \times 10^{-14}$ M.

*Step 2:* Calculate $[Co^{2+}]$ when $[S^{2-}] = 5.1 \times 10^{-13}$ M.

$[Co^{2+}] = K_{sp} / [S^{2-}] = 5.9 \times 10^{-21} / 5.1 \times 10^{-14} = 1.2 \times 10^{-7}$ M

19-72. (a) Which will precipitate first as sodium sulfate is added, $CaSO_4$ or $PbSO_4$?

$[SO_4^{2-}]$ required to ppt. $CaSO_4$ = $K_{sp} / [Ca^{2+}] = 2.4 \times 10^{-5} / 0.010 = 2.4 \times 10^{-3}$ M

$[SO_4^{2-}]$ required to ppt. $PbSO_4$ = $K_{sp} / [Pb^{2+}]$ = $1.8 \times 10^{-8} / 0.010$ = $1.8 \times 10^{-6}$ M

Therefore, $PbSO_4$ will precipitate before $CaSO_4$.

(b) Just before the more soluble salts begins to precipitate, what molarity of the first metal ion remains in solution?

$CaSO_4$ will ppt. when $[SO_4^{2-}]$ = $2.4 \times 10^{-3}$ M. This point is reached when

$[Pb^{2+}]$ = $K_{sp} / [SO_4^{2-}]$ = $1.8 \times 10^{-8} / 2.4 \times 10^{-3}$ = $7.5 \times 10^{-6}$ M

19-74. (a) $Cu^{2+}$ and $Ag^{+}$ can be separated by adding HCl to the solution. The $Cl^{-}$ ion leads to insoluble AgCl, while $CuCl_2$ is water-soluble.

(b) $Pb^{2+}$ and $Sn^{2+}$ can be separated as their hydroxides, since about $10^5$ more $OH^{-}$ is required to begin the precipitation of $Pb(OH)_2$ than to begin precipitating $Sn(OH)_2$. Alternatively, add HCl to precipitate $PbCl_2$ and leave $Sn^{2+}$ in solution.

(c) $Al^{3+}$ and $Fe^{3+}$ can be separated by precipitating $Fe_2S_3$ and leaving $Al^{3+}$ in the solution. Alternatively, add excess NaOH. Both $Al(OH)_3$ and $Fe(OH)_3$ will precipitate. However, $Al(OH)_3$ is amphoteric and so will dissolve in excess base to give $[Al(OH)_4]^{-}$.

## SIMULTANEOUS EQUILIBRIA AND COMPLEX IONS

19-76. $AgBr(s) \rightleftarrows Ag^{+}(aq) + Br^{-}(aq)$ $K_{sp} = 3.3 \times 10^{-13}$

$Ag^{+}(aq) + 2\ S_2O_3^{2-}(aq) \rightleftarrows [Ag(S_2O_3)_2]^{3-}(aq)$ $K_{form} = 2.0 \times 10^{13}$

---

$AgBr(s) + 2\ S_2O_3^{2-}(aq) \rightleftarrows [Ag(S_2O_3)_2]^{3-}(aq) + Br^{-}(aq)$ $K_{net} = 6.6$

19-78. $ZnS(s) \rightleftarrows Zn^{2+}(aq) + S^{2-}(aq)$ $K_{sp} = 1.1 \times 10^{-21}$

$Zn^{2+}(aq) + 2\ CN^{-}(aq) \rightleftarrows Zn(CN)_2(s)$ $1/K_{sp} = 1/8.0 \times 10^{-12}$

---

$ZnS(s) + 2\ CN^{-}(aq) \rightleftarrows Zn(CN)_2(s) + S^{2-}(aq)$ $K_{net} = 1.4 \times 10^{-10}$

The very small equilibrium constant for this reaction suggests that it would be difficult to transform ZnS into $Zn(CN)_2$.

19-80. $Au^+(aq) + 2\ CN^-(aq) \rightleftarrows [Au(CN)_2]^-(aq)$ $K_{form} = 2.0 \times 10^{38}$

$AuCl(s) \rightleftarrows Au^+(aq) + Cl^-(aq)$ $K_{sp} = 2.0 \times 10^{-13}$

--------------------------------------------------- -------------------------

$AuCl(s) + 2\ CN^-(aq) \rightleftarrows [Au(CN)_2]^-(aq) + Cl^-(aq)$ $K_{net} = 4.0 \times 10^{25}$

The equilibrium constant for the dissolving of AuCl is so large that it can be predicted that the solid will dissolve readily. To prove this, we shall calculate the moles of $CN^-$ required to dissolve the AuCl.

$0.015\ g\ AuCl \cdot 1\ mol / 232\ g = 6.5 \times 10^{-5}\ mol$

(There is more than enough NaCN present (0.090 mol) to provide the $CN^-$ to completely complex the Au+ ion.) If all AuCl were dissolved, the concentration of $[Au(CN)_2]^-$ in the solution would be $6.5 \times 10^{-5}$ mol/0.115 L = $5.7 \times 10^{-4}$ M. In order to achieve this concentration of $[Au(CN)_2]^-$, the concentration of CN- required by the equilibrium is

$$K_{net} = 4.0 \times 10^{25} = \frac{\{[Au(CN)_2]^-\}[Cl^-]}{[CN^-]^2} = \frac{(5.7 \times 10^{-4})^2}{[CN^-]^2}$$

$[CN^-] = 8.9 \times 10^{-17}$ M

This concentration is easily achieved given the fact that 0.090 mol of NaCN is dissolved in 115 mL.

## ACID-BASE REACTIONS FOR DISSOLVING SOLIDS AND PREPARING SALTS

19-82. $Ba(OH)_2$ and $BaCO_3$.

19-84. Treat $NiCO_3$ with $H_2SO_4(aq)$

$$NiCO_3(s) + H_2SO_4(aq) \rightarrow NiSO_4(aq) + CO_2(g) + H_2O(\ell)$$

and then heat the resulting solution to dryness

19-86. $FeCl_3(aq) + 3\ NaOH(aq) \rightarrow Fe(OH)_3(s) + 3\ NaOH(aq)$

$Fe(OH)_3(s) + 3\ HClO_4(aq) \rightarrow Fe(ClO_4)_3(aq) + 3\ H_2O(\ell)$

19-88. Adding water-soluble $BaCl_2$ to a solution that may contain $SO_4^{2-}$ or $CO_3^{2-}$ will precipitate $BaSO_4$ or $BaCO_3$. However, only $BaCO_3$ will dissolve in aqueous HCl.

19-90. $$\frac{K_{dissolve} \text{ for CuS}}{K_{dissolve} \text{ for ZnS}} = \frac{K_{sp} \text{ for CuS}}{K_{sp} \text{ for ZnS}} = \frac{[Cu^{2+}]}{[Zn^{2+}]}$$

$[Cu^{2+}]/[Zn^{2+}] = 7.9 \times 10^{-15}$

This means ZnS(s) is $1.3 \times 10^{14}$ times as soluble in HCl as CuS(s).

## GENERAL QUESTIONS

19-92. $K_{sp} = [Ba^{2+}][SO_4^{2-}] = 1.1 \times 10^{-10}$

Solubility of $BaSO_4 = [Ba^{2+}] = [SO_4^{2-}] = 1.0 \times 10^{-5}$ M

$1.0 \times 10^{-5}$ mol/L • 1.0 L • 233 g/mol • 1000 mg/g = 2.4 mg $BaSO_4$

19-94. (a) If you have a 0.015 M solution of $Ba^{2+}$, what is the minimum concentration of $F^-$ ion necessary to begin precipitation of $BaF_2$?

$K_{sp}$ for $BaF_2 = 1.7 \times 10^{-6} = [Ba^{2+}][F^-]^2$

$[F^-] = \{K_{sp}/[Ba^{2+}]\}^{1/2} = 0.011$ M

(b) If you started with 100.0 mL of the 0.015 M $Ba^{2+}$-containing solution, how many milligrams of NaF would be required to just begin precipitation?

0.011 mol $F^-$/L • 0.1000 L = 0.0011 mol $F^-$ req'd

0.0011 mol $F^-$ req'd • 42.0 g/mol • 1000 mg/g = 45 mg NaF

(c) 0.50 g NaF • 1 mol/42.0 g NaF = 0.012 mol

0.012 mol/0.1000 L = 0.12 M $F^-$

$[Ba^{2+}] = K_{sp}/[F^-]^2 = 1.7 \times 10^{-6}/(0.12)^2 = 1.2 \times 10^{-4}$ M

19-96. $AgBr(s) + 2\,NH_3(aq) \rightleftarrows [Ag(NH_3)_2]^+(aq) + Br^-(aq)$

Based on Example 19.11, we know the equilibrium constant for the reaction is $5.3 \times 10^{-6}$. If we dissolve $1.00 \times 10^{-4}$ moles of AgBr, it will furnish $1.00 \times 10^{-4}$ mol/L of $Br^-$ and $1.00 \times 10^{-4}$ mol/L of $[Ag(NH_3)_2]^+$. We shall use these concentrations to find the concentration of $NH_3$ that must exist at equilibrium.

$$K = 5.3 \times 10^{-6} = \frac{\{[Ag(NH_3)_2]^+\}\,[Br^-]}{[NH_3]^2} = \frac{(1.00 \times 10^{-4})^2}{[NH_3]^2}$$

$[NH_3] = 0.043$ M

Taking 5.0 mL of 2.5 M $NH_3$, we have 0.0125 mol of $NH_3$ available. After using $1.00 \times 10^{-4}$ mol of this to dissolve the AgBr, we have 0.0124 mol of the $NH_3$ left. This is dissolved in 1005 mL of solution (or 1.0 L to two significant figures). Therefore, the concentration of $NH_3$ remaining is 0.012 M. This is not sufficient to satisfy the equilibrium condition, so the *AgBr does not completely dissolve.*

19-98. Metal sulfides with moderately small $K_{sp}$ values can undergo the reaction

$$MS(s) + 2\,H_3O^+(aq) \rightarrow M^{2+}(aq) + H_2S(aq) + 2\,H_2O(\ell)$$

19-100. (a) Add 25 mg of $Zn(CN)_2$ to 1.0 L of pure water.

$K_{sp} = [Zn^{2+}]\,[CN^-]^2 = 8.0 \times 10^{-12}$

Solubility of $Zn(CN)_2$ = S = $[Zn^{2+}]$, so $K_{sp} = (S)\,(2S)^2$

$S = 1.3 \times 10^{-4}$ mol/L

$1.3 \times 10^{-4}$ mol/L $\cdot$ 117 g/mol = 0.015 g $Zn(CN)_2$/L

According to our calculation, $Zn(CN)_2$ should dissolve only to the extent of 15 mg/L Therefore, 10 mg of $Zn(CN)_2$ is left undissolved. If additional $Zn(CN)_2$ is added, none of it will dissolve. The concentrations of $Zn^{2+}$ and $CN^-$ in solution will not change.

(b) Add another 1.0 L of water.

$0.015 \text{ g } Zn(CN)_2/\text{L} \cdot 2.0 \text{ L} = 0.030 \text{ g } Zn(CN)_2$

When the total volume of solution is 2.0 L, 30 mg of $Zn(CN)_2$ should dissolve. Therefore, all of the original 25 mg of $Zn(CN)_2$ will dissolve.

(c) Add 100. mg KCN to 1.0 L of water containing 25 mg of $Zn(CN)_2$

$100. \text{ mg KCN} \cdot 1 \text{ g}/1000 \text{ mg} \cdot 1 \text{ mol}/65.12 \text{ g} = 1.5 \times 10^{-3} \text{ mol KCN}$

Solubility of $Zn(CN)_2 = S = [Zn^{2+}]$, so $K_{sp} = (S)(0.0015 + 2S)^2$

$S = 3.4 \times 10^{-6}$ M (neglecting 2S in $(0.0015 + 2S)$)

$3.4 \times 10^{-6} \text{ mol/L} \cdot 117 \text{ g/mol} = 4.0 \times 10^{-4}$ g (or 0.4 mg) of $Zn(CN)_2$ can dissolve. Of the original 25 mg of $Zn(CN)_2$ in the beaker, essentially all remains undissolved.

19-102. $[H_3O^+] = 10^{-pH} = 10^{-9.00} = 1.0 \times 10^{-9}$

$[OH^-] = 1.0 \times 10^{-5}$ M, and so $[Ce^{3+}] = 1/3\,(1.0 \times 10^{-5} \text{ M}) = 3.3 \times 10^{-6}$ M

$K_{sp} = [Ce^{3+}][OH^-]^3 = 3.3 \times 10^{-21}$

19-104. Calcite: $CaCO_3(s) \rightleftarrows Ca^{2+}(aq) + CO_3^{2-}(aq)$ $\quad K = K_{sp} = 3.8 \times 10^{-9}$

Aragonite: $Ca^{2+}(aq) + CO_3^{2-}(aq) \rightleftarrows CaCO_3(s)$ $\quad K = 1/K_{sp} = 1/6.0 \times 10^{-9}$

Calcite $\rightleftarrows$ Aragonite $\quad K = 0.63$

The equilibrium constant for the transformation of calcite to aragonite is less than one, so the conversion is not spontaneous.

19-106. $K_{sp} = [Ca^{2+}][CO_3^{2-}] = 3.8 \times 10^{-9}$

Calculate the $[Ca^{2+}]$ that can exist in solution when $[CO_3^{2-}] = 0.050$ M

$[Ca^{2+}] = K_{sp}/[CO_3^{2-}] = 7.6 \times 10^{-8}$ M

$$\frac{\text{Original } [Ca^{2+}] - \text{Equilibrium } [Ca^{2+}]}{\text{Original } [Ca^{2+}]} \cdot 100 \approx 100\%$$

19-108. (a) $[HS^-] = [OH^-] = 4.1 \times 10^{-10}$ M

(b) $K_{total}$ for CdS = ($K_{sp}$ for CdS) • ($K_h$ for $S^{2-}$) = $(3.6 \times 10^{-29}) \cdot (0.077)$

$= 2.8 \times 10^{-30}$

$K_{total} = [Cd^{2+}]\,[HS^-]\,[OH^-]$

$2.8 \times 10^{-30} = [Cd^{2+}]\,(4.1 \times 10^{-10})^2$

$[Cd^{2+}] = 1.6 \times 10^{-11}$

(c) $K_{sp}$ for CdS = $[Cd^{2+}]\,[S^{2-}]$

$[Cd^{2+}] = [S^{2-}] = K_{sp}^{1/2} = 6.0 \times 10^{-15}$

Therefore, hydrolysis of the sulfide ion raises the solubility of CdS from $6.0 \times 10^{-15}$ mol/L to $1.6 \times 10^{-11}$ mol/L, a factor of about 3000.

# CHAPTER 20

# SPONTANEITY OF REACTIONS: ENTROPY AND FREE ENERGY

## INSTRUCTOR'S NOTES

This subject is treated, in an admittedly superficial way, in 2-3 lectures. However, at the introductory level our goal is relatively limited: to convey the idea of reaction spontaneity and to differentiate this from reaction speed. Although we usually cover thermodynamics following our discussion of equilibria, we have sometimes reversed these topics and find no particular problems in doing so. When thermodynamics is discussed before equilibrium we make the connection between ΔG and K very early in the lectures on general equilibria

## LIST OF TRANSPARENCIES FOR THIS CHAPTER

| No. | Figure | Page | Subject |
|---|---|---|---|
| 93 | 20.6 | 828 | Entropy values and the periodic table |
| 94 | 20.7 | 833 | ΔG as a functions of the sign and magnitude of ΔS |

## SUGGESTED DEMONSTRATIONS

We usually begin these lectures with a series of reactions designed to show that some spontaneous reactions (for example, the combustion of $H_2$) proceed with the evolution of heat, while others proceed spontaneously while requiring heat (the melting of ice, the evaporation of liquid nitrogen, or the dissolution of some solids). There are many suitable demonstrations in Volume 1 of Shakhashiri's "Chemical Demonstrations." For example, the thermite reaction is outlined on page 85 (see textbook page 837), and the reaction of Mg with $CO_2$ is given on page 90 (see textbook page 102).

The following are additional demonstrations on thermodynamics that were obtained from the Project SERAPHIM database (see the introduction to this *Manual*):

- Bozzelli, J.W.; Barat, R.B. "The Thermite Lecture Demonstration" *J. Chem. Educ.* **1979**, *56*, 675.

- Dole, M. "Lecture Table Demonstration on Entropy" *ibid.* **1977**, *56*, 754.
- McNaught, I.J. "Thermodynamic versus Kinetic Control: A Lecture Demonstration" *ibid.* **1978**, *55*, 722.
- Miller, F.L; Fog, H.M. "Reversible Formation of Aluminum Xylenol Orange by Temperature Variation: An Experiment Demonstrating the Entropy Effect" *ibid.* **1973**, *50*, 147.

# SOLUTIONS TO SELECTED STUDY QUESTIONS

## ENTROPY

20-7. (a) Silicon with a trace of impurities.

(b) Liquid water

(c) Iodine vapor

20-9. (a) $MgCl_2(s)$

(b) $P_4(g)$

(c) $(CH_3)_2NH$

(d) $Hg(\ell)$

20-11. (a) $NH_4Cl(s) \rightarrow NH_4Cl(aq)$

$\Delta S^\circ$ = 1 mol (169.9 J/K • mol) - 1 mol (94.6 J/K • mol) = 75.3 J/K

(b) $S(g) \rightarrow S(s)$

$\Delta S^\circ$ = 1 mol (31.80 J/K • mol) - 1 mol (167.821 J/K • mol) = -136.0 J/K

(c) $CCl_4(g) \rightarrow CCl_4(\ell)$

$\Delta S^\circ$ = 1 mol (216.40 J/K • mol) - 1 mol (309.85 J/K • mol) = -93.45 J/K

20-13. $\Delta S = q/T$ = ($39.3 \times 10^3$ J/mol) / 351.2 K = 112 J/K • mol

## ENTROPY OF REACTIONS

20-15. $2\,Al(s) + 3\,Cl_2(g) \rightarrow 2\,AlCl_3(s)$

$\Delta S^\circ$ = 2 mol (110.67 J/K • mol) - [2 mol (28.3 J/K • mol) + 3 mol (223.066 J/K • mol)]

= -504.5 J/K

20-17. (a) $\frac{1}{2}\,I_2(g) + \frac{1}{2}\,Cl_2(g) \rightarrow ICl(g)$

$\Delta S^\circ$ = 1 mol (247.551 J/K • mol)

- [$\frac{1}{2}$ mol (116.135 J/K • mol) + $\frac{1}{2}$ mol (223.066 J/K • mol)]

= 77.951 J/K

(b) $C(s) + \frac{1}{2} O_2(g) + Cl_2(g) \rightarrow COCl_2(g)$

$\Delta S^\circ$ = 1 mol (283.53 J/K • mol) - [1 mol (5.740 J/K • mol)

+ ½ mol (205.138 J/K • mol) + 1 mol (223.066 J/K • mol)]

= -47.85 J/K

(c) $Ca(s) + C(s) + 3/2\ O_2(g) \rightarrow CaCO_3(s)$

$\Delta S^\circ$ = 1 mol (92.9 J/K • mol) - [1 mol (41.42 J/K • mol)

+ 1 mol (5.740 J/K • mol) + 3/2 mol (205.138 J/K • mol)]

= -262.0 J/K

(d) $C(s) + \frac{1}{2} H_2(g) + 3/2\ Cl_2(g) \rightarrow CHCl_3(g)$

$\Delta S^\circ$ = 1 mol (295.71J/K • mol) - [1 mol (5.740 J/K • mol)

+ ½ mol (130.684 J/K • mol) + 1.5 mol (223.066 J/K • mol)]

= -109.97 J/K

20-19. (a) $2\ Fe(s) + 3\ O_2(g) \rightarrow 2\ Fe_2O_3(s)$

$\Delta S^\circ$ = 2 mol (87.40 J/K • mol)

- [2 mol (27.78 J/K • mol) + 3 mol (205.138 J/K • mol)]

= -496.17 J/K

(b) $Ca(s) + 2\ H_2O(\ell) \rightarrow Ca(OH)_2(aq) + H_2(g)$

$\Delta S^\circ$ = 1 mol(-74.5 J/K • mol) + 1 mol (130.684 J/K • mol)

- [1 mol (41.42 J/K • mol) + 2 mol (69.91 J/K • mol)]

= -125.1 J/K

(c)* $Na_2CO_3(s) + 2\ HCl(aq) \rightarrow 2\ NaCl(aq) + H_2O(\ell) + CO_2(g)$

$\Delta S^\circ$ = 2 mol (115.5 J/K • mol) + 1 mol (69.91 J/K • mol)

+ 1 mol (213.74 J/K • mol)

- [1 mol (134.98 J/K • mol) + 2 mol (56.5 J/K • mol)]

= 266.7 J/K

* **NOTE:** In the first printing of the book, sodium carbonate in SQ 20-19 was specified as $Na_2CO_3(aq)$. It should be $Na_2CO_3(s)$, and is so indicated in the second and subsequent printings of the book.

## FREE ENERGY

20-21. (a) $Br_2(\ell) + 3\ F_2(g) \rightarrow 2\ BrF_3(g)$

$\Delta H^\circ = 2\ \Delta H^\circ_f\ [BrF_3(g)] = 2\ \text{mol}\ (-255.60\ \text{kJ/mol}) = -511.20\ \text{kJ}$

$\Delta S^\circ = 2\ \text{mol}\ (292.53\ \text{J/K} \cdot \text{mol})$

$- [1\ \text{mol}\ (152.2\ \text{J/K} \cdot \text{mol}) + 3\ \text{mol}\ (202.78\ \text{J/K} \cdot \text{mol})]$

$= -175.5\ \text{J/K}$

$\Delta G^\circ = -511.20\ \text{kJ} - (298\ \text{K})\ (-0.1755\ \text{kJ/K}) = -458.91\ \text{kJ}$

The value of $\Delta G^\circ$ calculated for this reaction is twice the value of $\Delta G^\circ_f$ for $BrF_3(g)$ (value from Appendix J = -229.43 kJ/mol).

(b) $Ca(s) + 2\ H_2O(\ell) \rightarrow Ca(OH)_2(aq) + 2\ H_2(g)$

$\Delta H^\circ = 1\ \text{mol}\ (-1002.82\ \text{kJ/mol}) + 2\ \text{mol}\ (0)$

$- [1\ \text{mol}\ (0) + 2\ \text{mol}\ (-285.83\ \text{kJ/mol})]$

$= -431.16\ \text{kJ}$

$\Delta S^\circ = -125.1\ \text{J/K}$ (from SQ 20-19)

$\Delta G^\circ = -431.16\ \text{kJ} - (298\ \text{K})\ (-0.1251\ \text{kJ/K}) = -393.89\ \text{kJ}$

The value of $\Delta G^\circ$ calculated here is NOT a molar free energy of formation.

(c) $6\ C(\text{graphite}) + 3\ H_2(g) \rightarrow C_6H_6(\ell)$

$\Delta H^\circ = 1\ \text{mol}\ (49.03\ \text{kJ/mol}) - [6\ \text{mol}\ (0) + 3\ \text{mol}\ (0)] = \Delta H^\circ_f[C_6H_6(\ell)]$

$\Delta S^\circ = 1\ \text{mol}\ (172.8\ \text{J/K} \cdot \text{mol})$

$- [6\ \text{mol}\ (5.740\ \text{J/K} \cdot \text{mol}) + 3\ \text{mol}\ (130.684\ \text{J/K} \cdot \text{mol})]$

$= -253.7\ \text{J/K}$

$\Delta G^\circ = 49.03\ \text{kJ} - (298)\ (-0.2537\ \text{kJ/K}) = 124.63\ \text{kJ}$

The reaction corresponds to the equation that defines the molar free energy of formation of benzene. Appendix J gives a value of 124.5 kJ/mol for $\Delta G^\circ_f$, very close to the value we have found.

Reactions (a) and (b) are predicted to be spontaneous.

20-23. (a) $Mg(s) + O_2(g) + H_2(g) \rightarrow Mg(OH)_2(s)$

$\Delta S^\circ = 1\ \text{mol}\ (63.18\ \text{J/K} \cdot \text{mol}) - [1\ \text{mol}\ (32.68\ \text{J/K} \cdot \text{mol})$

$+ 1\ \text{mol}\ (205.138\ \text{J/K} \cdot \text{mol}) + 1\ \text{mol}\ (130.684\ \text{J/K} \cdot \text{mol})]$

$= -305.32$ J/K

$\Delta G^\circ_f = -924.54\text{ kJ} - (298\text{ K})(-0.30532\text{ kJ/K}) = -833.55\text{ kJ}$

The value of $\Delta G^\circ_f$ for $Mg(OH)_2(s)$ in Appendix J is -833.51 kJ/mol.

(b) $\frac{1}{2}\, N_2(g) + \frac{1}{2}\, O_2(g) + \frac{1}{2}\, Cl_2(g) \rightarrow NOCl(g)$

$\Delta S^\circ = 1\text{ mol}\,(261.69\text{ J/K}\cdot\text{mol}) - [\frac{1}{2}\text{ mol}\,(191.61\text{ J/K}\cdot\text{mol})$

$+ \frac{1}{2}\text{ mol}\,(205.138\text{ J/K}\cdot\text{mol}) + \frac{1}{2}\text{ mol}\,(223.066\text{ J/K}\cdot\text{mol})]$

$= -48.22$ J/K

$\Delta G^\circ_f = 51.71\text{ kJ} - (298\text{ K})(-0.04822\text{ kJ/K}) = 66.08\text{ kJ}$

The value of $\Delta G^\circ_f$ for NOCl(g) in Appendix J is 66.08 kJ/mol.

(c) $2\,Na(s) + C(s) + 3/2\, O_2(g) \rightarrow Na_2CO_3(s)$

$\Delta S^\circ = 1\text{ mol}\,(134.98\text{ J/K}\cdot\text{mol}) - [2\text{ mol}\,(51.21\text{ J/K}\cdot\text{mol})$

$+ 1\text{ mol}\,(5.740\text{ J/K}\cdot\text{mol}) + 3/2\text{ mol}\,(205.138\text{ J/K}\cdot\text{mol})]$

$= -280.89$ J/K

$\Delta G^\circ_f = -1130.68\text{ kJ} - (298\text{ K})(-0.28089\text{ kJ/K}) = -1046.98\text{ kJ}$

The value of $\Delta G^\circ_f$ for $Na_2CO_3(s)$ in Appendix J is -1044.44 kJ/mol.

The formation of $Mg(OH)_2$ and $Na_2CO_3$ is predicted to be spontaneous.

20-25. $N_2H_4(\ell) + O_2(g) \rightarrow 2\,H_2O(\ell) + N_2(g)$

$\Delta G^\circ = 2\text{ mol}\,(-237.129\text{ kJ/mol}) - 1\text{ mol}\,(149.34\text{ kJ/mol}) = -623.60\text{ kJ}$

$-623.60\text{ kJ/mol} \cdot 1\text{ mol}/32.0453\text{ g} \cdot 454\text{ g/lb} = -8834.8\text{ kJ}$

20-27. (a) $HgS(s) + O_2(g) \rightarrow Hg(\ell) + SO_2(g)$

$\Delta G^\circ = 1\text{ mol}\,(-300.194\text{ kJ/mol}) - 1\text{ mol}\,(-50.6\text{ kJ}) = -249.6\text{ kJ}$

Reaction predicted to be spontaneous.

(b) $2\,H_2S(g) + 3\,O_2(g) \rightarrow 2\,H_2O(g) + 2\,SO_2(g)$

$\Delta G^\circ = 2\text{ mol}\,(-228.572\text{ kJ/mol}) + 2\text{ mol}\,(-300.194\text{ kJ/mol})$

$- 2\text{ mol}\,(-33.56\text{ kJ/mol}) = -990.41\text{ kJ}$

Reaction is predicted to be spontaneous.

(c) $SiCl_4(g) + 2\,Mg(s) \rightarrow 2\,MgCl_2(s) + Si(s)$

$\Delta G^\circ = 2\text{ mol}(-591.79\text{ kJ/mol}) - 1\text{ mol}(-616.98\text{ kJ/mol}) = -566.6\text{ kJ}$

Reaction is predicted to be spontaneous.

20-29. $H_2O(g) \rightarrow H_2(g) + \frac{1}{2}O_2(g)$

$\Delta H^\circ_{rxn} = -\Delta H^\circ_f = -(-241.818\text{ kJ/mol}) = +241.818\text{ kJ/mol}$

$\Delta S^\circ_{rxn} = 1\text{ mol}(130.684\text{ J/K}\cdot\text{mol}) + \frac{1}{2}\text{ mol}(205.138\text{ J/K}\cdot\text{mol})$

$- 1\text{ mol}(188.825\text{ J/K}\cdot\text{mol}) = 44.428\text{ J/K}$

$\Delta G^\circ_f = +241.818\text{ kJ} - (298\text{ K})(0.044428\text{ J/K}) = 228.578\text{ kJ}$

20-31. (a) $\Delta G^\circ_{rxn} = -272.8\text{ kJ for } TiCl_2(s) + Cl_2(g) \rightarrow TiCl_4(\ell)$

$\Delta G^\circ_{rxn} = -272.8\text{ kJ} = 1\text{ mol}(-737.2\text{ kJ/mol}) - \Delta G^\circ_f[TiCl_2(s)]$

$\Delta G^\circ_f[TiCl_2(s)] = -464.4\text{ kJ}$

(b) $\Delta G^\circ_{rxn} = -1830\text{ kJ for } 2\,CH_2CO(g) + 4\,O_2(g) \rightarrow 4\,CO_2(g) + 2\,H_2O(g)$

$\Delta G^\circ_{rxn} = -1830\text{ kJ}$

$= 4\text{ mol}(-394.359\text{ kJ/mol}) + 2(-228.572\text{ kJ/mol}) - 2\,\Delta G^\circ_f[CH_2CO(g)]$

$\Delta G^\circ_f[CH_2CO(g)] = -102.29\text{ kJ}$

20-33. $3\,H_2(g) + CO(g) \rightarrow CH_4(g) + H_2O(g)$

$\Delta H^\circ = 1\text{ mol}(-74.81\text{ kJ/mol}) + 1\text{ mol}(-241.818\text{ kJ/mol})$

$- [3\text{ mol}(0\text{ kJ/mol}) + 1\text{ mol}(-110.525\text{ kJ/mol})]$

$= -206.10\text{ kJ}$

$\Delta S^\circ = 1\text{ mol}(186.264\text{ J/K}\cdot\text{mol}) + 1\text{ mol}(188.825\text{ J/K}\cdot\text{mol}))$

$- [3\text{ mol}(130.684\text{ J/K}\cdot\text{mol})) + 1\text{ mol}(197.674\text{ J/K}\cdot\text{mol}))]$

$= -214.637\text{ J/K}$

$\Delta G^\circ = -206.10\text{ kJ} - (298\text{ K})(-0.214637\text{ J/K}) = -142.14\text{ kJ}$

The reaction is predicted to be spontaneous.

20-35. $CO(g) + 2\,H_2(g) \rightarrow CH_3OH(\ell)$

$$\Delta G^\circ_{rxn} = \Delta G^\circ_f[CH_3OH(\ell)] - \{\Delta G^\circ_f[CO(g)] + 2\,\Delta G^\circ_f[H_2(g)]\}$$

$$= 1\text{ mol}(-166.3\text{ kJ/mol}) - [1\text{ mol}(-137.168\text{ kJ/mol}) + 2\text{ mol}(0)]$$

$$= -29.1\text{ kJ}$$

The reaction is predicted to be spontaneous.

## THERMODYNAMICS AND EQUILIBRIUM CONSTANTS

20-37. $\Delta G^\circ = -RT \ln K_p$

$$-166.27\text{ kJ/mol} = -(8.314510 \times 10^{-3}\text{ kJ/K}\cdot\text{mol})(298\text{ K}) \ln K_p$$

$$\ln K_p = 67.106$$

$$K_p = 1.392 \times 10^{29}$$

The large, negative value of $\Delta G^\circ_{rxn}$ leads to a large value for $K_p$.

20-39. $K_p = 0.106$

$$\Delta G^\circ = -RT \ln K_p = -(8.314510 \times 10^{-3}\text{ kJ/K}\cdot\text{mol})(1800\text{ K}) \ln(0.106) = 33.59\text{ kJ/mol}$$

20-41. $H_2C{=}CH_2(g) + H_2(g) \rightarrow H_3C{-}CH_3(g)$

$$\Delta G^\circ_{rxn} = \Delta G^\circ_f[H_3C{-}CH_3(g)] - \{\Delta G^\circ_f[H_2C{=}CH_2(g)] + \Delta G^\circ_f[H_2(g)]\}$$

$$= 1\text{ mol}(-32.82\text{ kJ/mol}) - [1\text{ mol}(68.15\text{ kJ/mol}) + 1\text{ mol}(0)] = -100.97\text{ kJ}$$

$$\Delta G^\circ = -RT \ln K_p$$

$$-100.97\text{ kJ} = -(8.314510 \times 10^{-3}\text{ kJ/K}\cdot\text{mol})(298\text{ K}) \ln K_p$$

$$K_p = 4.99 \times 10^{17}$$

## GENERAL

20-43. What trend is observed in the entropies of the rare gases in Figure 20.6? Is this trend observed elsewhere in the periodic table?

Entropy generally increases with atomic mass. Another example of this is found in Group 4A (C, Si, Ge, Sn, and Pb).

20-45. (a) ethyne or acetylene $2\ C(s) + H_2(g) \rightarrow$ H–C≡C–H

$\Delta S^\circ = S^\circ(\text{ethyne}) - \{2\ S^\circ[C(s)] + S^\circ[H_2(g)]\}$

$= 1\ \text{mol}\ (200.94\ \text{J/K} \cdot \text{mol})$

$- [2\ \text{mol}\ (5.740\ \text{J/K} \cdot \text{mol}) + 1\ \text{mol}\ (130.684\ \text{J/K} \cdot \text{mol}]$

$= 58.78\ \text{J/K}$

(b) ethene or ethylene $2\ C(s) + 2\ H_2(g) \rightarrow$ 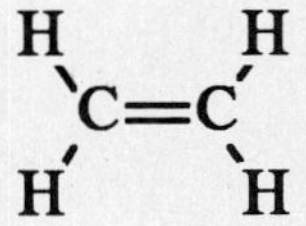

$\Delta S^\circ = S^\circ(\text{ethene}) - \{2\ S^\circ[C(s)] + 2\ S^\circ[H_2(g)]\}$

$= 1\ \text{mol}\ (219.56\ \text{J/K} \cdot \text{mol})$

$- [2\ \text{mol}\ (5.740\ \text{J/K} \cdot \text{mol}) + 2\ \text{mol}\ (130.684\ \text{J/K} \cdot \text{mol}]$

$= -53.29\ \text{J/K}$

(c) ethane $2\ C(s) + 3\ H_2(g) \rightarrow$ 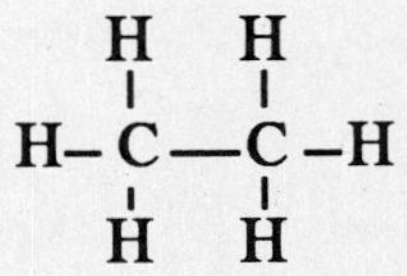

$\Delta S^\circ = S^\circ(\text{ethene}) - \{2\ S^\circ[C(s)] + 3\ S^\circ[H_2(g)]\}$

$= 1\ \text{mol}\ (229.60\ \text{J/K} \cdot \text{mol})$

$- [2\ \text{mol}\ (5.740\ \text{J/K} \cdot \text{mol}) + 3\ \text{mol}\ (130.684\ \text{J/K} \cdot \text{mol}]$

$= -173.93\ \text{J/K}$

There is a decrease in $\Delta S^\circ$ as more $H_2$ molecules are used as reactants and as they are added to the product. This is expected because, in general, the greater the reduction in the number of particles involved in a reaction, the greater the decrease in entropy. For ethyne, 3 mol of particles → 1 mol of product, whereas the change is from 4 mol to 1 mol for ethene, but 5 mol to 1 mol for ethane.

20-47. $BCl_3(g) + \frac{3}{2}\ H_2(g) \rightarrow B(s) + 3\ HCl(g)$

$\Delta H^\circ_{rxn} = \Delta H^\circ_f[B(s)] + 3\ \Delta H^\circ_f[HCl(g)] - \{\Delta H^\circ_f[BCl_3(g)] + 3/2\ \Delta H^\circ_f[H_2(g)]\}$

$= 1\ \text{mol}\ (0) + 3\ \text{mol}\ (-92.307\ \text{kJ/mol}) - [1\ \text{mol}\ (-403.8\ \text{kJ/mol}) + 3/2\ \text{mol}\ (0)]$

$= 126.9\ \text{kJ}$

$\Delta S^\circ = S^\circ[B(s)] + 3\ S^\circ[HCl(g)] - \{S^\circ[BCl_3(g)] + 3/2\ S^\circ[H_2(g)]\}$

$= 1 \text{ mol } (5.86 \text{ J/K} \cdot \text{mol}) + 3 \text{ mol } (186.908 \text{ J/K} \cdot \text{mol})$

$- [1 \text{ mol } (290 \text{ J/K} \cdot \text{mol}) + 3/2 \text{ mol } (130.684 \text{ J/K} \cdot \text{mol}]$

$= 81 \text{ J/K}$

$\Delta G^\circ rxn = 126.9 \text{ kJ} - (298 \text{ K}) (0.081 \text{ kJ/K}) = 103 \text{ kJ}$

The process is NOT spontaneous. The process is quite endothermic, and the slight increase in entropy is not sufficient to make the reaction spontaneous.

20-49. $FeO(s) + H_2(g) \rightarrow Fe(s) + H_2O(g)$

$\Delta G^\circ_{rxn} = -(8.314510 \times 10^{-3} \text{ kJ/K} \cdot \text{mol}) (700 + 273 \text{ K}) \ln (0.422)$

$= 6.98 \text{ kJ}$

20-51. Since the dissolution occurs readily, the sign of $\Delta G$ is expected to be negative. The increase in entropy drives the reaction.

20-53. *Step 1:* Calculate energy required.

20,300 ft • (1 m / 3.28 ft) • (2.1 kJ / 3.0 m) = $4.33 \times 10^3$ kJ required

*Step 2:* Calculate the energy supplied per mole of glucose

$C_6H_{12}O_6(aq) + 6\, O_2(g) \rightarrow 6\, CO_2(g) + 6\, H_2O(\ell)$

$\Delta H^\circ_{rxn} = 6\, \Delta H^\circ_f[CO_2(g)] + 6\, \Delta H^\circ_f[H_2O(\ell)] - \Delta H^\circ_f[C_6H_{12}O_6(aq)]$

$= 6 \text{ mol } (-393.509 \text{ kJ/mol}) + 6 \text{ mol } (-285.830 \text{ kJ/mol}) - 1 \text{ mol } (-1260 \text{ kJ/mol})$

$= -2816 \text{ kJ}$

*Step 3:* Calculate the grams of glucose required

$4.33 \times 10^3$ kJ required • (1 mol glucose / 2816 kJ) • (180.5 g / mol) = 277 g

20-55. $2\, SO_3(g) \rightarrow 2\, SO_2(g) + O_2(g)$

**Part (a)** — Calculate $\Delta G^\circ_{rxn}$

$\Delta G^\circ_{rxn} = 2\, \Delta G^\circ_f[SO_2(g)] + \Delta G^\circ_f[O_2(g)] - 2\, \Delta G^\circ_f[SO_3(g)]$

$= 2 \text{ mol } (-300.194 \text{ kJ/mol}) + 1 \text{ mol } (0) - 2 \text{ mol } (-371.06 \text{ kJ/mol})$

$= 141.93 \text{ kJ}$

The reaction is predicted to be *non*spontaneous at 25 °C.

**Part (b)** — Is the reaction spontaneous at a higher temperature? To answer this, we need to calculate $\Delta H^\circ_{rxn}$ and $\Delta S^\circ_{rxn}$.

$\Delta H^\circ_{rxn}$ = 2 $\Delta H^\circ_f[SO_2(g)]$ + $\Delta H^\circ_f[O_2(g)]$ - 2 $\Delta H^\circ_f[SO_3(g)]$

= 2 mol (-296.830 kJ/mol) + (0) - 2 mol (-395.72 kJ/mol) = 197.78 kJ

$\Delta S^\circ_{rxn}$ = 2 $S^\circ[SO_2(g)]$ + $S^\circ[O_2(g)]$ - 2 $S^\circ[SO_3(g)]$

= 2 mol (248.22 J/K • mol) + 1 mol (205.138 J/K • mol) - 2 mol (256.76 J/K • mol

= 188.06 J/K

T = $\Delta H^\circ_{rxn}$ / $\Delta S^\circ_{rxn}$ = 197.78 kJ / 0.18806 kJ/K = 1052 K or 779 °C

**Part (c)** — Calculate $K_p$ at 1500 °C

*Step 1:* Calculate $\Delta G^\circ_{rxn}$ at 1500 °C

$\Delta G^\circ_{rxn}$ = $\Delta H^\circ_{rxn}$ - $T\Delta S^\circ_{rxn}$ = 197.78 kJ - (1773 K) (0.18806 J/K)

= -135.7 kJ

*Step 2:* Calculate $K_p$ at 1500 °C

$\Delta G^\circ_{rxn}$ = -RT ln $K_p$

-135.7 kJ = - (8.314510 x $10^{-3}$ kJ/K • mol) (1500 + 273 K) ln $K_p$

$K_p$ = 9.92 x $10^3$

20-57. $N_2(g)$ + $O_2(g)$ → 2 NO(g)

**Part (a)** — Calculate $K_p$ at 25 °C

$\Delta G^\circ_{rxn}$ = 2 $\Delta G^\circ_f[NO(g)]$ = 2 mol (86.55 kJ/mol) = 173.1 kJ

$\Delta G^\circ_{rxn}$ = -RT ln $K_p$

173.1 kJ = - (8.314510 x $10^{-3}$ kJ/K • mol) (298 K) ln $K_p$

$K_p$ = 4.56 x $10^{-31}$

**Part (b)** — Estimate $K_p$ at 700. °C

Step 1: Calculate $\Delta H^\circ_{rxn}$ and $\Delta S^\circ_{rxn}$

$\Delta H^\circ_{rxn}$ = 2 $\Delta H^\circ_f[NO(g)]$ = 2 mol (90.25 kJ/mol) = 180.5 kJ

$\Delta S^\circ_{rxn}$ = 2 $S^\circ[NO(g)]$ - {$S^\circ[N_2(g)]$ + $S^\circ[O_2(g)]$}

$= 2\text{ mol }(210.76\text{ J/K}\cdot\text{mol})$

$- [1\text{ mol }(191.61\text{ J/K}\cdot\text{mol}) + 1\text{ mol }(205.138\text{ J/K}\cdot\text{mol})]$

$= 24.77\text{ J/K}$

*Step 2:* Calculate $\Delta G°rxn$ at 700. °C

$\Delta G°_{rxn} = \Delta H°_{rxn} - T\Delta S°_{rxn} = 180.5\text{ kJ} - (973\text{ K})(0.02477\text{ J/K})$

$= 156\text{ kJ}$

*Step 3:* Calculate $K_p$ at 700. °C

$\Delta G°_{rxn} = -RT \ln K_p$

$156\text{ kJ} = -(8.314510 \times 10^{-3}\text{ kJ/K}\cdot\text{mol})(973\text{ K}) \ln K_p$

$K_p = 4.02 \times 10^{-9}$

**Part (c)** — Find the equilibrium pressures of reactants and product

$$K_p = 4.02 \times 10^{-9} = \frac{[NO]^2}{[N_2][O_2]} = \frac{x^2}{(1.00 - 2x)^2} \quad \text{where } x = P(NO)$$

$x = P(NO) = 6.34 \times 10^{-5}\text{ atm}$

$P(N_2) = P(O_2) = 6.34 \times 10^{-5}\text{ atm}$

## SUMMARY QUESTION

20-59. **Part (a) — Stoichiometry**

150. g Ag • 1 mol / 107.9 g • 1 mol $H_2$ / 2 mol Ag = 0.695 mol $H_2$

0.695 mol • 2.016 g / mol = 1.40 g $H_2$

$$P = nRT/V = \frac{(0.695 \text{ mol})(0.082057 \text{ L} \cdot \text{atm/K} \cdot \text{mol})(296 \text{ K})}{15.0 \text{ L}}$$

P = 1.13 atm

**Part (b) — Solid State**

*Step 1:* Calculate the cell volume

Edge length for face-centered unit cell =

(1/√2) • diagonal distance across cube face = (1/√2) • (4 • atom radius)

Edge length = ((1/√2) • 4 (144.5 pm) = 408.7 pm

408.7 pm • (1 m / $10^{12}$ pm) • 100 cm / 1 m) = 4.087 x $10^{-8}$ cm

(4.087 x $10^{-8}$ cm)$^3$ = 6.827 x $10^{-23}$ cm$^3$

*Step 2:* Calculate the cell mass

(4 Ag atoms / cell) • (1 mol / 6.022 x $10^{23}$ atoms) • (107.9 g/mol)

= 7.167 x $10^{-22}$ g / cell

*Step 3:* Calculate the density of the silver

(7.167 x $10^{-22}$ g / cell) / (6.827 x $10^{-23}$ cm$^3$) = 10.50 g/cm$^3$

The *Periodic Table Stack* (see the section on Computer Software at the beginning of this *Manual*) gives 10.49 g/cm$^3$ as the density of silver.

**Part (c) — Thermodynamics**

For the reaction 2 Ag(s) + $H_2O$(g) → $Ag_2O$(s) + $H_2$(g)

$\Delta H^\circ_{rxn}$ = $\Delta H^\circ_f$[$Ag_2O$(s)] + $\Delta H^\circ_f$[$H_2$(g)] - {2$\Delta H^\circ_f$[Ag] + $\Delta H^\circ_f$[$H_2O$(g)]}

= 1 mol(-31.1 kJ/mol) + (0) - [(0) + 1 mol (-241.818 kJ/mol)]

= 210.7 kJ

$\Delta S^\circ_{rxn}$ = S°[$Ag_2O$(s)] + S°[$H_2$(g)] - {2S°[Ag] + S°[$H_2O$(g)]}

= 1 mol (121 J/K • mol) + 1 mol (130.684 J/K • mol)

- [2 mol(42.55 J/K • mol) + 1 mol (188.825 J/K • mol)]

$= -22 \text{ J/K}$

$\Delta G°_{rxn} = \Delta H°_{rxn} - T\Delta S°_{rxn}$

$= 210.7 \text{ kJ} - (298 \text{ K})(-0.022 \text{ kJ/K}) = 217.3 \text{ kJ}$

For the reaction $Ag_2O(s) \rightarrow 2\,Ag(s) + \frac{1}{2}\,O_2(g)$

$\Delta H°_{rxn} = -\Delta H°_f[Ag_2O(s)] = +31.1 \text{ kJ}$

$\Delta S°_{rxn} = 2\,S°[Ag(s)] + \frac{1}{2}\,S°[O_2(g)] - S°[Ag_2O(s)]$

$= 2 \text{ mol}\,(42.55 \text{ J/K} \cdot \text{mol}) + \frac{1}{2} \text{ mol}\,(205.138 \text{ J/K} \cdot \text{mol})$

$- 1 \text{ mol}\,(121 \text{ J/K} \cdot \text{mol})$

$= 67 \text{ J/K}$

$\Delta G°_{rxn} = \Delta H°_{rxn} - T\Delta S°_{rxn}$

$= +31.1 \text{ kJ} - (298 \text{ K})(0.067 \text{ kJ/K}) = 11.2 \text{ kJ}$

Overall reaction: $H_2O(g) \rightarrow H_2(g) + O_2(g)$

$\Delta H°_{rxn}$ for overall reaction = $-\Delta H°_f$ for $H_2O(g)$ = 241.818 kJ

$\Delta G°_{rxn}$ for overall reaction = $-\Delta G°_f$ for $H_2O(g)$ = 228.572 kJ

The overall process is NOT spontaneous at 25 °C.

Temperature at which the reaction $Ag_2O(s) \rightarrow 2\,Ag(s) + \frac{1}{2}\,O_2(g)$ becomes spontaneous.

$T = \Delta H°_{rxn} / \Delta S°_{rxn} = 31.1 \text{ kJ} / 0.067 \text{ kJ/K} = 460 \text{ K}$ or about 190 °C

# CHAPTER 21
# ELECTROCHEMISTRY

## INSTRUCTOR'S NOTES

About 4-5 lecture hours are reserved for a discussion of oxidation-reduction reactions. These lectures are usually placed after those on equilibrium and thermodynamics. However, there is no particular reason why it is not possible to place electrochemistry before these other subjects. In any event, electrochemistry should be covered before beginning any systematic descriptive chemistry.

## LIST OF TRANSPARENCIES FOR THIS CHAPTER

| No. | Figure | Page | Subject |
|---|---|---|---|
| 95 | 21.4 | 855 | A voltaic cell |
| 96 | 21.5 | 856 | Summary of terminology of voltaic cells |
| 97 | Table 21.1 | 865 | Standard electrode potentials |
| 98 | 21.8 | 872 | Dry cell |
| 99 | 21.10 | 875 | Lead storage battery |
| 100 | 21.20 | 886 | Downs cell for electrolysis of molten NaCl |
| 101 | 21.21 | 887 | Membrane cell for aqueous NaCl electrolysis |
| 102 | 21.22 | 888 | Iron corrosion |

## SUGGESTED DEMONSTRATIONS

This subject lends itself well to demonstrations, particularly to show the practical importance of redox reactions. A variety of redox reactions were described as demonstrations for Chapters 4 and 5, and the **Periodic Table Videodisc** contains many examples of such reactions. In addition, the videotape by Shakhashiri contains a number of suitable demonstrations.

1. To illustrate *redox reactions* in general we suggest the following possibilities.
   - The reaction of Al foil with concentrated $CuCl_2$ solution

$$3\,Cu^{2+}(aq) + 2\,Al(s) \rightarrow 2\,Al^{3+}(aq) + 3\,Cu(s) + \text{heat}$$

(A photo of this reaction is shown in Figure 1.8, page 13 of the text.) This reaction is interesting because it does not occur with other copper salts such as $Cu(NO_3)_2$ or $CuSO_4$. Apparently the reason is that the $Cl^-$ ion is able to diffuse through the layer of $Al_2O_3$ on the foil and open a pathway for electron transfer between $Cu^{2+}$ and Al. (See comments on page 890 of the text.) This is, in itself, a very useful point to make, since this shows the importance of $Cl^-$ in corroding systems; for example, one can note that aluminum objects are particularly sensitive to corrosion near the ocean.

• A copper screen is placed in a silver nitrate solution in a petri dish on an overhead projector (see Figure 5.24 on page 173).

$$2\,Ag^+(aq) + Cu(s) \rightarrow 2\,Ag(s) + Cu^{2+}(aq)$$

In a short time silver needles are clearly seen growing off of the copper wire, and the solution visibly takes on a blue color from aqueous copper(II) ion.

2. Demonstrations of *galvanic cells.*

• We like to begin this topic by first showing that oxidation-reduction can occur simply by putting zinc foil in a copper solution (Figure 5.23, page 169). This is then contrasted with a true galvanic cell such as that illustrated in Figure 21.4 (text page 855) where a current is obtained. (If one is reasonably careful in setting up this cell, the E° value of 1.10 V for this reaction can be obtained. Incidentally, although one should use 1.0 M solutions for the sake of "scientific purity," we find that 0.10 M solutions are much more practical and convenient.)

• To show that a fancy cell is not necessary to obtain a current and a reasonably large voltage difference, and that all one has to do is separate the electrodes and allow for ion flow between them, we regularly show the "grapefruit cell" pictured at the bottom of page 855. As noted there, a voltage of about 0.9 V is easily obtained with copper and zinc electrodes. (It is a student myth that the maraschino cherry enhances the voltage!)

• When discussing the various cells on pages 855-861 of the text, we place Zn metal in 1 M HCl and show that there is a rapid reaction while no reaction occurs when Cu metal is placed in the acid.

3. Demonstrations of *electrolysis*

• The electrolytic process in Figure 21.15 on page 878 can be done in a cell such as that shown there or in a petri dish on the overhead projector. Fill the dish with dilute KI and place platinum electrodes in the dish; add a few drops of phenolphthalein. When the cell is attached to a battery, $I_2$ is clearly seen at the anode and $H_2$ gas (as well as a red color from $OH^-$ ions) at the cathode. (This demonstration is also very useful for Chapter 5 to introduce the idea that one can write half-reactions for redox processes.)

• $H_2$ and $O_2$ can be obtained by the electrolysis of aqueous acid or base. See the photo on page 909, Figure 22.2.

4. Demonstrations of *corrosion*

• The demonstration on page 735 (Figure 19.20) is interesting and relatively simple to set up. Detailed directions are as follows:

> Add 3 g of agar to 100 mL of water and bring to a boil. Cool, and then stir in 1 mL of phenolphthalein solution and 2 mL of 1% $K_3Fe(CN)_6$ . Cover the bottom of a clean petri dish with the mixture and let it harden. Keep the remainder of the agar mixture liquid by keeping it warm on a water bath. Place an iron nail on the hardened agar and then pour on some more of the agar mixture. After an hour or so the blue color of Prussian blue is visible at the tips of the nail or where the nail was bent. Red phenolphthalein will then be visible along other regions of the nail.

A variation of this is suitable as a demonstration of electrolysis. Place two pieces of iron in the agar with one end sticking up through the surface of the hardened agar at least 1/2 inch. If these electrodes are attached to a battery (1.5-3 V), one soon sees a blue color around one electrode (the anode) due to the formation of $Fe^{2+}$. At the other electrode (the cathode) the surroundings turn red due to $OH^-$ ion from the reduction of water.

5. Additional suggestions

(a) Scharlin, P. "The Human Salt Bridge," *J. Chem. Educ.* **1990**, *67*, 156.

(b) Shearer, E.C. "The Construction and Use of Commercial Cell Displays in Freshman Chemistry," *ibid.*, **1990**, *67*, 158.

(c) Manjkow, J.; Levine, D. "Electrodeposition of Ni on Cu" *ibid.*, **1986**, *63*, 809.

(d) Heldeman, S. "Electrolysis of Water" *ibid.*, **1986**, *63*, 809.

# SOLUTIONS TO SELECTED STUDY QUESTIONS

## CELLS AND CELL POTENTIALS

21-13. $Zn(s) + Cl_2(g) \rightarrow ZnCl_2(aq)$

$Zn(s) \rightarrow Zn^{2+}(aq) + 2e-$ oxidation, anode compartment

$Cl_2(g) + 2e- \rightarrow 2\,Cl^{-}(aq)$ reduction, cathode compartment

21-15. $Fe(s) \rightarrow Fe^{2+}(aq) + 2e-$ oxidation, anode compartment

$O_2(g) + 4\,H_3O^{+}(aq) + 4e- \rightarrow 6\,H_2O(\ell)$ reduction, cathode compartment

$Fe \mid Fe^{2+} \parallel O_2 \mid H_2O$

21-17. $\Delta G^\circ = -nFE^\circ = -(2\text{ mol e-})(9.65 \times 10^4\text{ J / V} \cdot \text{mol})(2.12\text{ V})(1\text{ kJ}/10^3\text{ J})$

$= -409\text{ kJ}$

21-19. (a) $I_2(aq) + 2e- \rightarrow 2\,I^{-}(aq)$ $E^\circ = +0.535\text{ V}$

$Mg(s) \rightarrow Mg^{2+}(aq) + 2\,e^{-}$ $E^\circ = +2.37\text{ V}$

$I_2(aq) + Mg(s) \rightarrow Mg^{2+}(aq) + 2\,I^{-}(aq)$ $E^\circ_{net} = +2.91\text{ V}$

Spontaneous as written.

(b) $Ag(s) \rightarrow Ag^{+}(aq) + 2\,e-$ $E^\circ = -0.80\text{ V}$

$Fe^{3+}(aq)\ e- \rightarrow Fe^{2+}(aq)$ $E^\circ = +0.771\text{ V}$

$Ag(s) + Fe^{3+}(aq) \rightarrow Ag^{+}(aq) + Fe^{2+}(aq)$ $E^\circ_{net} = -0.03\text{ V}$

Not spontaneous as written.

(c) $2\,Ag^{+}(aq) + 2e- \rightarrow 2\,Ag(s)$ $E^\circ = +0.80\text{ V}$

$Sn^{2+}(aq) \rightarrow Sn^{4+}(aq) + 2\,2e-$ $E^\circ = -0.15\text{ V}$

$Sn^{2+}(aq) + 2\,Ag^{+}(aq) \rightarrow Sn^{4+}(aq) + 2\,Ag(s)$ $E^\circ_{net} = +0.65\text{ V}$

Spontaneous as written.

(d) $Zn(s) \rightarrow Zn^{2+}(aq) + 2e^-$ $E° = +0.763$ V

$O_2(g) + 2\,H_2O(\ell) + 4e^- \rightarrow 4\,OH^-(aq)$ $E° = +0.40$ V

$2\,Zn(s) + O_2(g) + 2\,H_2O(\ell) \rightarrow 2\,Zn^{2+}(aq) + 4\,OH^-(aq)$ $E°_{net} = +1.16$ V

Spontaneous as written.

21-21.(a) $Ce^{4+}(aq) + e^- \rightarrow Ce^{3+}(aq)$ $E° = +1.61$ V

$2\,Cl^-(aq) \rightarrow Cl_2(g) + 2e^-$ $E° = -1.360$ V

$2\,Ce^{4+}(aq) + 2\,Cl^-(aq) \rightarrow 2\,Ce^{3+}(aq) + Cl_2(g)$ $E°_{net} = +0.25$ V

Spontaneous as written.

(b) $Cu(s) \rightarrow Cu^{2+}(aq) + 2e^-$ $E° = -0.337$ V

$NO_3^-(aq) + 4\,H_3O^+(aq) + 3e^- \rightarrow NO(g) + 6\,H_2O(\ell)$ $E° = 0.96$ V

$3\,Cu(s) + 2\,NO_3^-(aq) + 8\,H_3O^+(aq) \rightarrow 3\,Cu^{2+}(aq) + 2\,NO(g) + 12\,H_2O(\ell)$

$E°_{net} = +0.62$

Spontaneous as written. (See Figure 4.7 on page 110.)

(c) $Fe^{2+}(aq) \rightarrow Fe^{3+}(aq) + e^-$ $E° = -0.771$ V

$Cr_2O_7^{2-}(aq) + 14\,H_3O^+(aq) + 6e^- \rightarrow 2\,Cr^{3+}(aq) + 21\,H_2O(\ell)$ $E° = 1.33$ V

$6\,Fe^{2+}(aq) + Cr_2O_7^{2-}(aq) + 14\,H_3O^+(aq)$

$\rightarrow 6\,Fe^{3+}(aq) + 2\,Cr^{3+}(aq) + 21\,H_2O(\ell)$ $E°_{net} = 0.56$ V

Spontaneous as written.

21-23. (a) The weakest oxidizing agent is $Al^{3+}$.

(b) The strongest oxidizing agent is $Ce^{4+}$.

(c) The strongest reducing agent is Al(s)

(d) The weakest reducing agent is $Ce^{3+}$.

(e) Sn will reduce $Ag^+$ to Ag ($E°_{net} = 0.94$ V)

(f) $Hg(\ell)$ cannot reduce $Sn^{2+}$ to Sn

(g) Sn will reduce $Hg_2^{2+}$, $Ag^+$, and $Ce^{4+}$.

(h) In principle, $Ag^+$ can oxidize Hg(ℓ), Sn(s), Ni(s), and Al(s).

21-25. (a) Reaction with maximum, positive potential is

$$3\ Ce^{4+}(aq) + Al(s) \rightarrow 3\ Ce^{3+}(aq) + Al^{3+}(aq) \qquad E^\circ_{net} = 3.27\ V$$

(b) Ni is a stronger reducing agent than Hg(ℓ). Therefore,

$$Ni(s) + Hg_2^{2+}(aq) \rightarrow 2\ Hg(\ell) + Ni^{2+}(aq) \qquad E^\circ_{net} = 1.04\ V$$

(c) Ag is a stronger reducing agent than $Ce^{3+}$. Therefore,

$$Ag(s) + Ce^{4+}(aq) \rightarrow Ce^{3+}(aq) + Ag^+(aq) \qquad E^\circ_{net} = 0.81\ V$$

21-27. (a) $Al(s) + 3\ Ag^+(aq) \rightarrow 3\ Ag(s) + Al^{3+}(aq) \qquad E^\circ_{net} = 2.46\ V$

(b) Al is a better reducing agent than Ag.

(c) The oxidation of Al(s) occurs in the anode compartment, and the reduction of $Ag^+(aq)$ occurs in the cathode compartment. The Al(s) electrode is negative, and the Ag(s) electrode is positive.

(d) $NO_3^-$ ions flow from the Ag | Ag+ compartment to the Al | $Al^{3+}$ compartment.

21-29. (a) Ni is a better reducing agent than Pb.

(b) $Ni(s) \rightarrow Ni^{2+}(aq) + 2e^- \qquad E^\circ = +0.25\ V$

$Pb^{2+}(aq) + 2e^- \rightarrow Pb(s) \qquad E^\circ = -0.126\ V$

$Ni(s) + Pb^{2+}(aq) \rightarrow Pb(s) + Ni^{2+}(aq) \qquad E^\circ_{net} = 0.12\ V$

(c) The Pb electrode is the cathode (reduction occurs).

(d) Electrons flow from Ni(s) to Pb(s)

(e) $NO_3^-$ ions flow from the $Pb^{2+}$ | Pb (cathode) compartment to the Ni | $Ni^{2+}$ compartment (anode).

21-31. (a) $Zn(s) + Cl_2(g) \rightarrow ZnCl_2(aq)$

(b) $Zn(s) \rightarrow Zn^{2+}(aq) + 2e^-$ oxidation, anode compartment

$Cl_2(g) + 2e^- \rightarrow 2\ Cl^-(aq)$ reduction, cathode compartment

(c) Zn electrode is negative.

(d) $Cl_2$ is at a pressure of 1 atm, and $ZnCl_2(aq)$ must have a concentration of 1 M.

(e) $E^\circ_{net} = 2.12$ V

(f) $E^\circ_{net} = 3.63$ V when $F_2$ replaces $Cl_2$.

## CELLS UNDER NONSTANDARD CONDITIONS AND THE CONNECTION BETWEEN E° AND K

21-33.

$$E = E^\circ - \frac{0.059}{2} \log \frac{[Fe^{3+}]^2}{[Fe^{2+}]^2\,[H_2O_2]\,[H_3O^+]^2}$$

If $E^\circ = 1.00$ V and all dissolved species have a concentration of 0.015 M, E = 0.84 V. The potential has decreased relative to $E^\circ_{net}$.

21-35. (a) $Zn(s) + 2\ Ag^+(aq) \rightarrow Zn^{2+}(aq) + 2\ Ag(s)$

(b) $E^\circ_{net} = 1.56$ V

(c) The Zn electrode is negative (source of electrons).

(d) Electrons flow from Zn to Ag.

(e)

$$E = E^\circ - \frac{0.0592}{2} \log \frac{[Zn^{2+}]}{[Ag^+]^2} = 1.54 \text{ V}$$

The potential of the cell has dropped by 0.02 V owing to the fact that the concentration of $Ag^+$ is less than that of $Zn^{2+}$.

21-37. (a) $Zn^{2+}(aq) + Ni(s) \rightarrow Zn(s) + Ni^{2+}(aq)$ $E^\circ = -0.51$ V

$\log K = nE^\circ/0.0592 = (2 \text{ mol})(-0.51 \text{ V})/0.0592 \text{ V} \cdot \text{mol} = -17$

$K = 5.9 \times 10^{-17}$

(b) $Cu(s) + 2\ Ag^+(aq) \rightarrow Cu^{2+}(aq) + 2\ Ag(s)$ $E^\circ = 0.46$ V

$\log K = nE^\circ / 0.0592 = (2)(0.46\ V)/0.0592 = 16$

$K = 3.4 \times 10^{15}$

## ELECTROLYSIS, ELECTRICAL ENERGY, AND POWER

21-39. 14.5 min • 60 sec / min = 870. sec

$1.0 \times 10^{-3}$ A • 870. sec = 0.87 C

0.87 C • 1 mol e- / 96500 C = $9.0 \times 10^{-6}$ mol e-

1 mol e- produce 1 mol Ag, Therefore,

$9.0 \times 10^{-6}$ mol Ag • 108 g Ag / mol = $9.7 \times 10^{-4}$ g Ag (or 0.97 mg)

21-41. 0.0125 amp • 2.00 hr • 3600 sec/hr = 90.0 C

90.0 C • 1 mol e- / 96,500 C • 1 mol Cu / 2 mol e- • 63.55 g Cu / mol = 0.0296 g Cu

90.0 C • 1 mol e- / 96,500 C • 1 mol $Cl_2$ / 2 mol e- • 70.9 g $Cl_2$ / mol = 0.0331 g $Cl_2$

21-43. 0.089 g Ag • 1 mol Ag / 108 g = $8.2 \times 10^{-4}$ mol Ag

$8.2 \times 10^{-4}$ mol Ag • 1 mol e- / mol Ag = $8.2 \times 10^{-4}$ mol e-

$8.2 \times 10^{-4}$ mol e- • 1 mol Au / 3 mol e- = $2.7 \times 10^{-4}$ mol Au

2.7 x 10-4 mol Au • 197 g/mol = 0.054 g Au

21-45. 0.0125 L • 0.015 mol / L = $1.9 \times 10^{-3}$ mol $V^{3+}$

$1.9 \times 10^{-3}$ mol $V^{3+}$ • 1 mol e- / 1 mol $V^{3+}$ • 96500 C / mol e- = 180 C

180 C = 0.268 amps • X sec

X sec = 680 sec (or a little more than an hour)

21-47. 84 g Al • 1 mol Al / 27.0 g • 3 mol e- / 1 mol Al • 96500 C / mole- = $9.0 \times 10^5$ C

$9.0 \times 10^5$ C = 1.0 amp • X sec

X sec = $9.0 \times 10^5$ sec

$9.0 \times 10^5$ sec • 1 hr / 3600 sec = 250 hours

21-49. $2.0 \times 10^3$ lb • 454 g / lb • 1 mol / 27.0 g = $3.4 \times 10^4$ mol Al

$3.4 \times 10^4$ mol Al • 3 mol e- / mol Al • 96500 C / mol e- = $9.7 \times 10^9$ C

$9.7 \times 10^9$ C • 5.0 V = $4.9 \times 10^{10}$ J

$4.9 \times 10^{10}$ J • 1 kwh / $3.60 \times 10^6$ J = $1.4 \times 10^4$ kwh

21-51. 24 hr • 3600 sec / hr = $8.6 \times 10^4$ sec

$8.6 \times 10^4$ sec • $4.0 \times 10^4$ amps • (1 C / amp • sec) = $3.5 \times 10^9$ C

The amount of *sodium produced* per day is:

$3.5 \times 10^9$ C • 1 mol e- / 96500 C • 1 mol Na / 1 mol e- • 23.0 g Na / mol = $8.2 \times 10^5$ g

The amount of *chlorine gas produced* per day is:

$3.5 \times 10^9$ C • 1 mol e- / 96500 C • 1 mol $Cl_2$ / 2 mol e- • 70.9 g $Cl_2$ / mol = $1.3 \times 10^6$ g

Number of *kilowatt-hours:*

$3.5 \times 10^9$ C • 7.0 V = $2.4 \times 10^{10}$ J

$2.4 \times 10^{10}$ J • 1 kwh / $3.60 \times 10^6$ J = 6700 kwh

21-53. (a) Anode reaction: 2 $Cl^-$(aq) → $Cl_2$(g) + 2e-

Electrode in anode compartment is positively charged.

(b) Cathode compartment: $Ni^{2+}$(aq) + 2e- → Ni(s)

Electrode in cathode compartment is negatively charged.

(c) 0.0250 amps • 1.25 hr • 3600 sec / hr • 1 C / amp•sec = 113 C

113 C • 1 mol e- / 96500 C = $1.17 \times 10^{-3}$ mol e-

$1.17 \times 10^{-3}$ mol e- • 1 mol Ni / 2 mol e- • 58.7 g Ni/mol = 0.0342 g Ni

$1.17 \times 10^{-3}$ mol e- • 1 mol $Cl_2$ / 2 mole- • 70.9 g/mol = 0.0413 g $Cl_2$

21-55. (a) $NiBr_2$(aq): Ni(s) will be produced at the cathode and $Br_2$($\ell$) at the anode.

(b) NaI(aq): $I^-$ is oxidized to $I_2$ the anode and $H_2O$ is reduced to $H_2$ (and $OH^-$) at the cathode.

(c) $CdCl_2$(aq): $Cd^{2+}$ is reduced to Cd(s) at the cathode, and $Cl^-$ is oxidized to $Cl_2$ at the anode.

## GENERAL QUESTIONS

21-57. (a) A and C are stronger reducing agents than $H_2$.

(b) C is the strongest reducing agent of the series.

(c) D is a stronger reducing agent than B.

Order of increasing strength as reducing agents: $B < D < H_2 < A < C$

21-59. (a) Balanced equation: $Ni^{2+}(aq) + Cd(s) \rightarrow Ni(s) + Cd^{2+}(aq)$

(b) Cd is the reducing agent and $Ni^{2+}(aq)$ is the oxidizing agent. $Ni^{2+}(aq)$ is reduced to Ni(s) and Cd(s) is oxidized to $Cd^{2+}(aq)$.

(c) *Cd is the anode* (oxidation occurs in the Cd-containing compartment) and *Ni is the cathode* (reduction of $Ni^{2+}$ occurs in the Ni-containing compartment). The *Cd electrode is negative* as this electrode supplies electrons.

(d) $E^\circ_{net} = 0.15$ V

(e) Electrons flow from the Cd electrode (anode) to the Ni electrode (cathode).

(f) $NO_3^-$ ions migrate from the $Ni^{2+}$/Ni compartment (the cathode) to the Cd/$Cd^{2+}$ compartment (the anode).

(g) $\log K = -nE^\circ/0.0592 = -(2.00 \text{ mol})(0.15 \text{ V})/(0.0592 \text{ V} \cdot \text{mol})$

$\log K = 5.07$, which gives $K = 1.17 \times 10^5$

(h) For $[Cd^{2+}] = 0.010$ M and $[Ni^{2+}] = 1.0$ M, and $E^\circ = 0.15$ V, we have

$$E = E^\circ - \frac{0.0592}{2} \log \frac{[Cd^{2+}]}{[Ni^{2+}]} = 0.21 \text{ V}$$

The reaction is still spontaneous in the direction written.

(i) The battery has, in the beginning, 1.00 mol each of $Ni^{2+}$ and $Cd^{2+}$. In addition,

50.0 g Ni • 1 mol / 58.69 g = 0.852 mol Ni

50.0 g Cd • 1 mol / 112.4 g = 0.445 mol Cd

The limiting reagent here is 0.445 mol Cd. Therefore, we use this to find the lifetime of the battery.

0.445 mol Cd • 2 mol e- / 1 mol Cd • 96500 C/ mol e- = $8.59 \times 10^4$ C

$8.59 \times 10^4$ C • 1 amp • sec / C • 1 / 0.050 amp = $1.7 \times 10^6$ sec (about 20 days)

21-61. $Pb(s) + PbO_2(s) + 2\,H_2SO_4(aq) \rightarrow PbSO_4(s) + 2\,H_2O(\ell)$ $E^\circ = 2.041$ V

$$E = E^\circ - \frac{0.0592}{2} \log \frac{1}{[H_2SO_4]^2} = 2.087 \text{ V}$$

21-63. $\Delta G^\circ = -nFE^\circ$

$-6.07 \times 10^5$ J = -(4 mol e-) • (96500 J/V • mol) • E°

$E^\circ = 1.57$ V

21-65. $Cu^{2+}(aq) + Sn(s) \rightarrow Cu(s) + Sn^{2+}(aq)$ $E^\circ_{net} = 0.48$ V

*Step 1:* Calculate the number of coulombs and then moles of electrons that passed through the cell.

0.400 amp • 48.0 hr • 3600 sec/hr = $6.91 \times 10^4$ C

$6.91 \times 10^4$ C • 1 mol e- / 96500 C = 0.716 mol e-

*Step 2:* Find the number of moles of $Cu^{2+}$ reduced and $Sn^{2+}$ produced.

0.716 mol e- • 1 mol $Cu^{2+}$ reduced / 2 mol e- = 0.358 mol $Cu^{2+}$/L reduced

Therefore, we know that 0.358 mol/L of $Sn^{2+}$ is also produced.

*Step 3:* Find the concentrations of $Cu^{2+}$ and $Sn^{2+}$ after 48.0 hours.

$[Cu^{2+}]$ = 1.00 M - 0.358 M = 0.64 M

$[Sn^{2+}]$ = 1.00 M + 0.358 M = 1.36 M

*Step 4:* Calculate the cell potential after 48.0 hours.

$$E = E^\circ - \frac{0.0592}{2} \log \frac{[Sn^{2+}]}{[Cu^{2+}]} = 0.47 \text{ V}$$

The cell potential has changed by only 0.01 V after 48.0 hours.

21-67. (a) $2\,Al(s) + 3\,Cl_2(g) \rightarrow 2\,Al^{3+}(aq) + 6\,Cl^-(aq)$

(b) Al is the anode and the $Cl_2|Cl^-$ compartment is the cathode.

(c) $E^\circ_{net} = 3.02$ V

(d) If the pressure of $Cl_2$ decreases, the cell potential should also decrease.

$$E = E^\circ - \frac{0.0592}{2} \log \frac{[Al^{3+}]^2 [Cl^-]^2}{P(Cl_2)^3} = 2.99 \text{ V}$$

(e) 30.0 g Al • 1 mol / 26.98 g = 1.12 mol Al

1.12 mol Al • 3 mol e- / mol Al • 96500 C / mol e- = $3.22 \times 10^5$ C

$3.22 \times 10^5$ C • 1 amp • sec/C • 1 / 0.75 amp = $4.29 \times 10^5$ sec (or 119 hours)

21-69. (a) $5.0 \times 10^{-3}$ g $C_{10}H_{10}Fe$ • 1 mol / 186 g = $2.7 \times 10^{-5}$ mol $C_{10}H_{10}Fe$

15.0 min • 60 sec/min • 0.00288 amp • 1 C/amp • sec = 2.59 C

2.59 C • 1 mol e- / 96500 C = $2.69 \times 10^{-5}$ mol e-

Therefore, we see that 1 mol of $C_{10}H_{10}Fe$ transfers 1 mol of electrons

$2.69 \times 10^{-5}$ mol e- / $2.7 \times 10^{-5}$ mol $C_{10}H_{10}Fe$ = 1.0 mol e- / mol $C_{10}H_{10}Fe$

21-71. (a) Anode $N_2H_4(aq) + 4\ OH^-(aq) \rightarrow N_2(g) + 4\ H_2O(\ell) + 4\ e^-$

Cathode $O_2(g) + 2\ H_2O(\ell) + 4\ e^- \rightarrow 4\ OH^-(aq)$

(b) Net cell reaction:

$O_2(g) + N_2H_4(aq) \rightarrow N_2(g) + 2\ H_2O(\ell)$

(c) 50.0 hr • 3600 sec/hr • 0.50 amp • 1 C/amp • sec = $9.00 \times 10^4$ C

$9.00 \times 10^4$ C • 1 mol e- / 96500 C • 1 mol $N_2H_4$ / 4 mol e- • 32.05 g $N_2H_4$ / mol = 7.47 g

(d) 7.47 g $N_2H_4$ • 1 mol $N_2H_4$ / 32.05 g • 1 mol $O_2$ / 1 mol $N_2H_4$ • 32.00 g $O_2$/mol = 7.46 g $O_2$

## SUMMARY QUESTIONS

21-73. *Step 1:* Calculate the number of moles of $H_2$ present

$$n = PV/RT = \frac{(200.\text{ atm})(1.0\text{ L})}{(0.082057\text{ L} \cdot \text{atm/K} \cdot \text{mol})(298\text{ K})} = 8.2 \text{ mol } H_2$$

*Step 2:* Calculate the number of coulombs of electricity that may be generated by the $H_2$

$8.2 \text{ mol } H_2 \cdot 2 \text{ mol e-} / \text{mol } H_2 \cdot 96500 \text{ C} / \text{mol e-} = 1.6 \times 10^6 \text{ C}$

Step 3: Calculate the time the cell can operate.

$1.6 \times 10^6 \text{ C} \cdot 1 \text{ amp} \cdot \text{sec} / \text{C} \cdot 1 / 1.5 \text{ amp} = 1.1 \times 10^6 \text{ sec}$ (or 290 hours)

21-75. (a) Electron dot structure of $CH_3SO_2F$. The molecular geometry of this 32-valence electron molecule is tetrahedral.

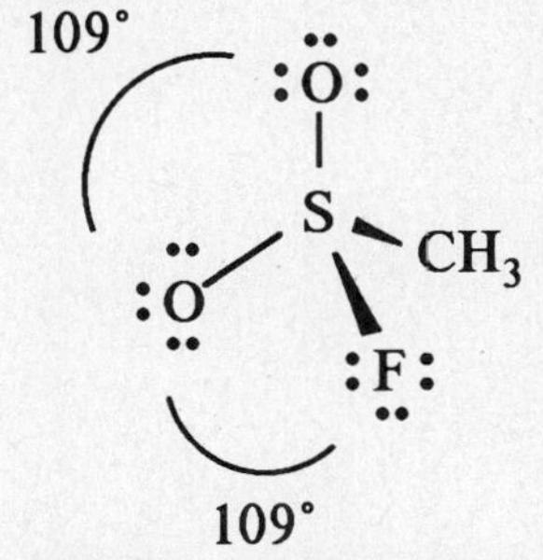


(b) Grams of HF required and grams product produced.

$CH_3SO_2F + 3 \text{ HF} \rightarrow CF_3SO_2F + 3 H_2$

$150 \text{ g } CH_3SO_2F \cdot 1 \text{ mol } CH_3SO_2F / 98.1 \text{ g} = 1.5 \text{ mol } CH_3SO_2F$

$1.5 \text{ mol } CH_3SO_2F \cdot 3 \text{ mol HF} / 1 \text{ mol } CH_3SO_2F \cdot 20. \text{ g HF/mol} = 92 \text{ g HF}$

$1.5 \text{ mol } CH_3SO_2F \cdot 1 \text{ mol } CF_3SO_2F / 1 \text{ mol } CH_3SO_2F$

$\cdot\ 152 \text{ g } CF_3SO_2F / \text{mol} = 232 \text{ g } CF_3SO_2F$

$1.5 \text{ mol } CH_3SO_2F \cdot 3 \text{ mol } H_2 / 1 \text{ mol } CH_3SO_2F \cdot 2.00 \text{ g } H_2/\text{mol} = 9.2 \text{ g } H_2$

(c) The $H_2$ is produced at the cathode (because the reaction is effectively the reduction of $H^+$ to $H_2$).

(d) $250 \text{ amps} \cdot 24 \text{ hr} \cdot 3600 \text{ sec/hr} \cdot 1 \text{ C} / \text{amp} \cdot \text{sec} = 2.2 \times 10^7 \text{ C}$

$2.2 \times 10^7 \text{ C} \cdot 8.0 \text{ V} \cdot 1 \text{ J/V} \cdot \text{C} = 1.7 \times 10^8 \text{ J}$

$1.7 \times 10^8 \text{ J} \cdot 1 \text{ kwh} / 3.6 \times 10^6 \text{ J} = 48 \text{ kwh}$

# PART V

# THE CHEMISTRY OF THE ELEMENTS AND THEIR COMPOUNDS

# CHAPTER 22
# THE CHEMISTRY OF HYDROGEN AND THE s-BLOCK ELEMENTS

## INSTRUCTOR'S NOTES

Beginning with this chapter, the material in the remainder of the book is covered to the extent that time is available at the end of the course. We have, regrettably, little time at the end of the course for a discussion of descriptive chemistry. Indeed, this was one motivating factor in placing as much chemistry as possible early in the text.

Normally we hope to have at least six to nine lectures on descriptive chemistry at the end of the second term. This enables us to discuss a few aspects of the chemistry of the following elements: H; Li and Na; Mg and Ca; B and Al; C and Si; N and P; O and S; and F and Cl. Since many of our students at SUNY-Oneonta go on to various courses in organic chemistry, we do not feel compelled to cover the organic and biochemistry chapters.

To rectify the lack of in-depth coverage of descriptive chemistry in the introductory course, the Chemistry Department at SUNY-Oneonta has added a 2-hour course in that subject, to be taken by chemistry majors or other interested students. It may be taken by any student after having General Chemistry, but it is taught with the assumption that the students only have a General Chemistry background. There are currently no separate texts available for such a course. However, the text that this *Manual* accompanies could be used for such a course, as long as some additional material is added. However, our experience with the descriptive course is that the material in Chapters 22-25 would not need much amplification to serve aa the basis of a 2-credit hour course.

## LIST OF TRANSPARENCIES FOR THIS CHAPTER

| No. | Figure | Page | Subject |
|---|---|---|---|
| 103 | 22.4 | 912 | Interstitial metal hydride |
| 104 | 22.9 | 921 | The Solvay process |
| 105 | 22.12 | 929 | Magnesium from seawater |

## SUGGESTED DEMONSTRATIONS

An excellent demonstration to use at the beginning of the discussion of hydrogen chemistry is to blow up a hydrogen-filled balloon. This can be made more interesting by filling a balloon with He and contrasting their behavior. The students can readily figure out which balloon contains $H_2$ gas. The **Periodic Table Videodisc** also has an excellent real-time example of the detonation of a hydrogen-filled balloon.

One can demonstrate the various ways of making $H_2$, such as electrolysis (Figures 21.15 and 22.2), a metal + acid (Figures 12.9, 17.1, 20.8, and 22.8), Al + base (Figures 12.10 and 15.1), or a metal hydride + water (Figure 17.2).

For more information on hydrogen, you will find it interesting and useful to read the book *Hydrogen Power* by L.O. Williams, Pergamon, 1980. Also useful are chapters in *The Modern Inorganic Chemicals Industry* (R. Thompson, ed., Royal Society, 1977) and *Industrial Inorganic Chemistry* by W. Büchner, R. Schliebs, G. Winter, and K.H. Büchel (VCH, 1989). (The latter is an especially useful, albeit expensive, book for teaching descriptive inorganic chemistry.) A very interesting article on the "hydrogen economy" was done by G. Graff in *High Technology* magazine, May, 1983. Finally, *ChemMatters* for October, 1985 was devoted to hydrogen and helium chemistry.

For the alkali metals, the following demonstrations are worthwhile:

- Flame colors of Na- and K-containing salts.
- The reactions of the alkali metals with water. Caution! The reaction of K with water should be done behind a shield. In both cases, place an indicator in the water to show the formation of metal hydroxide.
- The ease with which Na and K can be cut with a knife. See Figure 2.4 for Na.

Properties of the alkaline earth elements can be demonstrated as follows:

- Minerals containing alkaline earth metals are readily available: calcite, marble, shells, and fluorite, among others.
- Flame colors of the alkaline earth elements.
- Reactions of Mg and Ca to compare with Group 1 A elements. Magnesium will not react with water at room temperature, whereas Ca will react readily.
- Reactions of limestone. Among these reactions is one used on page 807. Here we

have illustrated the equilibria involved when $CO_2$ is added to a solution of $Ca(OH)_2$. The demonstration is discussed in detail in Shakhashiri, Volume 1.

- Calcium carbide is an important chemical. Its hydrolysis to give acetylene is readily illustrated [see Figures 22.13 (page 934) and 10.9 (page 381)].

The **Periodic Table Videodisc** has excellent images of the reactions of the alkali and alkaline earth metals with air, water, and acids and bases. The advantage in using the videodisc is that one can quickly scan through the various types of reactions and the students can get a feel for the periodic trends of the reactions. In addition, the **Videodisc** contains images of the uses of these elements.

To see flame colors in a large lecture room we have found the following setup to work very well. (The following is an edited description of this technique by George McKelvy that appeared in *ChemComments*, November, 1989.) The set-up for this demonstrations is illustrated below.

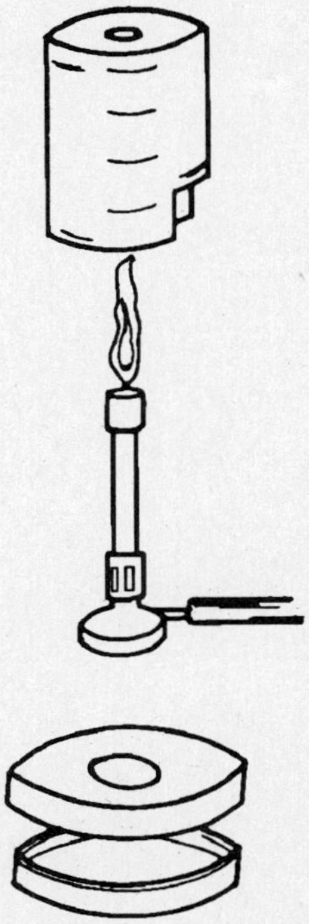

The material you need for this demonstration is the following:

- A one-pound coffee can, one end open with a 4 cm x 4 cm notch cut into the side and a 4-cm round hole cut into the other end.

- A laboratory burner, set for medium hot flame.
- A 14-cm-diameter plastic petri dish, with cover; cut a 5- to 6-cm round hole into cover.
- 25 mL 1 M HCl (aq)
- 5-6 pieces mossy zinc
- Metal chloride salts, 3-4 g each. Possible salts are LiCl, NaCl, $SrCl_2$, $BaCl_2$, KCl, and $CaCl_2$.

To perform the demonstration, place 25 mL of 1 M HCl in the bottom of the petri dish. Add 3-4 g of one of the metal chloride salts. Mix well so that a majority of the salt dissolves, and then cover with the top of petri dish. Drop 5-6 pieces of mossy zinc through the hole in the top of the petri dish, into the acid solution, and quickly place the lit burner over the hole. Place the coffee can over the burner, allowing it to rest on the petri dish.

The zinc reacts with the acid to evolve hydrogen gas. As the gas escapes, it carries with it some of the metal ions. The gas and ions are drawn into the air ports of the burner, and the representative color of the ion is imparted to the flame. The effect will last several minutes and allows viewing through a diffraction grating, student spectroscope, or cobalt glass.

We have also found that it works very well to simply place a piece of alkali or alkaline earth metal (such as Na, K, or Ca) in water in the petri dish. The flame colors are even more intense and longer-lasting.

## SOLUTIONS TO SELECTED STUDY QUESTIONS

22-1. (a) $2\ K(s) + H_2(g) \rightarrow 2\ KH(s)$

(b) $Cl_2(g) + H_2(g) \rightarrow 2\ HCl(g)$

(c) $S_8(s) + 8\ H_2(g) \rightarrow 8\ H_2S(g)$

(d) $C_3H_8(g) + 3\ H_2O(g) \rightarrow 7\ H_2(g) + 3\ CO(g)$

(e) $2\ K(s) + 2\ H_2O(\ell) \rightarrow 2\ KOH(aq) + H_2(g)$

22-2. Formation of sulfuric acid and hydrogen iodide

$$SO_2 + 2\ H_2O + I_2 \rightarrow 2\ HI + H_2SO_4$$

Decomposition of sulfuric acid

$$H_2SO_4 + \text{heat} \rightarrow H_2O + SO_2 + \tfrac{1}{2}\ O_2$$

Decomposition of hydrogen iodide

$$2\ HI + \text{heat} \rightarrow H_2 + I_2$$

Net reaction

$$H_2O \rightarrow H_2 + \tfrac{1}{2}\ O_2$$

22-5. (a) $2\ NaCl(aq) + 2\ H_2O(\ell) \rightarrow 2\ NaOH(aq) + Cl_2(g) + H_2(g)$

(b) The expected mass ratio of products is

$$\frac{1\text{ mol } Cl_2}{2\text{ mol NaOH}} = \frac{71\text{ g } Cl_2}{80.\text{ g NaOH}} = 0.89$$

The actual mass ratio is

$$\frac{22.66\text{ billion pounds } Cl_2}{23.97\text{ billion pounds NaOH}} = 0.95$$

The fact that the $Cl_2$/NaOH ratio is larger than expected from the electrolysis of NaCl means that $Cl_2$ is manufactured in some additional way. Among these is the "Kel-Chlor" process, which uses waste HCl from other industrial processes:

$$4\,HCl(g) + 2\,NOCl(g) + O_2(g) \rightarrow 2\,NO(g) + 2\,H_2O(g) + 3\,Cl_2(g)$$

22-7. Fr has the configuration $[Rn]7s^1$ and that of Ra is $[Rn]7s^2$.

22-9. Magnesium is more readily oxidized than the other parts of the ship's hull. Therefore, Mg becomes the anode in an electrochemical cell, and the ship's hull is the cathode.

22-11. Water is of course not be to used on a sodium fire. An inert blanketing gas such as $N_2$ could be used.

22-12. Magnesium and beryllium find their greatest use in alloys. Mg in particular is used in lightweight castings in automobile engines and for aircraft parts. Ca is alloyed with Al to make bearing metal, and it is used as a reducing agent.

22-14. The aquated beryllium(II) ion.

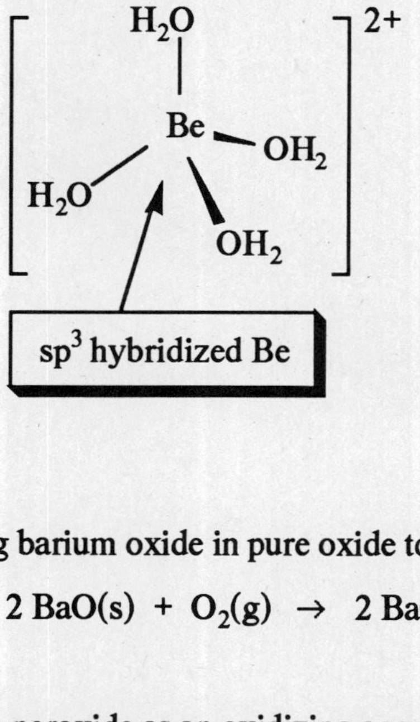


22-16. Heating barium oxide in pure oxide to give the peroxide:

$$2\,BaO(s) + O_2(g) \rightarrow 2\,BaO_2(s)$$

22-17. Barium peroxide as an oxidizing agent:

$$3\,BaO_2(s) + 2\,Fe(s) \rightarrow Fe_2O_3(s) + 3\,BaO(s)$$

22-18. Reaction of the Group 1A elements with oxygen:

*Formation of oxide:*

$$2\ H_2(g) + O_2(g) \rightarrow 2\ H_2O(g)$$

$$4\ Li(s) + O_2(g) \rightarrow 2\ Li_2O(s)$$

*Formation of Peroxide:*

$$2\ Na(s) + O_2(g) \rightarrow Na_2O_2(s)$$

*Formation of Superoxides:*

$$K(s) + O_2(g) \rightarrow KO_2(s)$$

$$Rb(s) + O_2(g) \rightarrow RbO_2(s)$$

$$Cs(s) + O_2(g) \rightarrow CsO_2(s)$$

22-19. $2\ MgO \cdot CaO(s) + Si(s) \rightarrow 2\ Mg(s) + Ca_2SiO_4(s)$

22-20. (a) $SrCO_3(s) + 2\ HCl(aq) \rightarrow SrCl_2(aq) + H_2O(\ell) + CO_2(g)$

(b) $BaCO_3(s) + 2\ HNO_3(aq) \rightarrow Ba(NO_3)_2(aq) + H_2O(\ell) + CO_2(g)$

(c) $3\ BaO(s) + 2\ Al(s) \rightarrow 3\ Ba(s) + Al_2O_3(s)$

(d) $LiH(s) + H_2O(\ell) \rightarrow LiOH(aq) + H_2(g)$

(e) $Na_2CO_3(aq) + H_2O(\ell) + CO_2(g) \rightarrow 2\ NaHCO_3(aq)$

22-22. (a) $Ca(s) + H_2(g) \rightarrow CaH_2(s)$

(b) $CaH_2(s) + 2\ H_2O(\ell) \rightarrow Ca(OH)_2(s) + H_2(g)$

22-25. All energies in the solution below are in kJ/mol.

| | $MgCO_3(s)$ | $\rightarrow$ | $MgO(s)$ | + | $CO_2(g)$ |
|---|---|---|---|---|---|
| $\Delta G^\circ_f$ | -1012.1 | | -569.43 | | -394.359 |

$$\Delta G^\circ_{rxn} = (-569.43\text{ kJ}) + (-394.359\text{ kJ}) - (-1012.1\text{ kJ}) = 48.3\text{ kJ}$$

$$CaCO_3(s) \rightarrow CaO(s) + CO_2(g)$$

$\Delta G^\circ_f$ -1128.79 -604.03 -394.359

$\Delta G^\circ_{rxn}$ = (-604.03 kJ) + (-394.359 kJ) - (-1128.79 kJ) = 130.40 kJ

$$BaCO_3(s) \rightarrow BaO(s) + CO_2(g)$$

$\Delta G^\circ_f$ -1137.6 -525.1 -394.359

$\Delta G^\circ_{rxn}$ = (-525.1 kJ) + (-394.359 kJ) - (-1137.6 kJ) = 218.1 kJ

The tendency for the Group 2A metal carbonates to decompose decreases as one descends the group.

22-27. Limestone and lime are used in steelmaking, to make calcium carbide, in water treatment, in pollution control, and in the paper industry.

$CaCO_3(s) + CO_2(g) + H_2O(\ell) \rightleftarrows Ca^{2+}(aq) + 2\ HCO_3^-(aq)$

Some uses of this mineral are described on page 112 of the text and are outlined by K. W. Watkins, *J. Chem. Educ.* **1983**, *60*, 60.

22-28. Hard water contains $Ca^{2+}$ and $Mg^{2+}$ ions and perhaps $CO_2$. One method for softening hard water is outlined on page 993.

22-29. $Na^+$, $K^+$, and $Ca^{2+}$ are important in controlling nerve impulses. $Mg^{2+}$ is the central element in chlorophyll, and $Ca^{2+}$ is found in teeth and bones. Lithium salts are used to treat manic depression.

# CHAPTER 23

# METALS, METALLOIDS, AND NONMETALS

## INSTRUCTOR'S NOTES

See the comments in this section in Chapter 22. A very useful source of information concerning the chemistry of this group is also found in *Industrial Inorganic Chemistry* by W. Büchner, R. Schliebs, G. Winter, and K.H. Büchel (VCH, 1989).

## LIST OF TRANSPARENCIES FOR THIS CHAPTER

| No. | Figure | Page | Subject |
|---|---|---|---|
| 106 | 23.2 | 943 | Boron icosahedron |
| 107 | 23.6 | 946 | Solid boric acid |
| 108 | 23.11 | 952 | Bonding in $B_2H_6$ |
| 109 | 23.20 | 959 | Cristobalite |
| 110 | 23.22 | 963 | Zeolite structure |

## SUGGESTED DEMONSTRATIONS

The following demonstrations of the properties of Group 3A elements have been used in our descriptive inorganic chemistry course:

- When boron was discovered in the Nevada and California deserts, prospectors were instructed that they could confirm a mineral containing boron if it "burned green." This can be duplicated in the following way: Place some sulfuric acid on a few grams of borax in an evaporating dish; the product is boric acid. Next, add methyl alcohol to convert the acid to trimethylborate, $B(OCH_3)_3$. Ignite the methyl alcohol, and the alcohol will burn with a green flame. Alternatively, just pour methyl alcohol onto boric acid and ignite it to observe the green flame.

- There are a number of consumer products containing boron that can be exhibited in class: borax and boric acid; Pyrex or borosilicate glass; or a detergent containing a perborate oxidizing agent.

- The formation of $Al(OH)_3$ and its amphoterism (Figure 17.9, page 737).
- The absorption of dyes by aluminum hydroxide and alumina (Figure 23.9, page 949).
- The nonreactivity of aluminum with an oxidizing acid (Figure 23.5, page 945). This can be contrasted with the reaction of Al with $CuCl_2$. See Figure 1.8, page 13, and the comments for Chapter 21 in this *Manual*.
- The thermite reaction (see Figure 20.9, page 837). Directions for this reaction are given in Shakhashiri, Volume 1. We do the reaction by piling a small quantity of the reactants on a bed of sand on the demonstration table. The students are kept back about 20 feet.
- Aluminum chloride is a very important industrial chemical; almost 25000 tons are produced annually in the United States. It is very useful to show that this halide is unstable toward water, in contrast with Group 1A and 2A halides. Use $AlCl_3$ from a newly opened bottle of $AlCl_3$ and add a spatula-full to 10-20 mL of water.
- The reaction of Al with $Br_2$ is spectacular (page 106). However, it can only be done in a lecture room that has an excellent hood.

Group 4A chemistry is very extensive, so there are many demonstrations that may be done.

- A major point to make is the difference between graphite and diamond. Models help here enormously, and graphite samples are easily obtained.
- The Special Section on "Lead Chemistry" (text page 964) was taken from a demonstration in Shakhashiri, Volume 1 (page 286).
- Silicon and silicon-based minerals are extremely important in our economy. It would be well to exhibit as many examples as possible. These include quartz, amethyst, various clays (including "kitty litter), zeolites, mica, talc, vermiculite, and (with proper precautions) asbestos. Other silicon-containing materials are silicon carbide, glass, and cement.
- The essay on *Asbestos Chemistry* on pages 968-969 reflects the current problems over the use of this mineral. An excellent discussion about the controversy that has arisen following the article by Mossman, to which we refer on page 969, appeared in the *New York Times Magazine*, November 25, 1990, page 38 ("The Asbestos Mess"). This article, which is highly recommended, is a very interesting discussing of the clash between social policy and science.
- Graphite fibers are being used in many consumer products. It would be interesting to show the students some of these products.

## SOLUTIONS TO SELECTED STUDY QUESTIONS

23-1. (a) $4\ Ga(s) + 3\ O_2(g) \rightarrow 2\ Ga_2O_3(s)$

(b) $2\ In(s) + 3\ Br_2(\ell) \rightarrow 2\ InBr_3(s)$

(c) $2\ BBr_3(g) + 3\ H_2(g) \rightarrow 2\ B(s) + 6\ HBr(g)$

(d) $B(OH)_3(s) + 3\ CH_3OH(\ell) \rightarrow B(OCH_3)_3(\ell) + 3\ H_2O(\ell)$

(e) $BF_3(g) + 3\ H_2O(\ell) \rightarrow B(OH)_3(s) + 3\ HF(g)$

(f) $B_2O_3(s) + 6\ HF(g) \rightarrow 2\ BF_3(g) + 3\ H_2O(\ell)$

(g) $2\ NaBH_4(s) + I_2(s) \rightarrow B_2H_6(g) + 2\ NaI(s) + H_2(g)$

(h) $2\ B_5H_9(g) + 12\ O_2(g) \rightarrow 5\ B_2O_3(s) + 9\ H_2O(g)$

23-2. (a) $SnCl_4(\ell) + 2\ H_2O(\ell) \rightarrow SnO_2(s) + 4\ HCl(g)$

(b) $2\ Pb(s) + O_2(g) \rightarrow 2\ PbO(s)$

(c) $Ge(s) + O_2(g) \rightarrow GeO_2(s)$

(d) $Sn(s) + Cl_2(g) \rightarrow SnCl_2(s)$

(e) $GeCl_4(\ell) + 4\ NaH(s) \rightarrow GeH_4(g) + 4\ NaCl(s)$

(f) $SiH_4(g) + 2\ O_2(g) \rightarrow SiO_2(s) + 2\ H_2O(g)$

23-3. (a) $SiO_2(s) + 3\ C(s) \rightarrow SiC(s) + 2\ CO(g)$

(b) $Si(s) + 2\ Cl_2(g) \rightarrow SiCl_4(\ell)$

(c) $2\ PbS(s) + 3\ O_2(g) \rightarrow 2\ PbO(s) + 2\ SO_2(g)$

(d) $PbO(s) + CO(g) \rightarrow Pb(s) + CO_2(g)$

(e) $CCl_4(\ell) + HF(\ell) + \text{catalyst} \rightarrow CFCl_3(g) + HCl(g)$

(f) $Si(s) + 2\ CH_3Cl(g) + \text{catalyst} \rightarrow (CH_3)_2SiCl_2(\ell)$

23-4. (a) $2\ NaBH_4(s) + I_2(s) \rightarrow B_2H_6(g) + 2\ NaI(s) + H_2(g)$

(b) $4\ In(s) + 3\ O_2(g) \rightarrow 2\ In_2O_3(s)$

(c) $B_2O_3(s) + 3\ CaF_2(g) + 3\ H_2SO_4(\ell) \rightarrow 2\ BF_3(g) + 3\ H_2O(\ell) + 3\ CaSO_4(s)$

(d) $Al_2O_3(s) + 3\ C(s) + 3\ Cl_2(g) \rightarrow 2\ AlCl_3(s) + 3\ CO(g)$

(e) $GeO_2(s) + 2\ C(s) \rightarrow Ge(s) + 2\ CO(g)$

$Ge(s) + 2\ Cl_2(g) \rightarrow GeCl_4(\ell)$

(f) $NiO(s) + H_2(g) + 4\ CO(g) \rightarrow Ni(CO)_4(g) + H_2O(g)$

(g) $SiO_2(s) + 3\ C(s) \rightarrow SiC(s) + 2\ CO(g)$

(h) $CaCO_3(s) \rightarrow CaO(s) + CO_2(g)$

$2\ CaO(s) + 5\ C(s) \rightarrow 2\ CaC_2(s) + CO_2(s)$

(i) $CaC_2(s) + N_2(g) \rightarrow CaCN_2(s) + C(s)$

$CaCN_2(s) + CO_2(g) + H_2O(\ell) \rightarrow CaCO_3(s) + NH_2CN(s)$

23-5.

| COMPOUND | USES |
|---|---|
| cyanamide | fertilizer |
| hydrogen cyanide | polymers, metal complexes |
| boron carbide | bullet proof armor |
| aluminum sulfate | paper industry, water purification |
| boric acid | antiseptic |
| sodium borohydride | reducing agent |
| carbon black | reinforcing agents in rubber, printing ink pigment |

23-6. $B_3O_6^{3-}$ ion $B_2O_5^{4-}$ ion

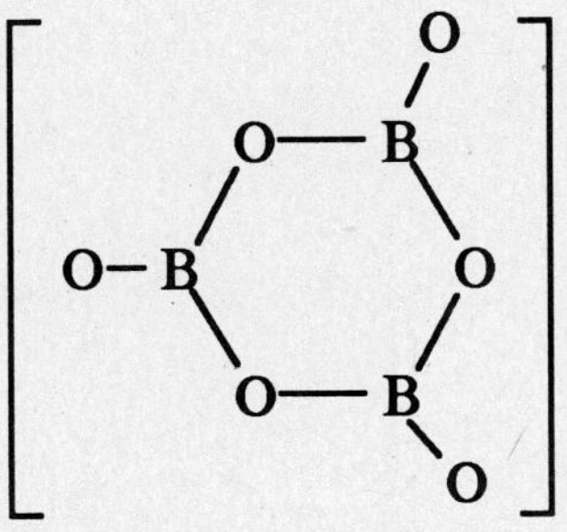


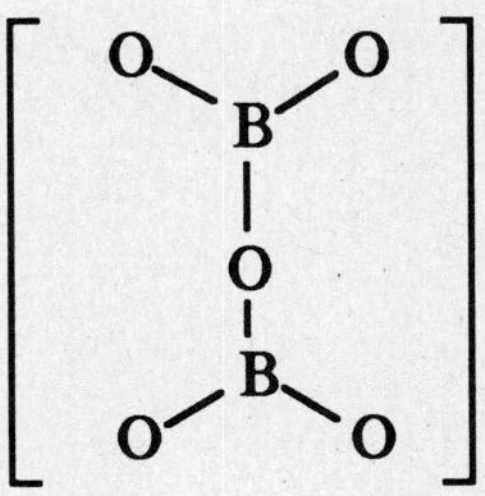


23-8. $2\ B_5H_9(g) + 12\ O_2(g) \rightarrow 5\ B_2O_3(s) + 9\ H_2O(g)$

$\Delta H^\circ_{rxn}$ = 5 mol (-1254.5 kJ/mol) + 9 mol (-241.818 kJ/mol) - [(2 mol)(54 kJ/mol)]

= -8557 kJ

Therefore, $\Delta H^\circ_{rxn}$ per mole of $B_5H_9(g)$ = -4278 kJ

| **COMPOUND** | $\Delta H^\circ_{rxn}$ |
|---|---|
| $B_5H_9(g)$ | -4278 kJ/mol |
| $B_2H_6(g)$ | -2016 kJ/mol |
| $C_2H_6(g)$ | -1428 kJ/mol |

The compound containing 5 boron atoms has a much higher enthalpy change for the combustion reaction than either diborane or ethane.

23-10. $2\ BF_3(g) + 6\ NaH(s) \rightarrow B_2H_6(g) + 6\ NaF(s)$

23-12. *Part (a)*

$2\ NaBH_4(s) + 6\ H_2O(\ell) + 2\ AgNO_3(aq)$

$\rightarrow 2\ Ag(s) + 7\ H_2(g) + 2\ B(OH)_3(aq) + 2\ NaNO_3(aq)$

*Part (b)*

0.575 L • 0.011 M $AgNO_3$ = 0.0063 mol $AgNO_3$

13.0 g $NaBH_4$ • 1 mol/ 37.83 g = 0.344 mol $NaBH_4$

Since the required ratio of moles of $AgNO_3$ to moles of $NaBH_4$ is 1:1, it is evident that $AgNO_3$ is the limiting reagent.

$$0.0063 \text{ mol } AgNO_3 \cdot 2 \text{ mol Ag}/2 \text{ mol } AgNO_3 \cdot 108 \text{ g/mol} = 0.68 \text{ g Ag}$$

23-14. In the reaction in which sand is converted to silicon carbide, the materials of interest may be represented as:

$$SiO_2(s) + 3\,C(s) \rightarrow SiC(s) + 2\,CO(g)$$

$$1.0 \times 10^5 \text{ T} \cdot 1 \text{ mol SiC}/40. \text{ T} \cdot 1 \text{ mol } SiO_2/1 \text{ mol SiC} \cdot 60. \text{ T } SiO_2/\text{mol}$$

$$= 1.5 \times 10^5 \text{ T } SiO_2$$

Only 70% of the $SiO_2$ is converted to SiC. Therefore, we need to increase the initial amount of $SiO_2$ by an appropriate amount.

$$1.5 \times 10^5 \text{ T } SiO_2 \cdot 100/70 = 2.1 \times 10^5 \text{ T } SiO_2$$

23-16. The surface of metallic aluminum is covered with an oxide coating that prevents further reaction. If the oxide coating is abraded, the metal readily reduces other metals (until the protective oxide coating is reformed).

23-18. If calcium ions occupy the sites of a face-centered lattice, the carbide ions are in *octahedral holes*, each being adjacent to six nearest neighbors. The unit cell would have:

| | | |
|---|---|---|
| $Ca^{2+}$ ions: | 8 corners • 1/8 ion per corner | = 1 net ion |
| | 6 faces • 1/2 ion per face | = 3 net ions |
| | ------------------------------------------------ | |
| | 4 net $Ca^{2+}$ ions per unit cell | |
| $C_2^{2-}$ ions: | 12 edges • 1/4 ion per edge | = 3 net ions |
| | 1 ion in the center | = 1 ion |
| | ------------------------------------------------ | |
| | 4 net carbide ions per unit cell | |

Since there are 4 net ions of each type in a unit cell, the lattice stoichiometry is 1:1 or $CaC_2$.

23-20. (a) $Al_2O_3(s) + 3\ SiO_2(s) \rightarrow Al_2(SiO_3)_3(s)$

(b) $Al_2O_3(s) + CaO(s) \rightarrow Ca(AlO_2)_2(s)$

23-22. (a) $CaC_2(s) + 2\ H_2O(\ell) \rightarrow Ca(OH)_2(s) + C_2H_2(g)$

(b) Dot structure of carbide ion $[\ :C \equiv C:\ ]^{2-}$

(c) $1.35\ g\ CaC_2 \cdot 1\ mol/64.10\ g = 0.0211\ mol\ CaC_2$

$0.0211\ mol\ CaC_2 \cdot 1\ mol\ C_2H_2 / 1\ mol\ CaC_2 = 0.0211\ mol\ C_2H_2$

$$P = nRT/V = \frac{(0.0211\ mol\ C_2H_2)(0.082057\ L \cdot atm/K \cdot mol)(294.7\ K)}{(0.865\ L)}$$

$P = 0.589\ atm\ (447\ mmHg)$

23-24. Reaction of $BF_3$ and $NH_3$

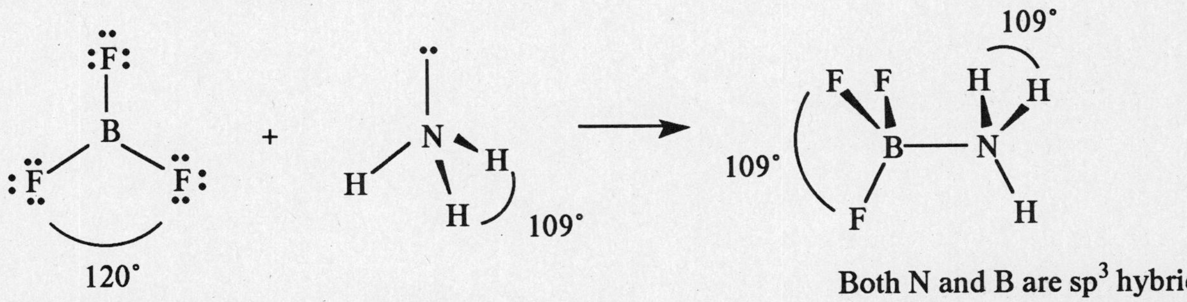


$sp^2$ hybridized B

$sp^3$ hybridized N

Both N and B are $sp^3$ hybridized.
LP electrons on F are not shown.

23-26. Given below is a plot of the standard reduction potentials for Group 3A elements. The diagram shows there is a general decline in the reducing ability of the metal on descending the group, just as was observed for the Group 1A metals (pp. 919-920). It is noteworthy, however, that boron does not conform to this trend. If we examine the factors used in analyzing the Group 1A values on pages 919-920, we find that the ionization energy of boron is considerably higher than that of the metals of Group 3A. Although other factors are surely involved (for example, a very high enthalpy of hydration of $Al^{3+}$), the ionization energy factor could provide the explanation for the seemingly anomalous behavior of boron.

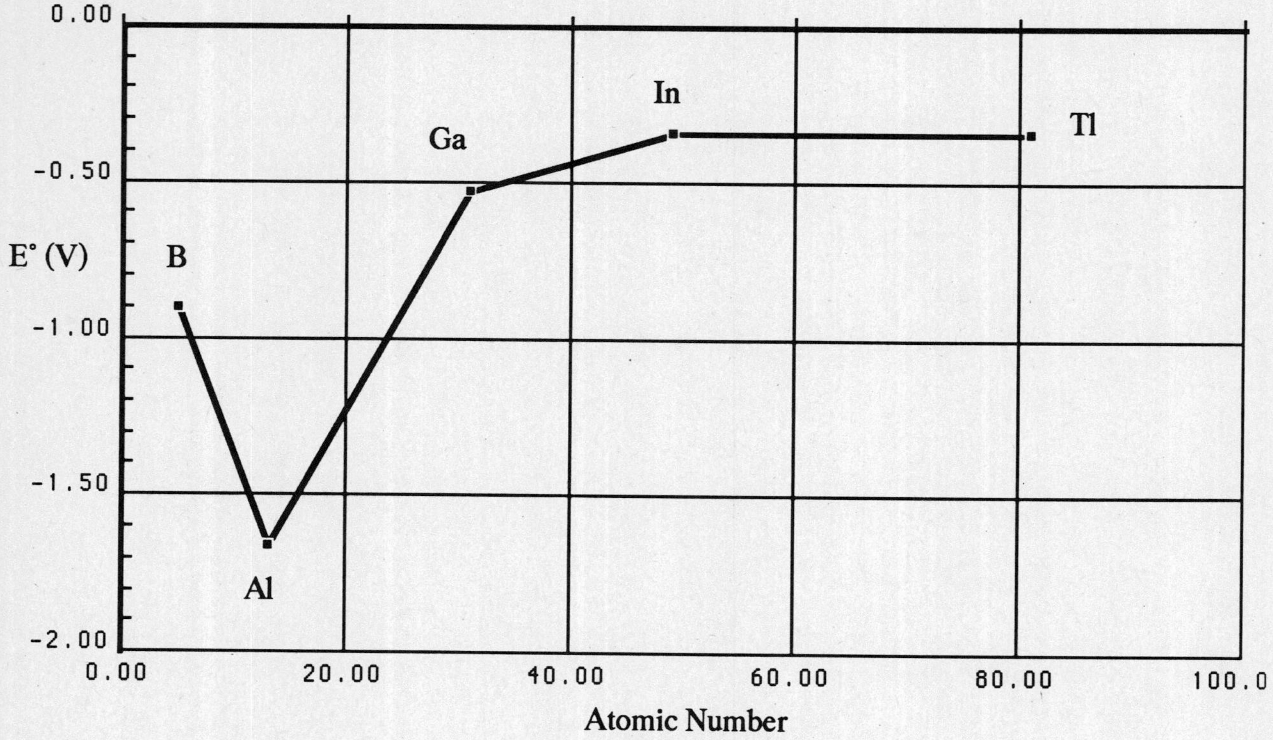


23-28. (a) $Sn^{4+}$ ions:

| | | |
|---|---|---|
| 8 corners • 1/8 ion per corner | = | 1 net ion |
| 1 ion in the center | = | 1 net ion |

2 net $Sn^{4+}$ ions per unit cell

$O^{2-}$ ions:

| | | |
|---|---|---|
| 2 faces • 1/2 ion per face | = | 2 net ions |
| 2 ions wholly in the lattice | = | 2 ion |

4 net oxide ions per unit cell

Since there are two net $Sn^{4+}$ ions and four net oxide ions in a unit cell, the lattice stoichiometry is 1:2 or $SnO_2$.

(b) In this calculation, we shall assume that water is produced as liquid water

$$SnO_2(s) + 4\,HCl(g) \rightarrow SnCl_4(\ell) + 2\,H_2O(\ell)$$

$\Delta G^\circ_f$ -519.6 -95.299 -440.1 -237.129 kJ/mol

$$\Delta G^\circ_{rxn} = 2\text{ mol }(-237.129\text{ kJ/mol}) + (-440.1\text{ kJ/mol}) - [1\text{ mol }(-519.6\text{ kJ/mol}) + 4\text{ mol }(-95.299\text{ kJ/mol})]$$

$$= -13.6\text{ kJ}$$

$\Delta G°_{rxn} = -RT \ln K$

$-13.6 \times 10^3 \text{ J} = -(8.314510 \text{ J/K} \cdot \text{mol})(298) \ln K$

$\ln K = 5.47$

$K = 240$

Yes, it should be feasible to convert $SnO_2$ to liquid $SnCl_4$.

23-30. Kilograms of tin required:

$$1.00 \text{ kg} \cdot \frac{1 \text{ mol } (C_4H_9)_2SnCl_2}{303.8 \text{ kg}} \cdot \frac{1 \text{ mol Sn}}{1 \text{ mol } (C_4H_9)_2SnCl_2} \cdot \frac{118.7 \text{ kg}}{1 \text{ mol Sn}} \cdot \frac{100}{40}$$

$= 0.977$ kg Sn

Kilograms of butyl chloride, $C_4H_9Cl$

$$1.00 \text{ kg} \cdot \frac{1 \text{ mol } (C_4H_9)_2SnCl_2}{303.8 \text{ kg}} \cdot \frac{2 \text{ mol } C_4H_9Cl}{1 \text{ mol } (C_4H_9)_2SnCl_2} \cdot \frac{92.57 \text{ kg}}{1 \text{ mol Sn}} \cdot \frac{100}{40}$$

$= 1.52$ kg $C_4H_9Cl$

23-32. $Al_2O_3(s) + 3\, H_2SO_4(\ell) \rightarrow Al_2(SO_4)_3(aq) + 3\, H_2O(\ell)$

Kilograms of aluminum oxide required:

$$1.00 \text{ kg } Al_2(SO_4)_3 \cdot \frac{1 \text{ mol } Al_2(SO_4)_3}{342.1 \text{ kg}} \cdot \frac{1 \text{ mol } Al_2O_3}{1 \text{ mol } Al_2(SO_4)_3} \cdot \frac{102.1 \text{ kg}}{1 \text{ mol } Al_2O_3}$$

$= 0.298$ kg $Al_2O_3$

Kilograms of sulfuric acid required:

$$1.00 \text{ kg } Al_2(SO_4)_3 \cdot \frac{1 \text{ mol } Al_2(SO_4)_3}{342.1 \text{ kg}} \cdot \frac{3 \text{ mol } H_2SO_4}{1 \text{ mol } Al_2(SO_4)_3} \cdot \frac{98.07 \text{ kg}}{1 \text{ mol } H_2SO_4}$$

$= 0.860$ kg $H_2SO_4$

23-34. Orthosilicate — olivine

Amphibole — asbestos

Sheet silicate — mica

Aluminosilicate — clays, feldspars, zeolites

23-36. The dissociation of the ion may be written:

$[Ag(CN)_2]^-(aq) \rightleftarrows Ag^+(aq) + 2\,CN^-(aq)$

The equilibrium constant for this dissociation, $K_{diss}$, is the reciprocal of the formation constant, $K_{form}$.

$$K_{diss} = \frac{[Ag^+]\,[CN^-]^2}{\{[Ag(CN)_2]^-\}} = \frac{1}{K_{form}} = 1.8 \times 10^{-19}$$

The magnitude of the constant suggests that very little of the complex ion dissociates. The equilibrium concentrations of the various species in solution are:

$[Ag+] = x$ $[CN-] = 2x$ $\{[Ag(CN)_2]^-\} = 0.010 - x$

$$K_{diss} = 1.8 \times 10^{-19} = \frac{(x)\,(2x)^2}{0.010 - x}$$

If we make the reasonable assumption that $x$ is very small relative to 0.010, then

$1.8 \times 10^{-19} = (x)\,(2x)^2 / 0.010$

and $x = 7.7 \times 10^{-8}$ M

Therefore, $[Ag+] = 7.7 \times 10^{-8}$ M, $[CN-] = 1.5 \times 10^{-8}$ M, and $\{[Ag(CN)_2]^-\}$ = 0.010 M.

# CHAPTER 24
# THE CHEMISTRY OF THE NONMETALS

## INSTRUCTOR'S NOTES

See comments on this section of the book at the beginning of Chapter 22.

## LIST OF TRANSPARENCIES FOR THIS CHAPTER

| No. | Figure | Page | Subject |
|---|---|---|---|
| 111 | 24.1 | 977 | 14 chemicals produced in the largest amounts |
| 112 | 24.2 | 978 | Furnace for phosphorus production |
| 113 | 24.3 | 978 | Uses of phosphate rock |
| 114 | --- | 989 | Phosphorus oxides and acids |
| 115 | Table 24.5 | 991 | Phosphorus oxyacids |
| 116 | 24.12 | 999 | Products made from sulfur |
| 117 | --- | 1004 | Fate of CFC's in the atmosphere; ozone chemistry |
| 118 | 24.15 | 1007 | Production of sulfuric acid |
| 119 | Table 24.7 | 1008 | Some oxyacids of sulfur |

## SUGGESTED DEMONSTRATIONS

There are a number of demonstrations in Shakhashiri's volumes that can be used to illustrate the chemistry of Groups 5A-7A. They are listed below according to the appropriate group of the periodic table.Experiments listed with an asterisk have been photographed for the book and comments are given below. Many of the others appear in the videotape of demonstrations done by Professor Shakhashiri and his colleagues.

| Experiment | Page/Volume |
|---|---|
| *Group 5A* | |
| Combustion under water ($H_2SO_4$ is added to $KClO_3$ and white P) | 40/1 |
| Reaction of Zn with $NH_4NO_3$ and $NH_4Cl$ | 51/1* |
| Reaction of Sb and $Cl_2$ | 64/1 |
| Reaction of white P and $Cl_2$ | 70/1 |
| Reaction of red P and $Br_2$ | 72/1* |
| Combustion of white P | 74/1 |
| Decomposition of $(NH_4)_2Cr_2O_7$ | 81/1* |
| Explosion of $NI_3$ | 96/1* |
| Explosive reactions of allotropes of P and $KClO_3$ | 99/1 |
| Explosion of NO and $CS_2$ | 117/1 |
| Reactions between Sb(III) and $Cl_2$ | 324/1 |
| Preparation and properties of NO | 163/2 |
| $NO_2/N_2O_4$ equilibrium | 180/2* |
| Preparation and properties of $NH_3$ | 202/2 |
| Catalytic oxidation of $NH_3$ | 214/2 |
| *Group 6A* | |
| Heat of Dilution of $H_2SO_4$ | 17/1 |
| Crystallization of $Na_2S_2O_3$ | 31/1 |
| Reaction of Zn and S | 53/1 |
| Reaction of Fe and S | 55/1 |
| Reaction of $Na_2O_2$ and S | 57/1 |
| Dehydration of sugar by $H_2SO_4$ | 77/1* |
| Preparation and properties of $O_2$ | 137/2 |
| $O_2$ and dextrose: the "blue-bottle reaction | 142/2 |
| Preparation and properties of liquid $O_2$ | 147/2 |
| Preparation and properties of $SO_2$ | 184/2 |

*Group 7A*

| | |
|---|---|
| Reaction of Zn and $I_2$ | 49/1* |
| Reaction of Na and $Cl_2$ | 61/1 |
| Reaction of Fe and $Cl_2$ | 66/1 |
| Reaction of Al and $Br_2$ (see page 106 of the text)* | |
| Photochemical reaction of $H_2$ and $Cl_2$ | 121/1 |
| Preparation and properties of HCl | 211/2 |
| Preparation and properties of $Cl_2$ | 220/2 |

Some of the reactions above have been photographed for this book and these and others have been used in classroom demonstrations. Some comments may be helpful.

- Reaction of Zn with $NH_4NO_3$ and $NH_4Cl$. Photographs of this reaction appear on pages 234 and 988 of the text. Although we have done this in the classroom, we find that copious fumes of ZnO are emitted and that it is probably not safe unless it is done in a hood.
- Reaction of P with $Br_2$. The photograph appears on page 70 of the text as with any reaction involving $Br_2$, this *must* be done in a hood.
- Decomposition of ammonium dichromate; the "dichromate volcano." This reaction gives of copious amounts of Cr203. However, the experimental design in Shakhashiri confines the powder very well and is shown in Figure 4.8 (page 111)
- Explosion of $NI_3$. Photograph appears on text page 995. This experiment deserves respect. When we were making the $NI_3$ for the photograph, we spread a small amount onto a filter paper and placed the beaker with the remainder of the moist, solid $NI_3$ in the sink. On spraying water into the beaker the $NI_3$ exploded noisily and spread $NI_3$ around the laboratory.
- $NO_2/N_2O_4$ equilibrium. Photograph appears on page 681.
- Dehydration of sugar with $H_2SO_4$. This is a good demonstration of this useful property of sulfuric acid. Photograph appears on page 1130.
- Reaction of Zn with $I_2$. A photograph of this reaction appears on text page 106. As is evident there, a large amount of $I_2$ vapor is evolved. Thus, it can only be done in a hood and not in

a lecture room.

- Reaction of Al with $Br_2$. As noted above, this can only be done in a very good hood. Photographs appear on page 106.

Other possible demonstrations are shown in photographs in Chapter 24. These include the hydrolysis of $P_2O_5$ (page 992), the properties of sulfur (page 1000), reaction of $Br_2$ and $I_2$ (page 1012), and the preparation of $Cl_2$ (page 1011) and $I_2$ (page 1013). All (except for the preparation of chlorine gas) can be performed in lecture with a minimum of preparation

## SOLUTIONS TO SELECTED STUDY QUESTIONS

24-1. *Haber:* $N_2(g) + 3\,H_2(g) \rightarrow 2\,NH_3(g)$ (450 °C, high pressure, catalyst)

*Marsh:* $Zn(s) + 2\,HCl(aq) \rightarrow Zn^{2+}(aq) + 2\,Cl^-(aq) + H_2(g)$

Arsenic containing material is reduced with $H_2$ to $AsH_3(g)$. When this is heated it decomposes to elemental As, which can then be deposited as a shiny mirror.

$2\,AsH_3(g) \rightarrow 2\,As(s) + 3\,H_2(g)$

*Raschig:* $2\,NH_3(aq) + NaOCl(aq) \rightarrow N_2H_4(aq) + NaCl(aq) + H_2O(\ell)$

Reaction is run under basic conditions and in the presence of gelatin.

*Contact Process:* $2\,SO_2(g) + O_2(g) \rightarrow 2\,SO_3(g)$ ($V_2O_5$ is used as a catalyst)

$SO_3(g) + H_2O(\ell) \rightarrow H_2SO_4(\ell)$

24-3. (a) $N_2(g) + 2\,O_2(g) \rightarrow 2\,NO_2(g)$

(b) $4\,HNO_3 + P_4O_{10} \rightarrow 2\,N_2O_5 + 4\,HPO_3$

(c) $NaNO_2(aq) + HCl(aq) \rightarrow NaCl(aq) + HNO_2(aq)$

(d) $H_3AsO_4(aq) + 4\,H_2(g) \rightarrow AsH_3(g) + 4\,H_2O(\ell)$

(e) $4\,NH_3(g) + 3\,F_2(g) + \text{catalyst} \rightarrow 3\,NH_4F(s) + NF_3(g)$

(f) $CCl_4(\ell) + F_2(g) \rightarrow CCl_2F_2(g) + Cl_2(g)$

(g) $NH_4NO_3(s) + \text{heat} \rightarrow N_2O(g) + 2\,H_2O(g)$

(h) $2\,NH_3(aq) + NaOCl(aq) \rightarrow N_2H_4(aq) + NaCl(aq) + H_2O(\ell)$

24-5.

$$3.00 \times 10^4\ \text{L} \cdot \frac{3.08\ \text{cm}^3\ O_2}{0.100\ \text{L}} \cdot \frac{1\ \text{mol}\ O_2}{22{,}400\ \text{cm}^3\ \text{at STP}} \cdot \frac{2\ \text{mol}\ Na_2SO_3}{1\ \text{mol}\ O_2} \cdot \frac{126\ \text{g}\ Na_2SO_3}{1\ \text{mol}\ Na_2SO_3} = 1.04 \times 10^4\ \text{g}\ Na_2SO_3$$

24-6. $NH_3(g) + 2\,O_2(g) \rightarrow HNO_3(aq) + H_2O(\ell)$

$\Delta G^\circ_f$ -16.45 0 -111.25 -237.129 kJ/mol

$$\Delta G^\circ_{rxn} = 1 \text{ mol} (-111.25 \text{ kJ/mol}) + 1 \text{ mol} (-237.129 \text{ kJ/mol}) - 1 \text{ mol} (-16.45 \text{ kJ/mol})$$

$$= -331.93 \text{ kJ/mol}$$

$$\Delta G^\circ_{rxn} = -331.93 \times 10^3 \text{ J} = -RT \ln K = -(8.314510 \text{ J/K} \cdot \text{mol})(298 \text{ K}) \ln K$$

$\ln K = 134$

$K = 1.52 \times 10^{58}$

24-9. Dot structure of the azide ion: 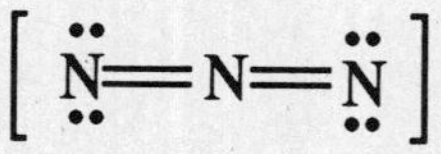

24-10. Cyanate ion, $NCO^-$, with formal charges indicated on each atom.

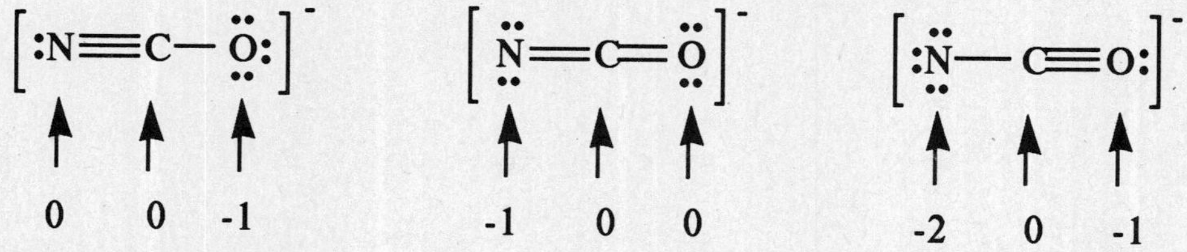


(most important resonance structure)

Fulminate ion, $CNO^-$, with formal charges indicated.

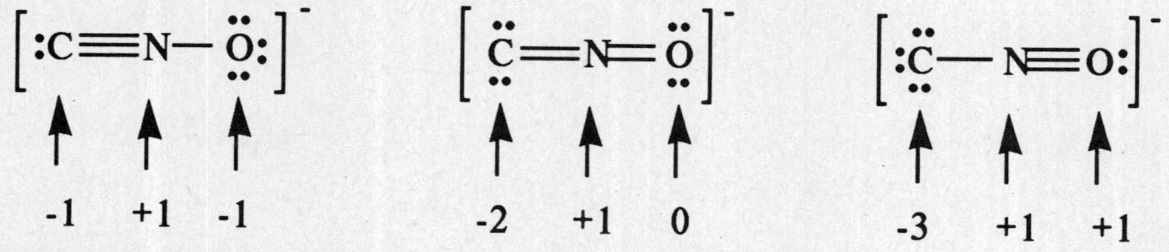


(most important resonance structure)

24-11. As the size and mass of the molecules increases, so does the boiling point (owing to increased dispersion forces). The anomalous boiling point of $NH_3$ is due to hydrogen bonding.

24-13. Nitric acid has three NO bonds.

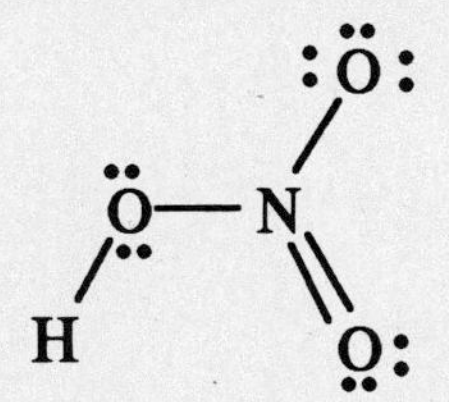

The terminal NO bonds have a bond order of $1\frac{1}{2}$ and are of the same length. The NO bond bearing the H atom is a single bond and is therefore expected to be longer than the other two bonds.

24-16. Diphosphonic acid

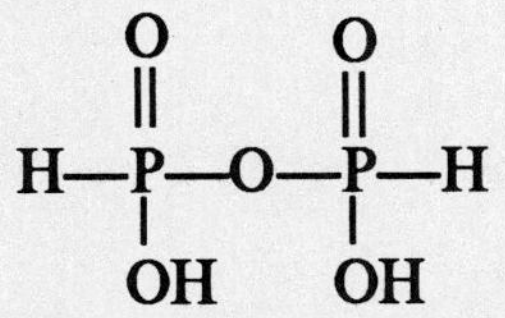

This has can lose *two* protons (the O—H hydrogens).

24-18. (a) Structures of $NH_3$ and $NF_3$. The direction of the molecular dipole is shown in each case.

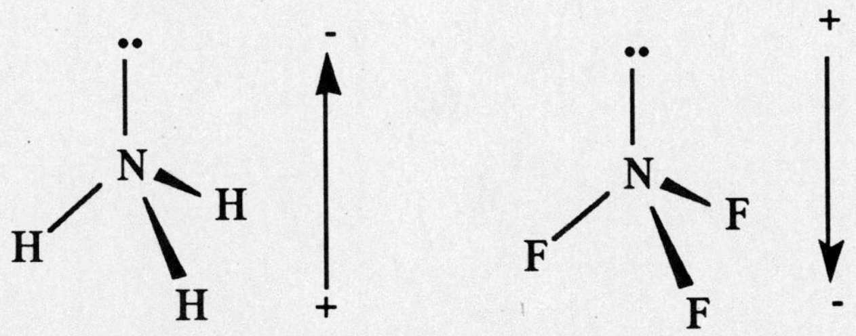

(b) The N—F bond dipole is in the opposite direction from the N—H bond dipole, and in the opposite direction from the lone pair dipole, so the dipole moment of $NF_3$ is considerably smaller than the $NH_3$ dipole.

24-20. $H_2PO_2^-$ as a reducing agent: **E° (V)**

$H_2PO_2^- + 3\,OH^- \rightarrow HPO_3^{2-} + 2\,H_2O + 2e^-$ -1.57

$Zn(OH)_2(s) + 2\,e^- \rightarrow Zn(s) + 2\,OH^-(aq)$ -1.22

$Cr(OH)_3(s) + 3\,e^- \rightarrow Cr(s) + 3\,OH^-(aq)$ -1.30

The $H_2PO_2^-$ ion should be capable of reducing salts of either $Cr^{3+}$ or $Zn^{2+}$ in basic solution. The net reactions would be

$3\,H_2PO_2^-(aq) + 2\,Cr(OH)_3(s) + 3\,OH^-(aq) \rightarrow$

$2\,Cr(s) + 3\,HPO_3^{2-}(aq) + 6\,H_2O(\ell)$ $E° = 0.27\ V$

$H_2PO_2^-(aq) + Zn(OH)_2(s) + OH^-(aq) \rightarrow$

$Zn(s) + HPO_3^{2-}(aq) + 2\,H_2O(\ell)$ $E° = 0.35\ V$

24-21. Structure of $As_4O_6$ and its hydrolysis product, $H_3AsO_3$.

$As_4O_6 + 6\,H_2O \rightarrow 4\,H_3AsO_3$.

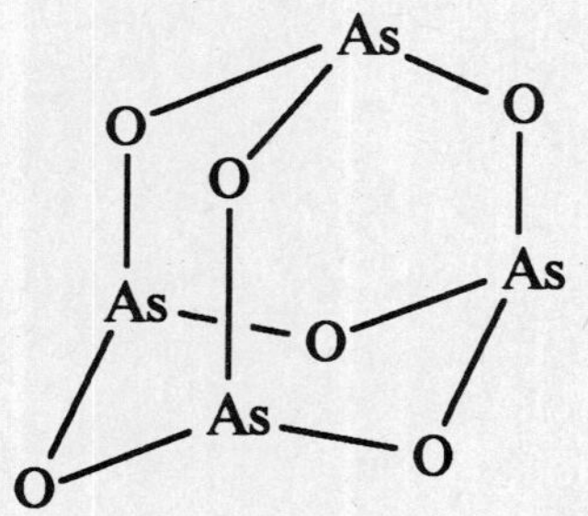


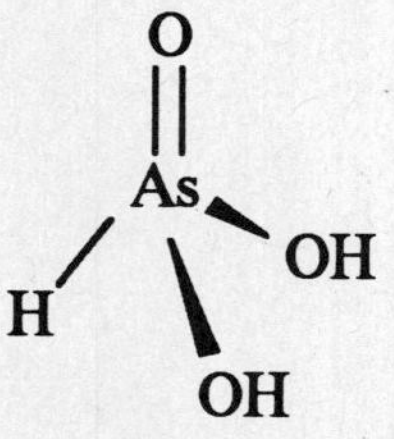


As oxidation number = +3

24-24. (a) $H_2S_2O_8 + 2\,H_2O \rightarrow 2\,H_2SO_4 + H_2O_2$

(b) $K(s) + O_2(g) \rightarrow KO_2(s)$

(c) $Na_2SO_3(s) + \frac{1}{8}\,S_8 \rightarrow Na_2S_2O_3(s)$

(d) See Study Question 24-1 above.

24-26. $O_2$ reacts with virtually all other elements largely because the energy of the bonds formed with O greatly outweighs the energy cost of breaking the O=O bond in $O_2$.

24-27. The $H_2O_2$ molecule is not planar owing to repulsions between OH bonds and the O atom lone pairs, as well as repulsions between the lone pairs on neighboring O atoms.

24-29. $\frac{1}{2}O_2(g) + F_2(g) \rightarrow OF_2(g)$ $\Delta H^\circ_{rxn} = \Delta H^\circ_f[OF_2(g)] = 24.5$ kJ/mol

$\Delta H^\circ_{rxn} = \frac{1}{2}D_{O=O} + D_{F—F} - 2D_{O—F}$

$24.5 \text{ kJ} = \frac{1}{2}(498 \text{ kJ}) + (159 \text{ kJ}) - 2D_{O—F}$

$D_{O—F} = 192$ kJ/mol

The reported O—F bond energy is 184 kJ/mol.

24-32. Dot structure of the 20-electron $S_3^{2-}$ ion:

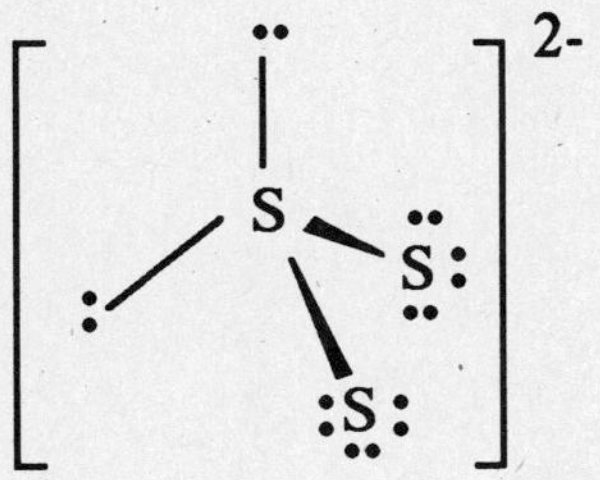


24-33. In all of the structures below, only S-atom lone pairs are shown.

(a) $S_2F_2$, 26 valence electrons. S oxidation number = +1

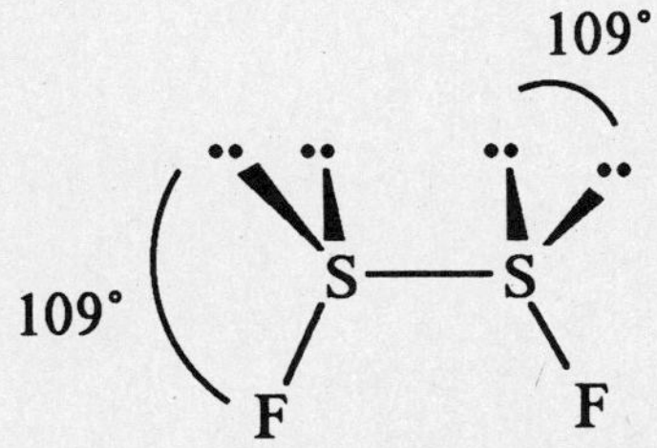


(b) $SF_2$, 20 valence electrons. S oxidation number = +2

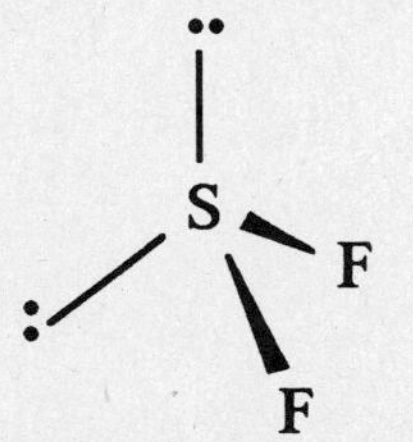


(c) $SF_4$, 34 valence electrons. S oxidation number = +4

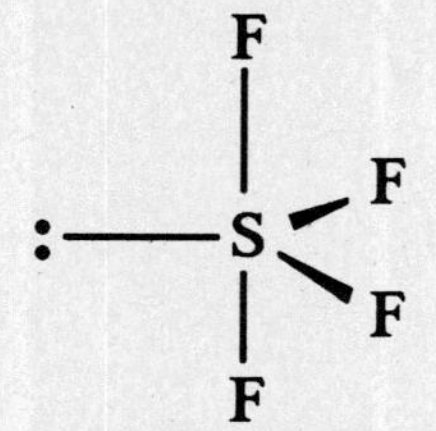

(d) $SF_6$, 48 valence electrons. S oxidation number = +6

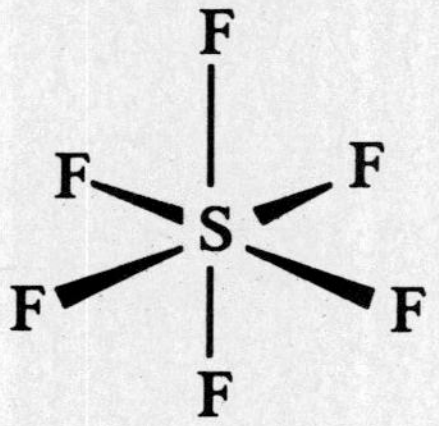

24-35. *Burning S to $SO_2$* $\Delta H^\circ_{rxn}$

$S(s) + O_2(g) \rightarrow SO_2(g)$ -296.830 kJ

*Oxidation of $SO_2$*

$SO_2(g) + \frac{1}{2} O_2(g) \rightarrow SO_3(g)$ -98.89 kJ

*Hydrolysis of $SO_3$*

$SO_3(g) + H_2O(\ell) \rightarrow H_2SO_4(\ell)$ -130 kJ

$\Sigma \Delta H^\circ_{rxn} = -526$ kJ

$$\frac{-526 \text{ kJ}}{1 \text{ mol } H_2SO_4} \cdot \frac{1 \text{ mol } H_2SO_4}{98.07 \text{ g}} \cdot \frac{9.08 \times 10^5 \text{ g}}{1.00 \text{ T}} = 4.87 \times 10^6 \text{ kJ/T}$$

24-37. (a) $CaF_2(s) + H_2SO_4(\ell) \rightarrow CaSO_4(s) + 2 HF(g)$

(b) $H_2(g) + Cl_2(g) \rightarrow 2 HCl(g)$

(c) $Xe(g) + 2 F_2(g) \rightarrow XeF_4(g)$

24-39. $Cl_2(aq) + 2 e^- \rightarrow 2 Cl^-(aq)$ $E^\circ = 1.36$ V

$2\,I^-(aq) \rightarrow I_2(s) + 2\,e^-$ $E^\circ = -0.535$ V

---

$Cl_2(aq) + 2\,I^-(aq) \rightarrow I_2(s) + 2\,Cl^-(aq)$ $E^\circ = 0.83$ V

24-40. (a) Mineral sources of fluoride:

| | |
|---|---|
| fluorspar | $CaF_2$ |
| fluoroapatite | $Ca_{10}(PO_4)_6F_2$ |
| cryolite | $Na_3AlF_6$ |

(b) Iodide ion course: brine wells

24-42. $5.00 \times 10^3$ T $F_2$ • 55 T $F_2$ used/100. T produced • 352 T $UF_6$ / 114 T $F_2$

= 8500 T $UF_6$

24-43. Since the F—F bond distance is so small, the nonbonding or lone pairs of electrons of one F atom repel those of the other atom and weaken the bond. The halogen-halogen bond in $Cl_2$, $Br_2$, and $I_2$ is longer, so this repulsive effect is lessened relative to $F_2$.

24-44. (a) The F—F bond is the weakest of the halogen-halogen bonds.

(b) Bond energies of fluorine attached to other elements are relatively large; that is, particularly stable bonds are formed.

24-45. The molecule is T-shaped. The bond angles will be reduced relative to their ideal values owing to the two lone pairs of electrons on the bromine atom.

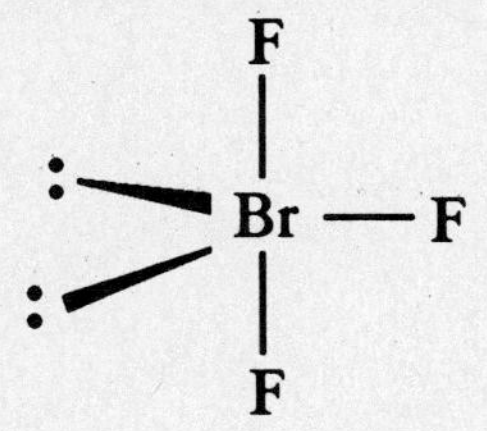


24-47. Lattice energy is one controlling factor in the melting points of metal halides and their solubilities (see 365-367 and 537-539). For a given metal ion, the lattice energy should decline as the halide ions become larger. (This is due to a weakened Coulombic force of attraction between $M^+$ and $X^-$.) Thus, the lattice is more easily disrupted as $X^-$ becomes larger. This means the solubility of the salt should increase, and the melting point should decrease, on going from chlorides to bromides to iodides.

24-49. $Mn^{2+}(aq) + 4\ H_2O(\ell) \rightarrow MnO_4^-(aq) + 8\ H^+(aq) + 5\ e^-$  $E° = -1.51\ V$

$BrO_3^-(aq) + 6\ H^+(aq) + 6\ e^- \rightarrow Br^-(aq) + 3\ H_2O(\ell)$  $E° = 1.44\ V$

------------------------------------------------------------------------ -------------------

$6\ Mn^{2+}(aq) + 9\ H_2O(\ell) + 5\ BrO_3^-(aq)$

$\rightarrow 6\ MnO_4^-(aq) + 18\ H^+(aq) + 5\ Br^-(aq)$  $E° = -0.07\ V$

The bromate ion should *not* spontaneously oxidize manganese(II) under standard conditions. (However, the equilibrium constant is not too small, so one can imagine that if the concentration of bromate is increased, the reaction can occur spontaneously.

24-50. $OCl^-(aq) + H_2O(\ell) \rightleftarrows HOCl(aq) + OH^-(aq)$

$$K_b = \frac{[HOCl]\,[OH^-]}{[OCl^-]} = \frac{1.0 \times 10^{-14}}{3.5 \times 10^{-8}} = 2.86 \times 10^{-7}$$

When $[OCl^-] = 0.10$ M, $[HOCl] = [OH^-] = 1.7 \times 10^{-4}$ M. Therefore, the pH = 10.23.

24-52. (a) $\Delta G°_{rxn} = \Delta G°_f\,[BaCl_2(s)] - 2\,\Delta G°_f\,[HCl(g)]$

$= (-810.4\ kJ) - (-190.598\ kJ)$

$= -619.8$ kJ, spontaneous

$\Delta G°_{rxn} = \Delta G°_f\,[PbCl_2(s)] - 2\,\Delta G°_f\,[HCl(g)]$

$= (-314.10\ kJ) - (-190.598\ kJ)$

$= -123.50$ kJ, spontaneous

$\Delta G°_{rxn} = \Delta G°_f\,[HgCl_2(s)] - 2\,\Delta G°_f\,[HCl(g)]$

$= (-178.6\ kJ) - (-190.598\ kJ)$

$= +12.0$ kJ, nonspontaneous

$\Delta G°_{rxn} = \Delta G°_f[TiCl_4(\ell)] - 4\,\Delta G°_f\,[HCl(g)]$

= (-737.2 kJ) - (-381.196 kJ)

= -356.0 kJ, spontaneous

(b) $\Delta G^\circ_{rxn} = \Delta G^\circ_f[CaF_2(s)] - 2\,\Delta G^\circ_f\,[HF(g)]$

= (-1167.3 kJ) - (-546.4 kJ)

= -620.9 kJ, HF should react spontaneously with Ca

(c) $\Delta G^\circ_{rxn} = \Delta G^\circ_f[NiF_2(s)] - 2\,\Delta G^\circ_f\,[HF(g)]$

= (-604 kJ) - (-546.4 kJ)

= -58 kJ, Ni should react spontaneously with HF

(d) $\Delta G^\circ_{rxn} = \Delta G^\circ_f[NiBr_2(s)] - 2\,\Delta G^\circ_f\,[HBr(g)]$

= (-253.6 kJ) - (-106.90 kJ)

= -146.7 kJ, Ni should react spontaneously with HBr

(e) $\Delta G^\circ_{rxn} = \Delta G^\circ_f[BI_3(g)] - 3\,\Delta G^\circ_f\,[HI(g)]$

= +20.7 kJ - (5.1 kJ)

= 15.6 kJ, B should not react spontaneously with HI

24-53. Structure of $XeF_4$

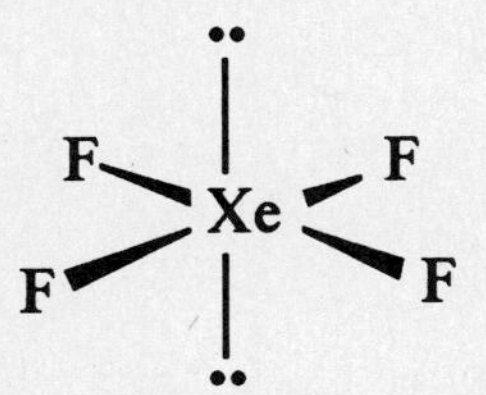


(In this and the following structures the LP electrons on the F atoms are not shown.)

With six groups attached to the Xe atom the geometry of the groups is octahedral. The lone pairs are located on the axial positions to minimize repulsion. The fluorine atoms are arranged in a square planar configuration.

Structure of $[XeF_3]^-$

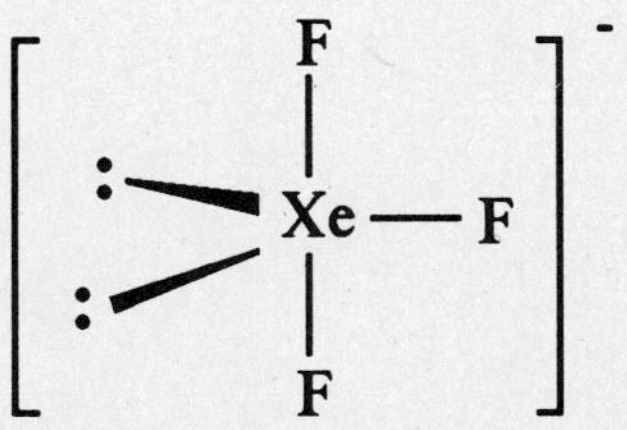


The ion is isoelectronic with $ClF_3$, which is also T-shaped with two lone pair of electrons in the equatorial plane.

24-54. (a) $XeF_6 + H_2O \rightarrow XeOF_4 + 2\,HF$

(b) The structural pair geometry of the compound is octahedral, but its molecular geometry is that of a square pyramid.

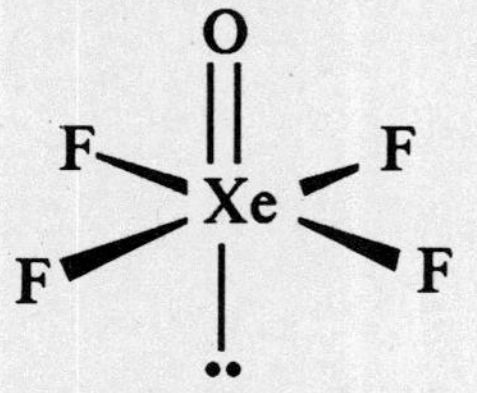


(The LP electrons on the F atoms are not shown.)

# CHAPTER 25
# THE TRANSITION METALS AND THEIR CHEMISTRY

## INSTRUCTOR'S NOTES

As this material ia usually covered at the end of the , there is not usually enough time to cover all of the chapter. However, enough of the basic ideas of crystal field theory can be introduced so that the color of complexes may be understand.

## LIST OF TRANSPARENCIES FOR THIS CHAPTER

| No. | Figure | Page | Subject |
|---|---|---|---|
| 120 | 25.2 | 1027 | Oxidation numbers of 4th period transition elements |
| 121 | 25.4 | 1029 | Atomic radii of transition metals |
| 122 | 25.5 | 1030 | The density of transition metals |
| 123 | 25.6 | 1031 | Sources of the elements |
| 124 | 25.7 | 1032 | Blast furnace chemistry |
| 125 | 25.13 | 1038 | Structure of $[Co(en)_3]^{3+}$ |
| 126 | 25.16 | 1045 | Isomers of $[Co(en)_2Cl_2]^+$ |
| 127 | 25.17 | 1046 | Stereoisomers of lactic acid |
| 128 | 25.18 | 1047 | Stereoisomers of lactic acid; mirror images |
| 129 | 25.50 | 1048 | Rotation of plane of polarized light |
| 130 | 25.23 | 1051 | The five *d* orbitals and their relation to ligand positions |
| 131 | 25.24 | 1051 | *d* orbital energy change as ligands approach; octahedron |
| 132 | 25.25 | 1052 | *d* orbital energy change as ligands approach; square plane |
| 133 | 25.27 | 1054 | Electron configurations for transition metal ions |
| 134 | 25.29 | 1057 | Color discs |
| 135 | 25.30 | 1059 | Spectrophotometer and spectrum of $Cu^{2+}$ |
| 136 | --- | 1066 | Unit cell of 1-2-3 superconductor |

## SUGGESTED DEMONSTRATIONS

The photographs in this chapter suggest many demonstrations, particularly of colored complexes. One that we have used to show the different colors of vanadium in its different oxidation states. (See Cornelius, R. *J. Chem. Educ.* **1980**, *57*, 316 for an experiment involving vanadium oxidation states.) This has already been outlined in Chapter 3 of this *Manual.* The chemistry shown in the photo essays on the chemistry of nickel and iron was taken from demonstrations outlined in Shakhashiri, Volume 1. (See also *Special Section: Silver Chemistry,* text page 809). In addition, there are several other demonstrations in Shakhashiri involving transition metal compounds and chemical equilibria.

The **Periodic Table Videodisc** of course has images of almost all of the transition elements, the uses of these elements, and their reactions with air, water, acids, and bases.

We have recently been working on a project for the REACTIVITY NETWORK (a group formed by Professor E. Mellon of Florida State University and sponsored by the National Science Foundation) to work out laboratory experiments and demonstrations involving chromium chemistry. The following are some ideas that have come from this work:

- Chromium salts are widely used in leather tanning, and an article on that subject appeared in *Chem Matters* magazine, April 1990. The Teacher's Guide to that issue contains a very interesting experiment on this subject.
- Chromium complexes are easy to prepare, and some references are —
  - (a) Gillard, P.D.; Mitchell, P.R. *J. Chem. Soc. A,* 1968, 2129.
  - (b) Pass, G.; Sutcliffe, H. *Practical Inorganic Chemistry,* 1974.
- We have purchased in a local supermarket an "alcohol breath tester." This tester uses the oxidation of alcohol in the breath by a dichromate salt (which is reduced to $Cr^{3+}$ or $Cr^{2+}$). The yellow salt in the tester turns green. This is a very simple demonstration that is certainly of interest to students.

The following list of additional demonstrations ia taken from the database by Bunce as published by Project SERAPHIM (see the introduction to this Manual).

- Greenaway, L.M.; Lancashire, R.J. "Cobalt(III) Ammines" *J. Chem. Educ.* **1982**, *59*, 419. (An undergraduate experiment. Could be used as such or to synthesize complexes for demonstration purposes.)
- Ophardt, C.E. "Cobalt Complexes in Equilibrium" *ibid.* **1980**, *57*, 453.

• Vitz, E. W. "The Ferrioxalate Actinometer: A Lecture Demonstration" *ibid.* **1981**, *58*, 655.

• Burke, B.A. "Chemical Equilibrium" *ibid.* **1977**, *54*, 29. (See also Shakhashiri, Volume 1, page 314).

• Arora, C.L "A Lecture Demonstration of the Various Oxidation States of Manganese" *ibid.* **1977**, *54*, 302.

• Ondrus, M.G. "Reaction Stoichiometry and Theoretical Yield" (the ammonium dichromate volcano) *ibid.* **1976**, *53*, 228. (See also Volume 1 of Shakhashiri's books on demonstrations, Figure 4.8 on page 111 of the text, and Chapter 4 in this *Manual.*)

• Walker, N. "Paramagnetic Properties of Fe(II) and Fe(III)" *ibid.* **1977**, *54*, 431.

• Bozzelli, J.W.; Barat, R.B. "The Thermite Lecture Demonstration" *ibid.* **1979**, *56*, 475. (See also Shakhashiri, Volume 1, page 85 . See textbook page 837.)

• Hill, J.W.; Foss, D.L.; Scott, L.W. "A Copper Mirror: Electroless Plating of Copper" *ibid.* **1979**, *56*, 752.

• Lau, O.W. "The Precipitation of Ferrous Hydroxide: A Lecture Demonstration" *ibid.* **1979**, *56*, 474.

• Long, R.H. "The Reducing Property of Ammonia" (Black CuO and white $(NH_4)_2CO_3$ react together to give red $Cu/Cu_2O$.)*ibid.* **1972**, *49*, A85.

# SOLUTIONS TO SELECTED STUDY QUESTIONS

## REVIEW QUESTIONS

25-1 Transition elements are characterized by having valence electrons assigned to d orbitals, while the inner transition elements, the lanthanides and actinides, have valence electrons assigned to f orbitals.

25-2. *Pyrometallurgy* uses high temperature methods to recover metals from their ores, as in the production of iron from iron ore. *Hydrometallurgy* uses aqueous solutions at relatively low temperatures for the same purposes. Copper, zinc, tungsten, and gold are recovered by the latter method.

25-3. Iron, nickel, and cobalt, among others,

25-4. (a) Transition elements have valence electrons assigned to d or f orbitals. When assigned to f orbitals, such elements are called the "inner transition elements," or lanthanides and actinides .

(b) A coordination compound is a grouping of a metal atom or ion with one or more Lewis bases functioning as ligands. For example, a $Pt^{2+}$ ion surrounded by two $NH_3$ molecules and two $Cl^-$ ions gives the coordination compound diamminedichloro-platinum(II), $Pt(NH_3)_2Cl_2$.

(c) Complex ions are a sub-set of coordination compounds. Here the coordination compound bears a charge, as in $[PtCl_4]^{2-}$.

(d) A ligand is a Lewis base (such as $Cl^-$ or $NH_3$) that can bind to a metal atom or ion functioning as a Lewis acid.

(e) A chelating ligand is one having two or more Lewis base sites that may be used to bind to a metal atom or ion.

(f) A bidentate ligand has two Lewis base sites involved in binding to one (or possibly two) metal atoms or ions.

25-5. **Common Coordination Numbers** **Possible Structure**

| Common Coordination Numbers | Possible Structure |
|---|---|
| 2 | linear |
| 4 | square planar or tetrahedral |
| 6 | octahedral |

25-6. Geometric isomers result when the ligand atoms bonded to the metal are sequenced in a different order about the metal.

*trans* isomer *cis* isomer

Optical isomers have the same atom-to-atom bonding sequence but they differ in the details of the arrangement of atoms in space. The result is that the isomers are non-superimposable mirror images of one another.

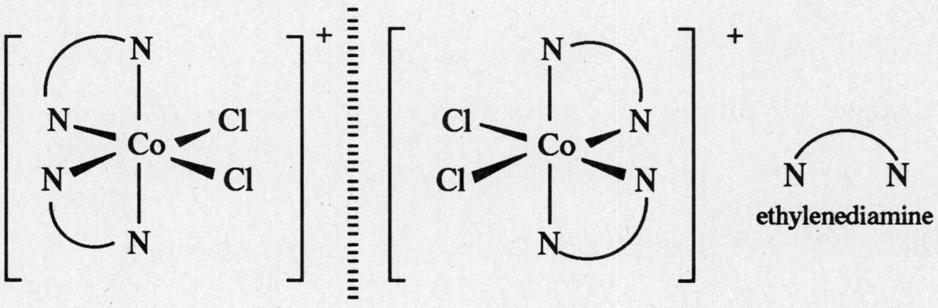


25-7. According to crystal field theory, the splitting of d orbitals into two sets arises from repulsions between electrons assigned to metal d orbitals and ligand electron pairs.

25-8. Factors that determine the spin of a coordination compound are (a) the energy difference between the two sets of d orbitals (Δ) and (b) the electron pairing energy. See page 1053.

## CONFIGURATIONS AND PHYSICAL PROPERTIES

25-9. $Cr^{3+}$ is a *paramagnetic ion* with three unpaired electrons.

| | | | | | | | |
|---|---|---|---|---|---|---|---|
| [Ar] | ↑ | ↑ | ↑ | | | | |
| | | | | | | | 4s |

$Cr^{6+}$ has no electrons remaining assigned to the 3d or 4s orbitals.

25-11. Figure 25.5 is a plot of the densities of the elements of the three transition series. In general, the elements increase in density toward the middle of the series, when the density again declines. Further, the density increases down a group.

(a) Fe is more dense than Ti

(b) Zr is more dense than Ti

(c) Os is more dense than Ti

(d) Hf is more dense than Zr

## METALLURGY

25-12. (a) $Cr_2O_3(s) + 2\ Al(s) \rightarrow Al_2O_3(s) + 2\ Cr(s)$

(b) $TiCl_4(\ell) + 2\ Mg(s) \rightarrow Ti(s) + 2\ MgCl_2(s)$

(c) $2\ [Ag(CN)_2]^-(aq) + Zn(s) \rightarrow 2\ Ag(s) + Zn^{2+}(aq) + 4\ CN^-(aq)$

25-14. *Step 1:* Calculate the liters of sulfuric acid required.

$1.00 \times 10^3$ g $FeTiO_3$ • 1 mol $FeTiO_3$ / 151.7 g = 6.59 mol $FeTiO_3$

6.59 mol $FeTiO_3$ • 3 mol $H_2SO_4$ / 1 mol $FeTiO_3$ • 1.00 L / 18.0 mol $H_2SO_4$ = 1.10 L

*Step 2:* Calculate the kilograms of $TiO_2$ produced.

6.59 mol $FeTiO_3$ • 1 mol $TiO_2$ / 1 mol $FeTiO_3$ • 79.90 g/mol $TiO_2$ = 527 g

527 g $TiO_2$ $\rightarrow$ 0.527 kg $TiO_2$

25-16. *Step 1:* Calculate the moles of Ni to be produced.

$908 \times 10^3$ g Ni • 1 mol / 58.69 g = $15.5 \times 10^3$ mol Ni

*Step 2:* Calculate the mass of NiO required.

$15.5 \times 10^3$ mol Ni • 1 mol NiO / 1 mol Ni • 74.70 g/mol = $1.16 \times 10^3$ g NiO

*Step 3:* Calculate the mass of $H_2$ required.

$15.5 \times 10^3$ mol Ni • 1 mol $H_2$ / 1 mol Ni • 2.016 g/mol = $3.12 \times 10^4$ g $H_2$

*Step 4:* Calculate the mass of CO required

$15.5 \times 10^3$ mol Ni • 4 mol CO / 1 mol Ni • 28.01 g/mol = $1.73 \times 10^6$ g CO

## LIGANDS

25-18.

| Compound | Metal Ion | Oxidation Number |
|---|---|---|
| $[Ni(NH_3)_6]SO_4$ | $Ni^{2+}$ | +2 |
| $[Co(NH_3)_4Cl_2]Cl$ | $Co^{3+}$ | +3 |
| $K_4[Fe(CN)_6]$ | $Fe^{2+}$ | +2 |
| $Ni(en)_2Cl_2$ | $Ni^{2+}$ | +2 |

25-20. $[Co(en)_2(H_2O)Cl]^{2+}$

## NAMING

25-22.

| Name of Compound | Formula |
|---|---|
| diamminetriaquahydroxochromium(II) nitrate | $[Cr(NH_3)_2(H_2O)_3(OH)]NO_3$ |
| hexaammineiron(III) nitrate | $[Fe(NH_3)_6](NO_3)_3$ |
| pentacarbonyliron(0) | $Fe(CO)_5$ |

25-23.

| **Ligand** | **Name** |
|---|---|
| $OH^-$ | hydroxo |
| $O^{2-}$ | oxo |
| $I^-$ | iodo |
| $C_2O_4^{2-}$ | oxalato |

25-25.

| **Formula** | **Name** |
|---|---|
| $[Ni(C_2O_4)_2(H_2O)_2]^{2-}$ | diaquadioxalatonickelate(II) |
| $[Co(en)_2(NH_3)Cl]^{2+}$ | amminechlorobis(ethylenediamine)cobalt(III) |
| $[Co(en)_2(NO_2)_2]^+$ | bis(ethylenediamine)dinitritocobalt(III) |
| $Pt(NH_3)_2(C_2O_4)$ | diammineoxalatoplatinum(II) |

25-27.

| **Formula** | **Name** |
|---|---|
| $[Cr(H_2O)_4(Cl)_2]Cl$ | dichlorotetraaquachromium(III) chloride |
| $[Cr(NH_3)_5SO_4]Cl$ | pentaamminesulfatochromium(III) chloride |
| sodium tetrachloropalladate(II) | $Na[PdCl_4]$ |

## ISOMERS

25-29. (a) $[Co(H_2O)_4Cl_2]^+$

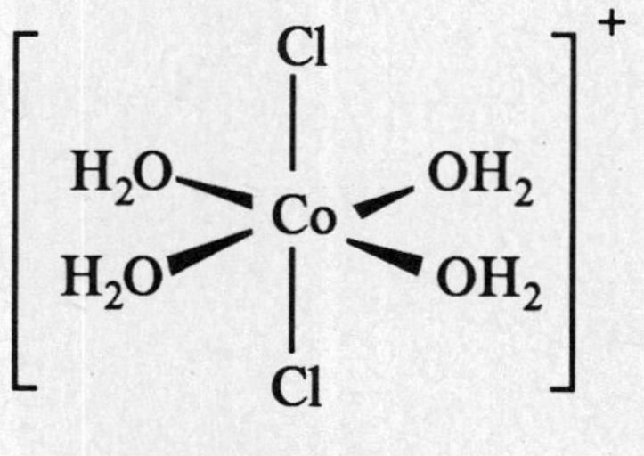


*trans* isomer

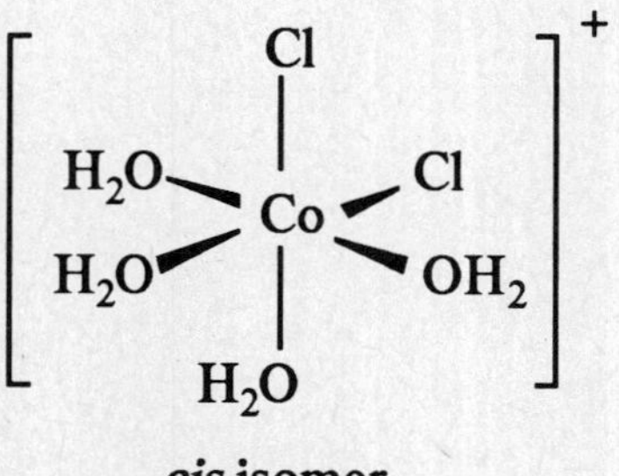


*cis* isomer

(b) $[Pt(NH_3)Br_3]^-$ , no isomers

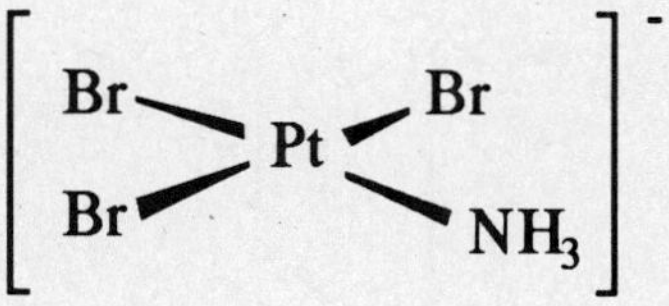

(c) $Co(H_2O)_3F_3$

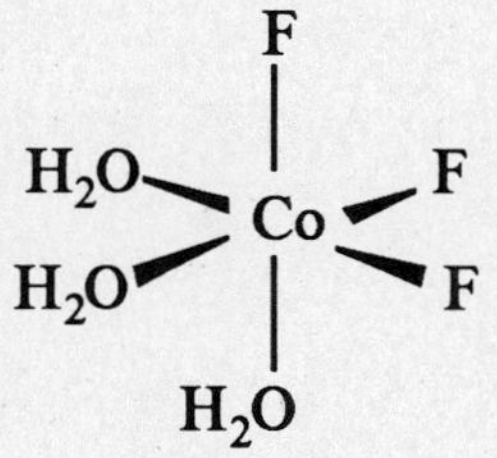
*fac* isomer

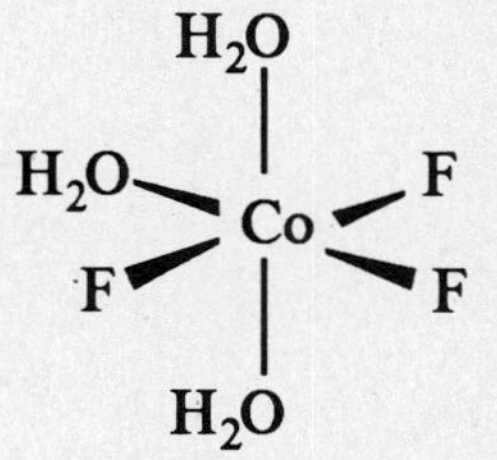
*mer* isomer

(d) $[Co(en)_2(NH_3)Cl]^{2+}$

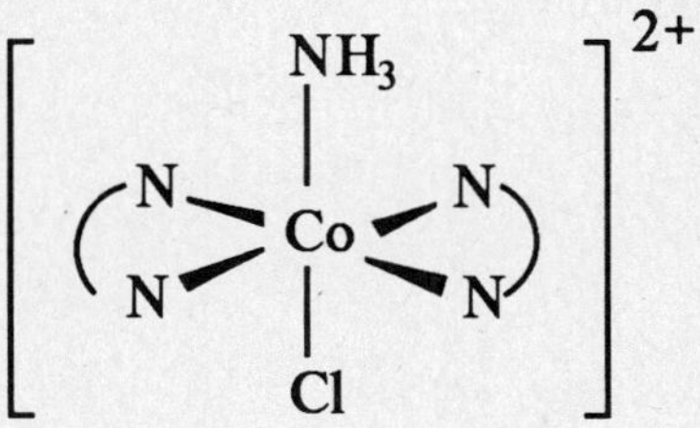
*trans* isomer

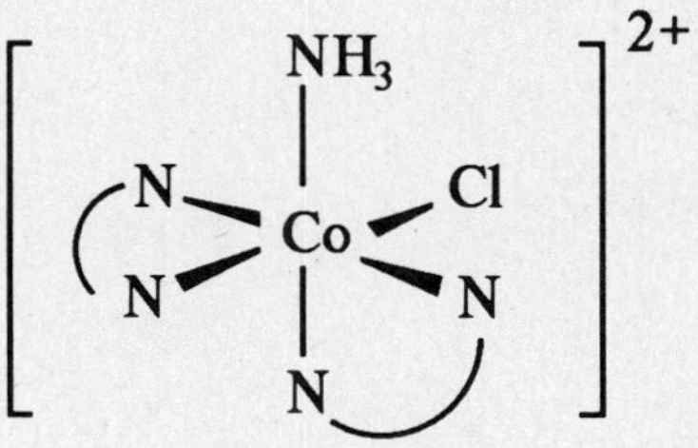
*cis* isomer

25-31. (a) This molecule does NOT have an enantiomer.

(b) This molecule has four different groups attached to the central atom and so has an enantiomer.

(c) This molecule does NOT have an enantiomer.

25-33. The isomers (geometric and optical) for the ion $[Cr(C_2O_4)_2(H_2O)_2]^-$.

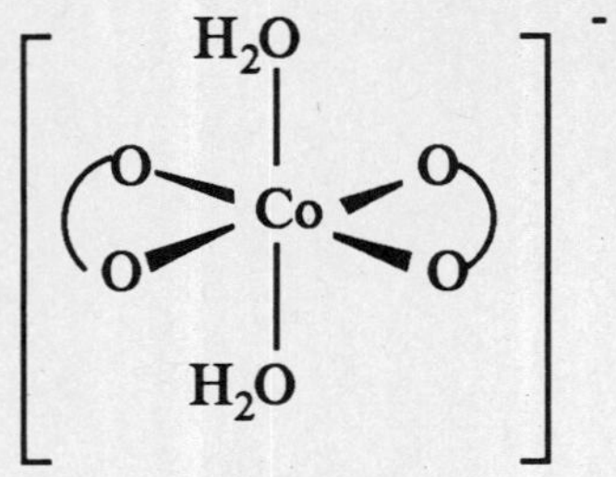


*trans* isomer

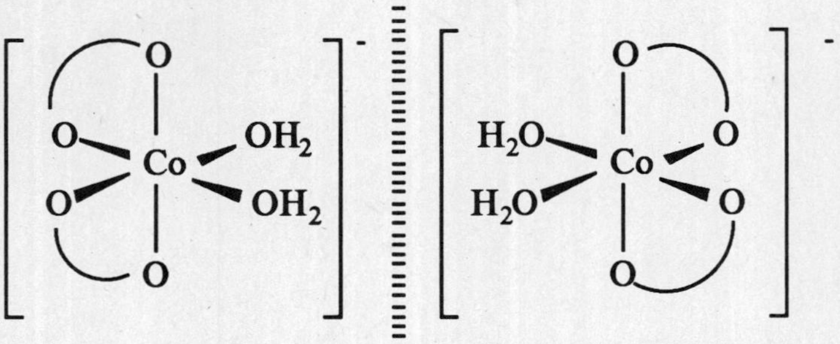


enantiometers of *cis* isomer

The bidentate, oxalato ion is represented by O—O.

## MAGNETISM OF COORDINATION COMPLEXES

25-35 (a) $[Fe(CN)_6]^{4-}$ contains $Fe^{2+}$, with the electron configuration $[Ar]3d^6$. Only the high spin ion is paramagnetic; the low spin ion is diamagnetic.

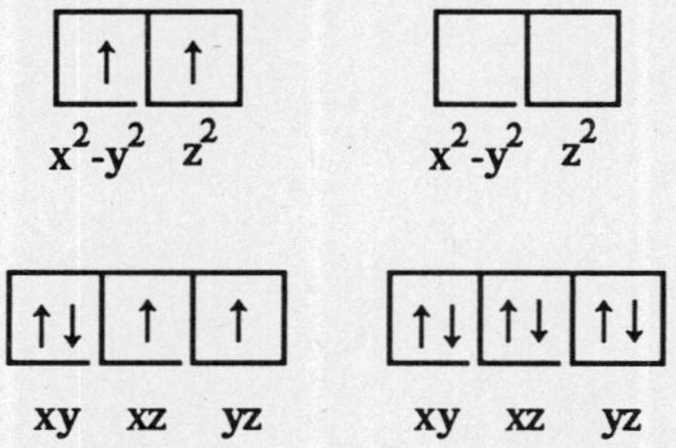


high spin low spin

(b) $[Co(NH_3)_6]^{3+}$ contains $Co^{3+}$, with the electron configuration $[Ar]3d^6$. The high and low spin configurations are the same as in part (a) above.

(c) $[Fe(H_2O)_6]^{3+}$ contains $Fe^{3+}$, an ion with the electron configuration $[Ar]3d^5$. Both the high and low spin ions are paramagnetic.

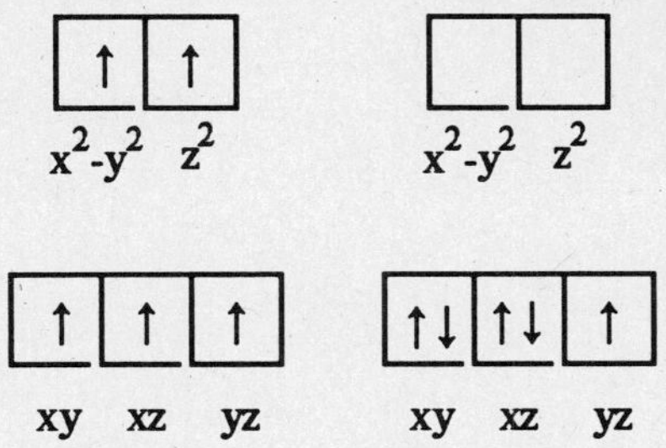


high spin low spin

25-37. The cyanide ion is a very strong field ligand and often leads to low spin complexes.

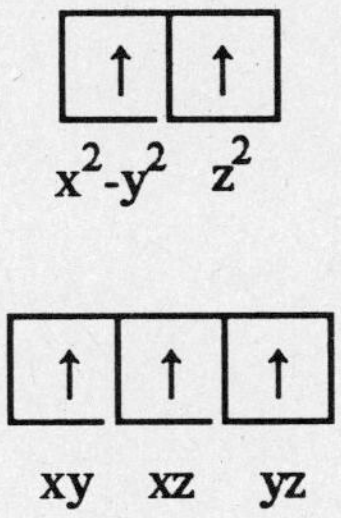


$[Mn(H_2O)_6]^{2+}$ is paramagnetic to the extent of 5 unpaired electrons.

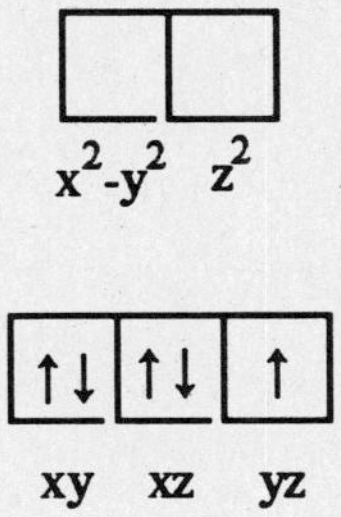


$[Mn(CN)_6]^{4-}$ is paramagnetic, but only to the extent of one unpaired electron.

## COLOR

25-38. $F^- < H_2O < NH_3 < CN^-$

25-40. Since the complex absorbs light in the red region of the spectrum, it transmits the complementary color, cyan. (See Figure 25.29).

## GENERAL QUESTIONS

25-42. (a) The central ion is $Pt^{2+}$.

(b) diammineoxalatoplatinum(II)

(c) The $NH_3$ ligands are monodentate and the oxalate ion is bidentate.

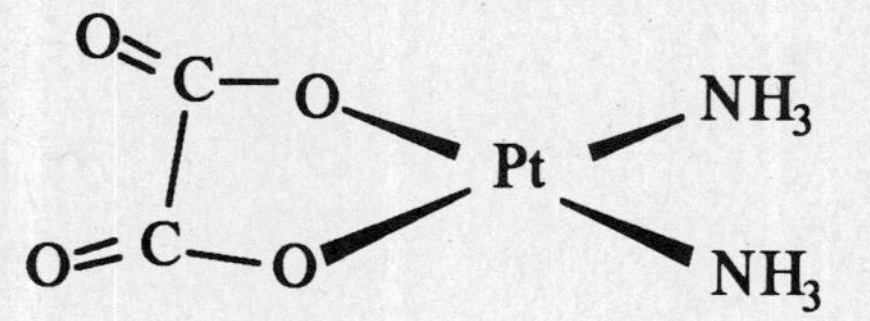


25-44. $[Pt(NH_3)_4][PtCl_4]$ — tetraammineplatinum(II) tetrachloroplatinate(II)

25-46. $Pt(NH_3)_2(NO_2)Cl$, diamminechloronitritoplatinum(II)

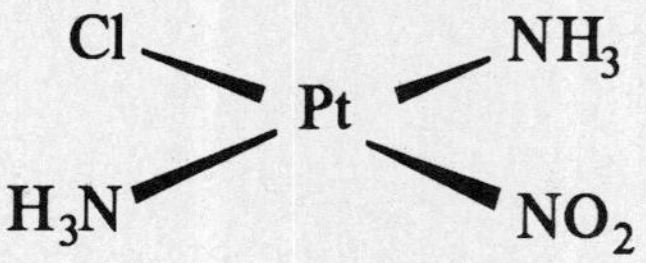


*trans* isomer

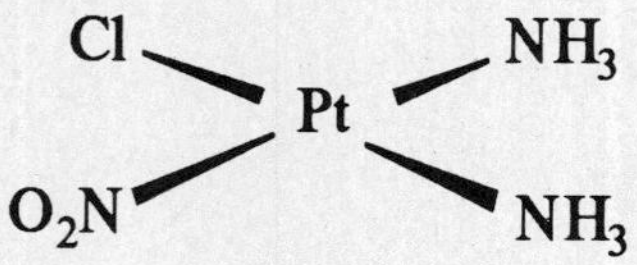


*cis* isomer

25-48. See Study Question 25-36. The aquated $Co^{3+}$ ion is diamagnetic owing to the fact that the ligand and the ion charge lead to a large value of $\Delta_0$. This leads to a low-spin, diamagnetic species, just as $[Co(NH_3)_6]^{3+}$ is diamagnetic. When $F^-$ ion is added in excess, most or all of the $H_2O$ ligands are displaced and the ion $[CoF_6]^{3-}$ is formed. Since $F^-$ is such a weak field ligand (and leads to a small value of $\Delta_0$), the complex is high-spin and paramagnetic.

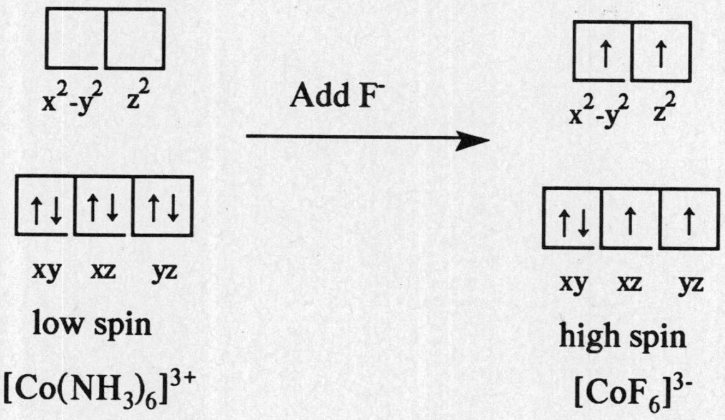

25-50. $K_{form}$ for the ammine complex = $10^8$, so $\Delta G$ = -45.6 kJ

$K_{form}$ for the en complex = $10^{18}$, so $\Delta G$ = -102.7 kJ

The difference in $\Delta G$ values for the two complexes is about 57 kJ, much greater than the difference in the enthalpies of the reactions. Thus, enthalpy changes alone cannot account for the great difference in $\Delta G$ values, and entropy must play a role in these reactions. At 298 K, $\Delta S^\circ$ for the hexammine complex is approximately -210 J/K, while that for the ethylenediamine reaction is about -48 J/K. This difference in entropy changes seems reasonable in view of the difference in the number of molecules reacting. For the ammine reaction, seven ions or molecules form one complex ion, while in the ethylenediamine reaction only four molecules or ions form one complex ion.

Another view, and one that provides more insight into chelation reactions, is obtained if we calculate $\Delta H^\circ$, $\Delta S^\circ$, and $\Delta G^\circ$ for the *exchange* of ligands.

$[Ni(NH_3)_6]^{2+}$ + 3 en → $[Ni(en)_3]^{2+}$ + 6 $NH_3$

$\Delta H^\circ$ = -8 kJ, $\Delta S^\circ$ = +165 J/K, and $\Delta G^\circ$ = -57.1 kJ

This shows that the chelated complex is much more stable than the complex ion with monodentate ligands and that the reaction is primarily *entropy driven.*

25-52. Cu,Al alloy + $HNO_3$ and HCl → $Cu^{2+}(aq)$ + $Al^{3+}(aq)$

$Cu^{2+}$ + $Al^{3+}$ + $NH_3(aq)$ → $Al(OH)_3(s)$ + $Cu^{2+}(aq)$

$Al(OH)_3(s)$ + heat → 3.8249 g $Al_2O_3$

3.8249 g $Al_2O_3$ • 1 mol $Al_2O_3$ / 101.96 g • 2 mol Al / 1 mol $Al_2O_3$

• 26.9815 g Al/1 mol Al = 2.0244 g Al

Weight percentage of Al = (2.0244 g Al / 2.1309 g alloy)100 = 95.00 % Al

Weight percentage of Cu = 5.00 % Cu

25-54.

| **Complex Ion** | **Formation Constant** |
|---|---|
| $[Co(NH_3)_6]^{2+}$ | $7.7 \times 10^4$ |
| $[Ni(NH_3)_6]^{2+}$ | $5.6 \times 10^8$ |
| $[Cu(NH_3)_4]^{2+}$ | $6.8 \times 10^{12}$ |
| $[Zn(NH_3)_4]^{2+}$ | $2.9 \times 10^9$ |

The formation constants are indeed in the order $Co^{2+} < Ni^{2+} < Cu^{2+} > Zn^{2+}$

# CHAPTER 26
# ORGANIC CHEMISTRY

## INSTRUCTOR'S NOTES

The major theme of this chapter is the reactivity of organic compounds. Our goal was to present only a very few reaction types and then encourage the student to see how these could be put together to synthesize more complex organic compounds. The idea of multistep reactions is a difficult concept for beginning students, but we believe it is a worthwhile goal.

As one can see from this chapter, we also believe it is important to use models when teaching the subject. The students needs to be trained to see molecular structures in three dimensions.

Since the first edition of this book was prepared, numerous, elegant computer programs have been developed that are sufficiently inexpensive that they can be used by students or in lectures. One example is "Beaker — An Expert System for the Organic Chemistry Student" by J. Brockwell, J. Werner, S. Townsend, and N. Tea. The program, which runs on a Macintosh, is available for about $30 from Brooks/Cole Publishing.

Many of the figures in the textbook and in this Manual were prepared using an excellent program, also for the Macintosh, "ChemDraw" from Cambridge Scientific Computing, 875 Massachusetts Avenue, Suite 41, Cambridge, MA 02139. Finally, we plan to begin experimenting with "Alchemy II", a molecular modeling program that runs on both IBM-compatible and Macintosh computers. It is available from TRIPOS Associates, 1699 S. Hanley Road, Suite 303, St. Louis, MO 63144.

## LIST OF TRANSPARENCIES FOR THIS CHAPTER

| No. | Figure | Page | Subject |
|---|---|---|---|
| 137 | --- | 1096 | $\pi$ orbital system of benzene |
| 138 | --- | 1106 | Fats and oils |
| 139 | 26.3 | 1115 | Petroleum refinery |

# SOLUTIONS TO SELECTED STUDY QUESTIONS

## ALKANES

26-2. Alkanes have the general formula $C_nH_{2n+2}$. Therefore, only $C_{14}H_{30}$ fits this criterion.

26-4. $C_nH_{2n+2} = C_{13}H_{28}$

26-6. As the molar mass of the compounds increases (as atoms are added) the weak dispersion forces between molecules become stronger, and the boiling points of the compounds increase.

26-8. (a) 2,3-dimethylpentane

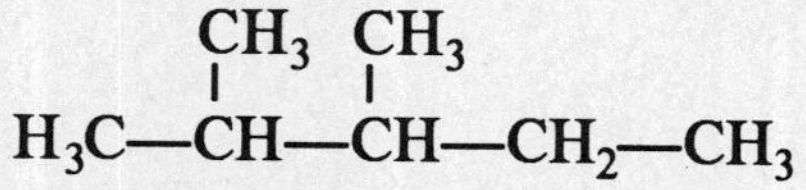


(b) 2,4-dimethyloctane

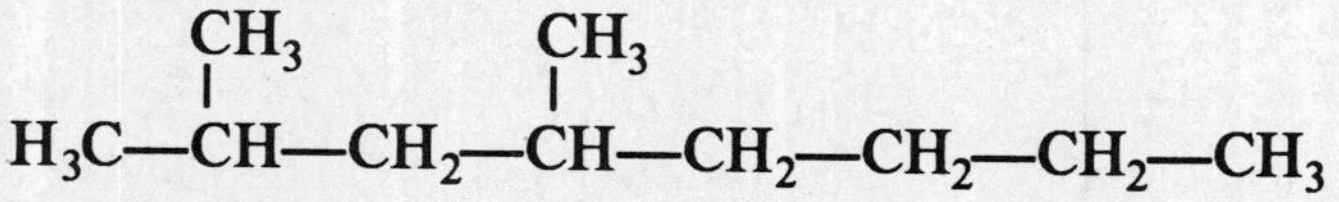


(c) 3-ethylhexane

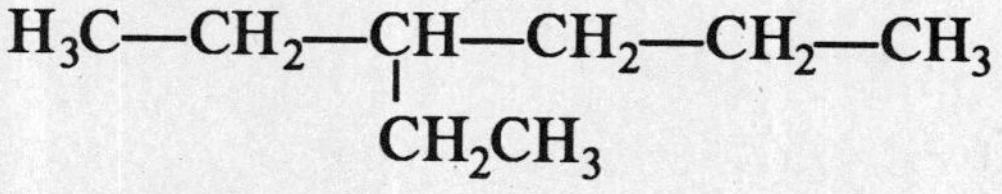


(d) 2-methyl-3-ethylhexane

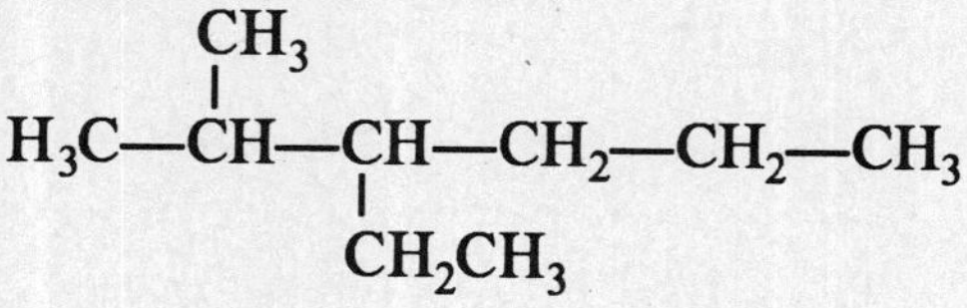


## FUNCTIONAL GROUPS

26-11. Structure and bonding in dimethyl ether

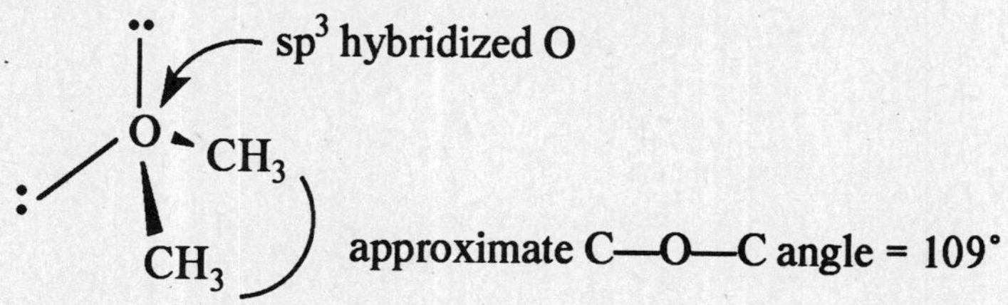

26-13. Structure and bonding in formic acid

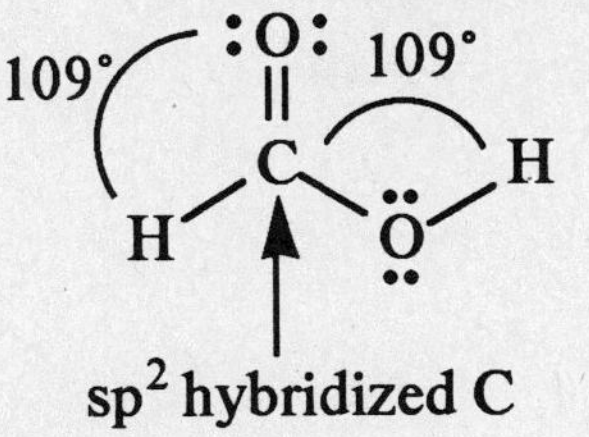


## ALCOHOLS

26-16. (a) 1-pentanol $H_3C—CH_2—CH_2—CH_2—CH_2—O—H$

2-pentanol 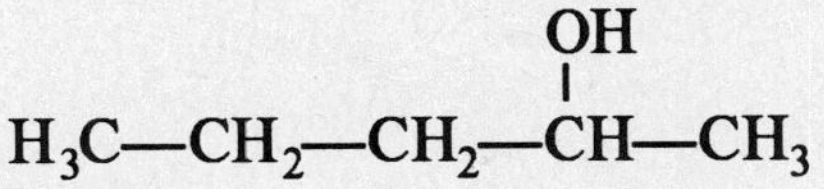

(b) 2-methyl-1-hexanol

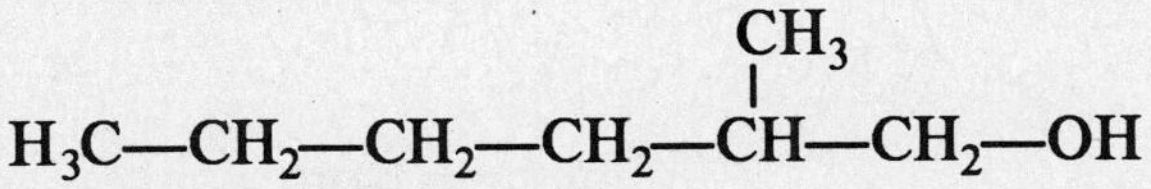

(c) 3,3-dimethyl-2-butanol

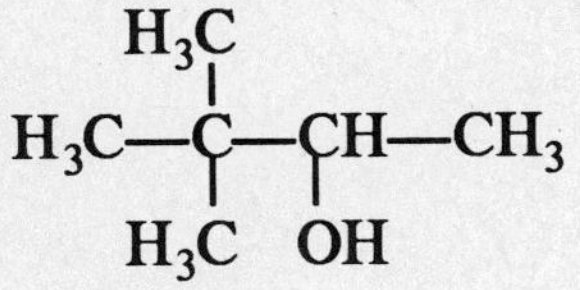

26-18. 1-butanol $H_3C—CH_2—CH_2—CH_2—O—H$

2-butanol 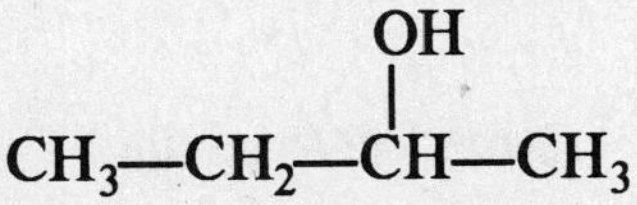

2-methyl-1-propanol 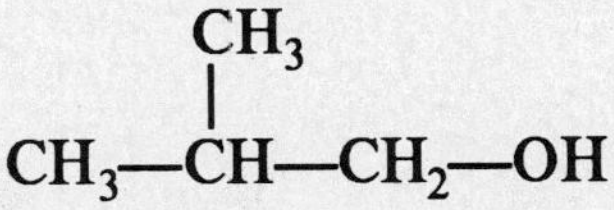

26-19. There are seven possible isomers for an alcohol with the formula $C_5H_{12}O$.

$H_3C-CH_2-CH_2-CH_2-CH_2-O-H$ 1-pentanol

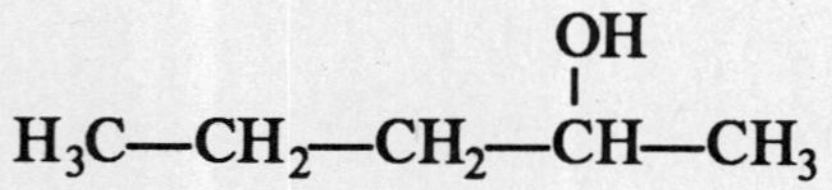

2-pentanol

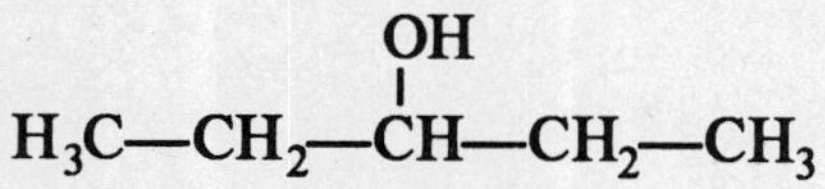

3-pentanol

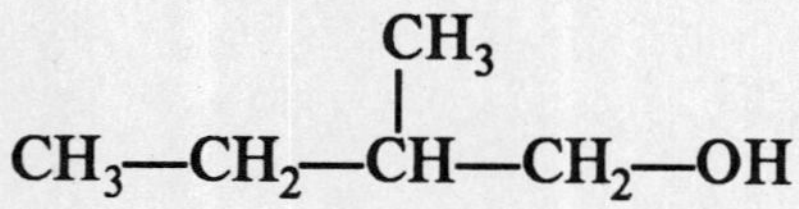

2-methyl-1-butanol

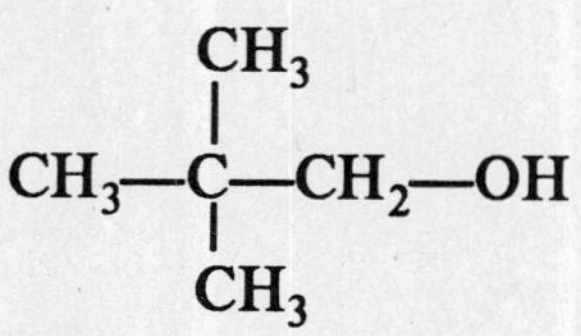

2,2-dimethyl-1-propanol

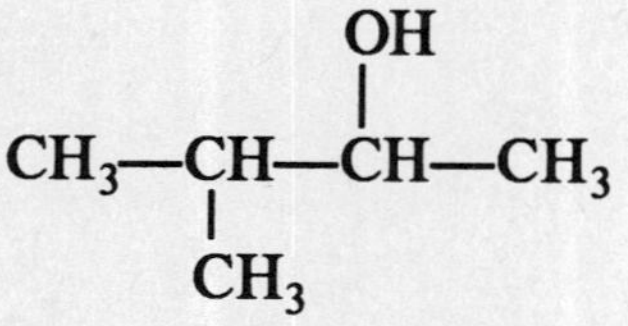

3-methyl-2-butanol

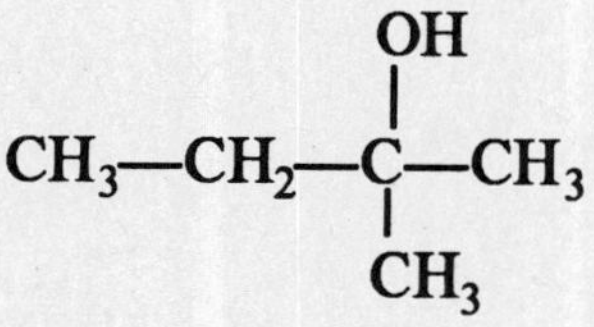

2-methyl-2-butanol

26-21. The ether will have the lower boiling point. The alcohol is capable of hydrogen bonding while the ether is not. The more strongly bound alcohol molecules would therefore boil at a higher temperature.

26-23. (a) *cis* and *trans* isomers of $C_2H_2Cl_2$

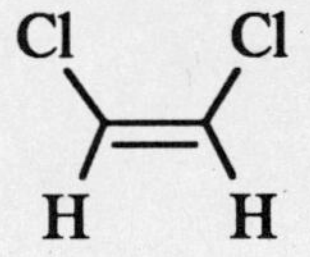

*cis* isomer

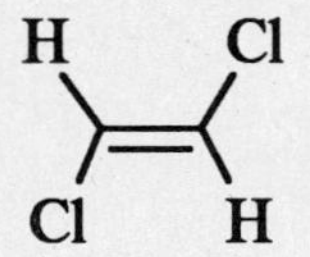

*trans* isomer

(b) *cis* and *trans* isomers of 3-methyl-3-hexene

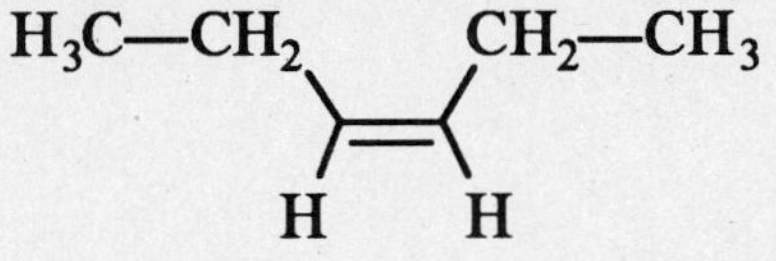

*cis* isomer

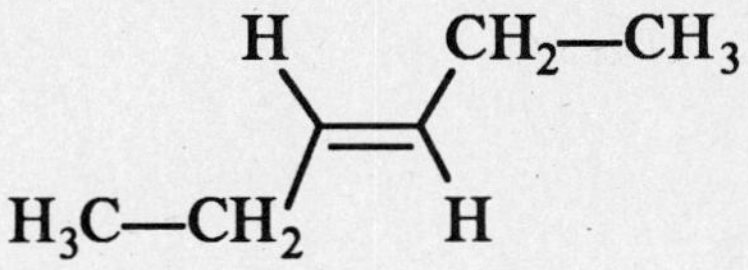

*trans* isomer

26-25. Reactions of 2-methyl-2-pentene

(a)

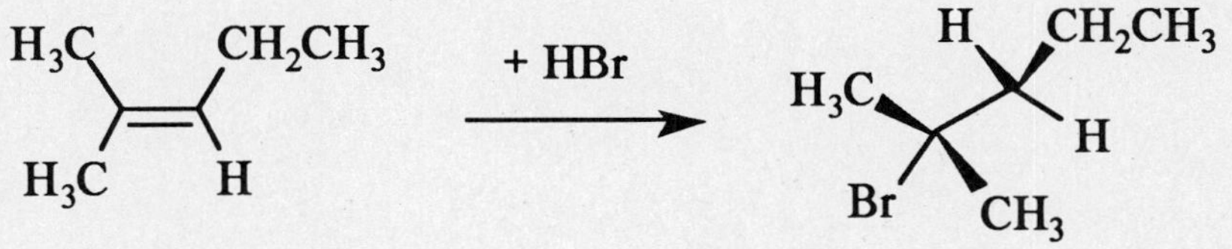

(b)

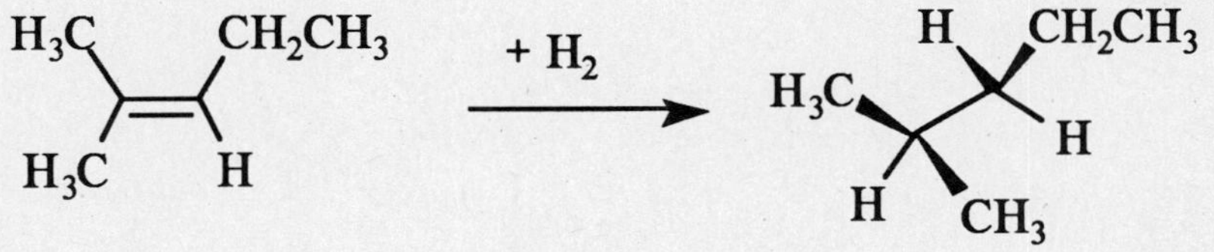

26-26. (a) 2-bromo-2-methylpentane

(b) 2-methylpentane

26-28. Synthesis of 2-butanol from 1-butanol

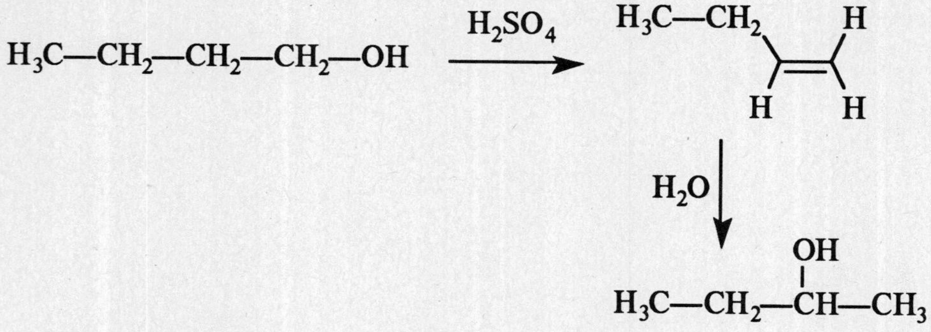

26-30. Structures of alkynes

(a) 3-hexyne $H_3C{-}CH_2{-}C{\equiv}C{-}CH_2{-}CH_3$

(b) 1-pentyne $H_3C{-}CH_2{-}CH_2{-}C{\equiv}CH$

(c) 4-methyl-2-pentyne

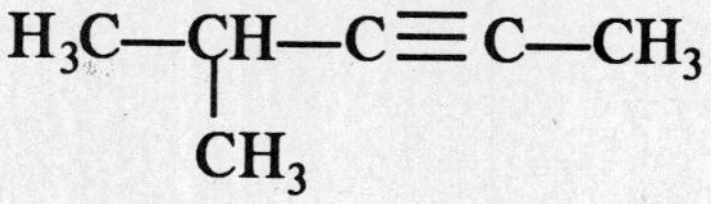

26-31. (a) $H_3C{-}CH_2{-}C{\equiv}CH \xrightarrow{+\,H_2} H_3C{-}CH_2{-}CH_2{-}CH_3$

(b) Reactions of 2-pentyne with HBr

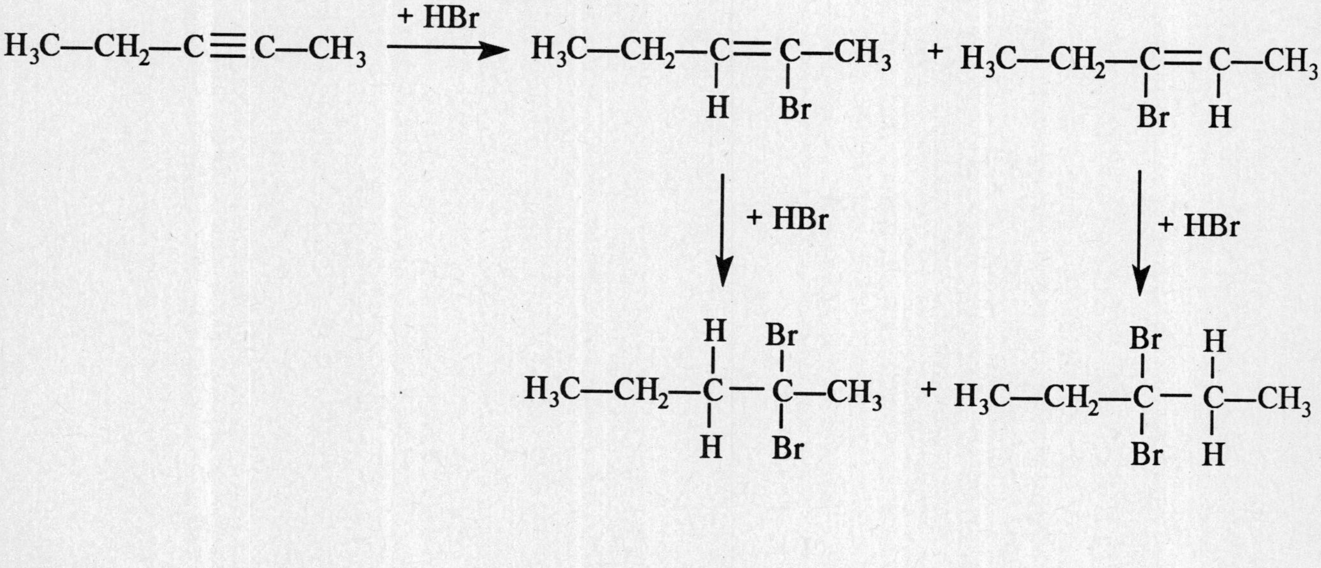

## BENZENE AND AROMATIC COMPOUNDS

26-32. Structures of aromatic compounds

(a) *o*-dichlorobenzene

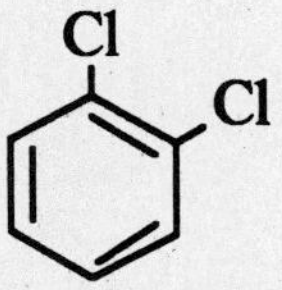

(b) *m*-chlorotoluene

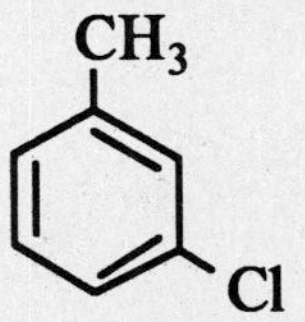

(c) *p*-diethylbenzene

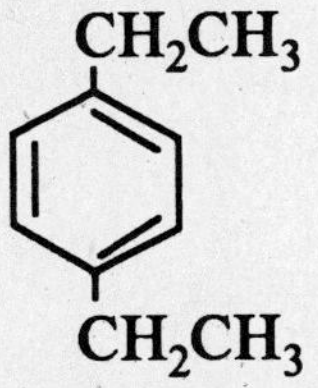

(d) *p*-chlorophenol

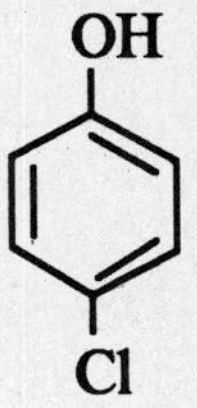

(e) *p*-chlorostyrene

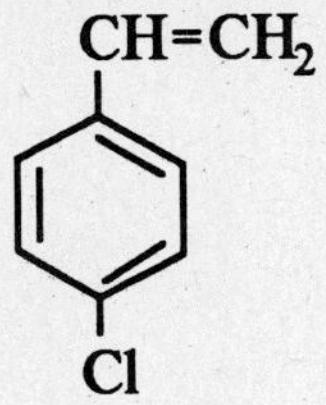

26-34. Synthesis of chlorobenzene

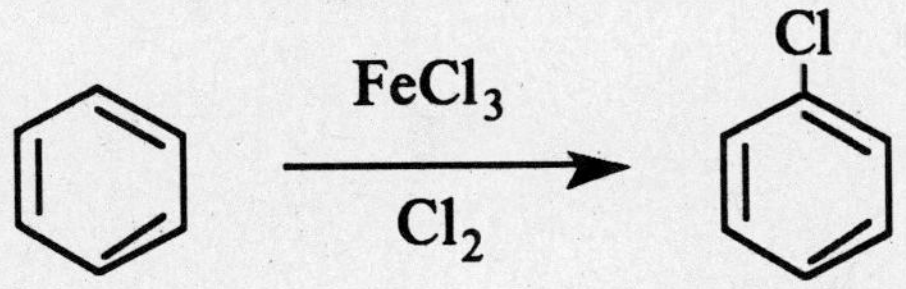

## ALKYL HALIDES

26-36. Elimination

$$H_3C-\underset{\displaystyle I}{CH}-CH_3 \xrightarrow[C_2H_5OH]{KOH} H_3C-CH=CH_3 + HI$$

Substitution

$$H_3C-\underset{\displaystyle I}{CH}-CH_3 \xrightarrow[H_2O]{NaOH} H_3C-\underset{\displaystyle OH}{CH}-CH_3 + NaI$$

Grignard reaction

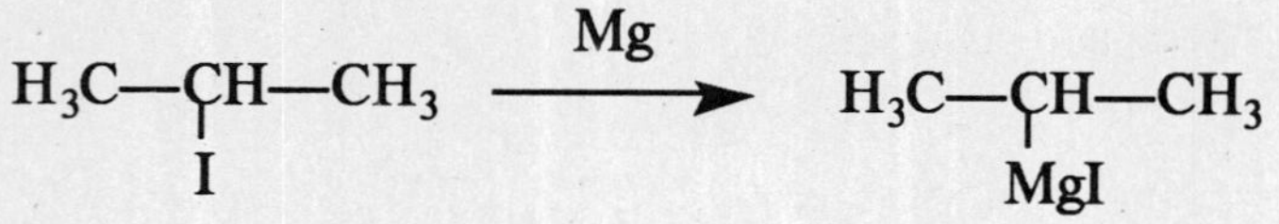

26-39. Synthesis of 2-deuteropropane

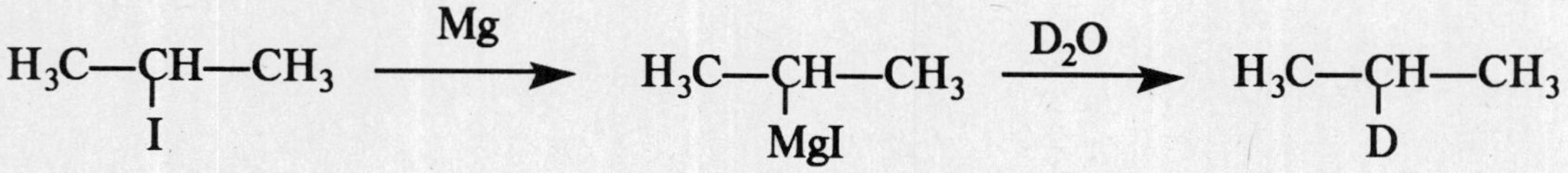

## CARBOXYLIC ACID AND ESTERS

26-41. (a)

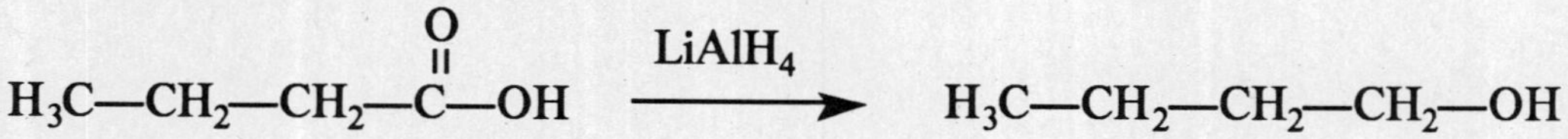

(b)

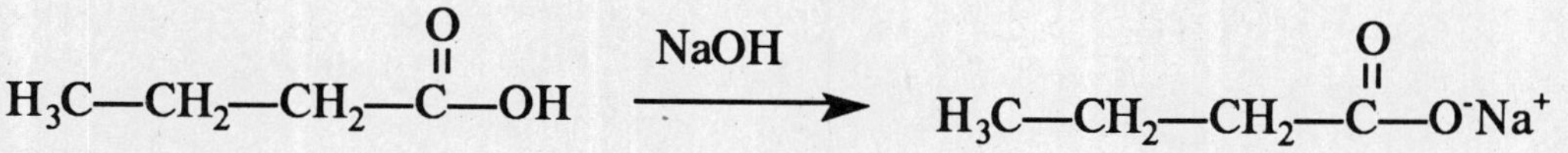

(c)

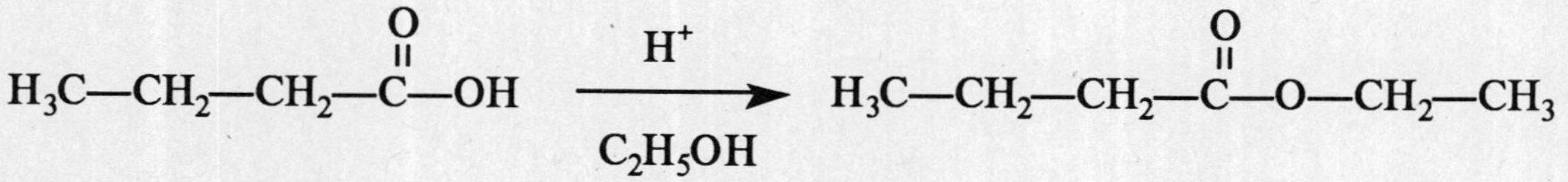

26-43. Benzoic acid and isopropyl alcohol → isopropyl benzoate

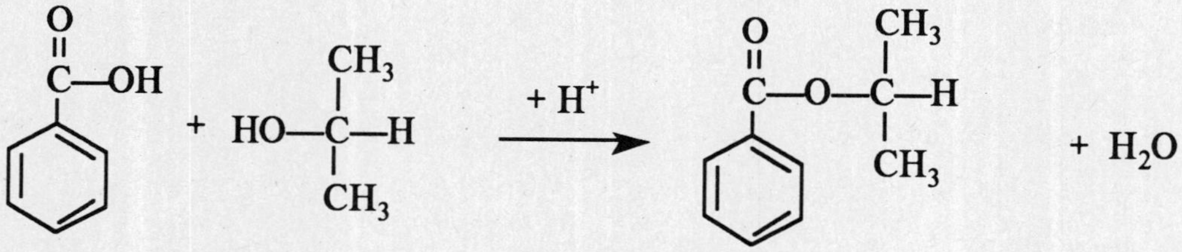

26-45. The saponification reaction is the hydrolysis of esters. Soaps (alkali metal salts — typically sodium salts of fatty acids) are produced when fats or oils are treated with an alkali metal hydroxide (e.g., NaOH).

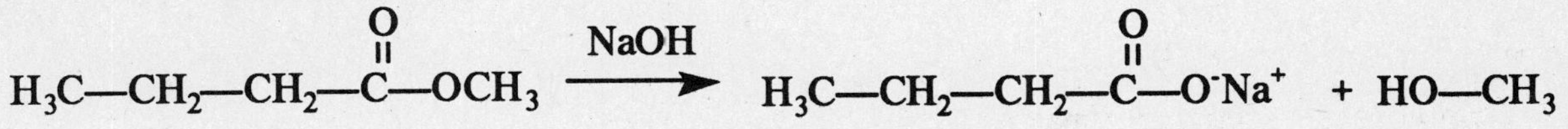


## ALDEHYDES AND KETONES

26-49. (a) propanone (common name: acetone)

(b) butanal

(c) 1-phenyl-2-propanone

26-51. Reactions of acetone

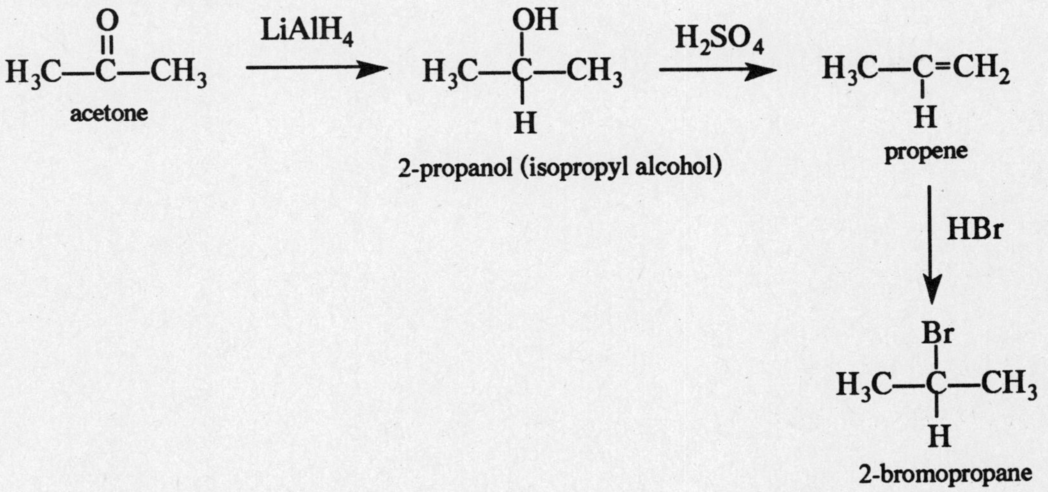


26-53. In the following reaction sequence, the first step is to prepare methyl magnesium bromide. Reacting this with 2-pentanone gives an intermediate that, when treated with acid and water, gives the final product.

$$H_3C\text{—}Br + Mg \longrightarrow H_3C\text{—}MgBr$$

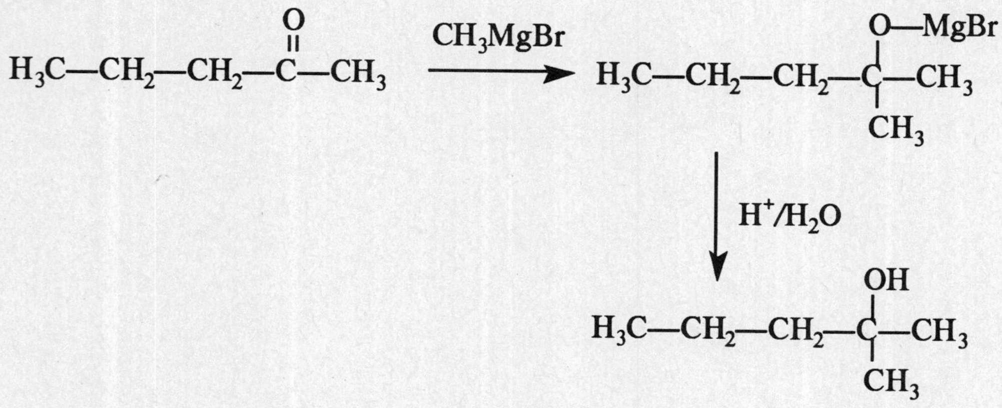


## AMINES AND AMIDES

26-55. (a) diphenylamine

(b) triethylamine

(c) 4-chloroaniline

26-56. $H_3CNH_2(aq) + H_2O(\ell) \rightleftarrows H_3CNH_3^+(aq) + OH^-(aq)$

$$K_b = 5.0 \times 10^{-4} = \frac{[H_3CNH_3^+][OH^-]}{H_3CNH_2} = \frac{[OH^-]^2}{0.10}$$

$[OH^-] = 7.1 \times 10^{-3}$ M

pH = 11.85

26-58. $(C_2H_5)_3N + HBr \rightarrow (C_2H_5)_3NH^+Br^-$ (triethylammonium bromide)

26-60. The condensation of an amine with acetyl chloride depends on the loss of a hydrogen atom bound to N as the amide bond is formed. Tri-*n*-propylamine has no such hydrogen atom, hence no reaction.

26-61. Refinery processes:

(a) Fractional distillation — separates the petroleum into fractions, depending on their relative boiling points.

(b) Catalytic cracking — reduces large alkanes to smaller, generally branched-chain alkanes.

(c) Reforming — catalytic process that converts alkanes to aromatic compounds.

(d) Steam reforming — reduces small alkanes (primarily ethane and propane) to their unsaturated analogs.

(e) Alkylation — catalytic process that transforms low molecular weight alkanes and alkenes into longer, branched-chain hydrocarbons.

## GENERAL QUESTIONS

26-63. Isomers of $C_4H_8Cl_2$

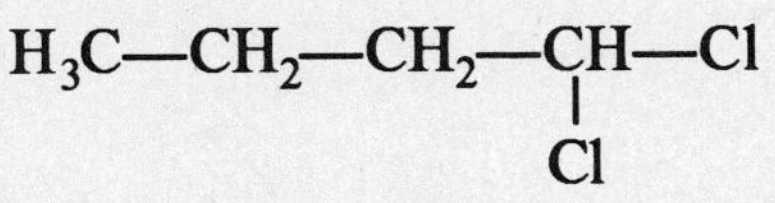

1,1-dichloropropane

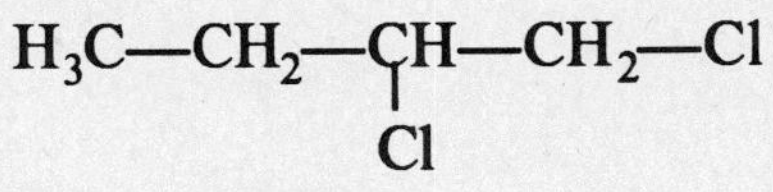

1,2-dichloropropane

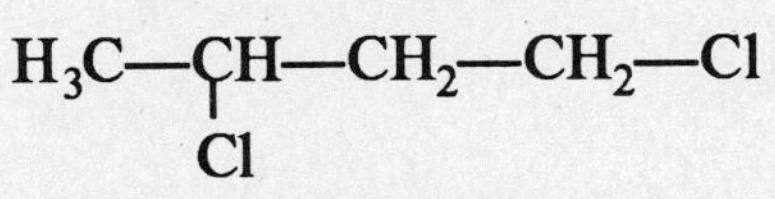

1,3-dichloropropane

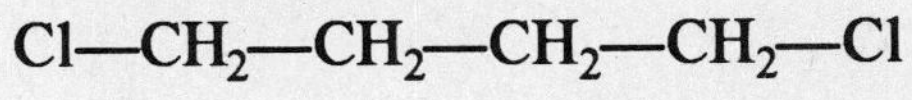

1,4-dichloropropane

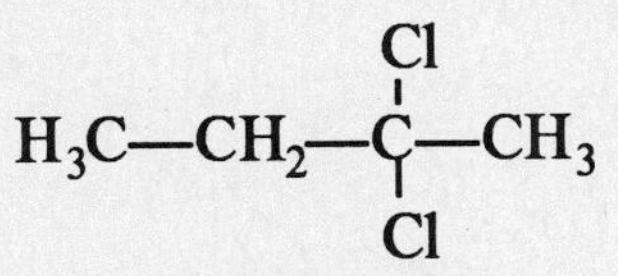

2,2-dichlorobutane

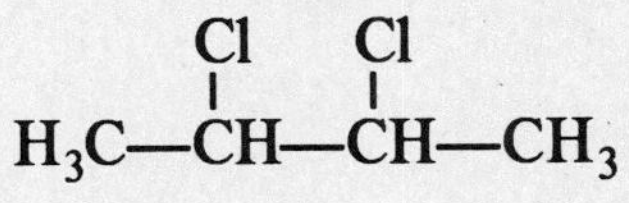

2,3-dichlorobutane

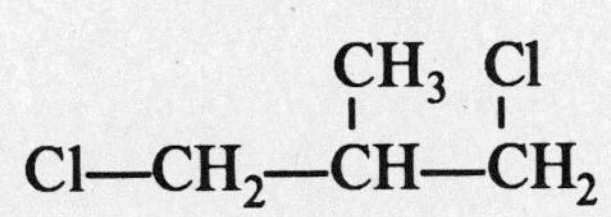

1,3-dichloro-2-methylpropane

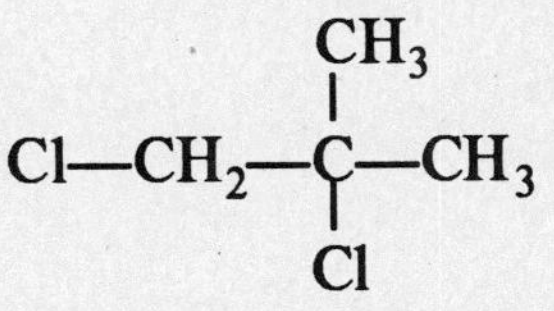

1,2-dichloro-2-methylpropane

26-65. Synthesis of 2-hexanol

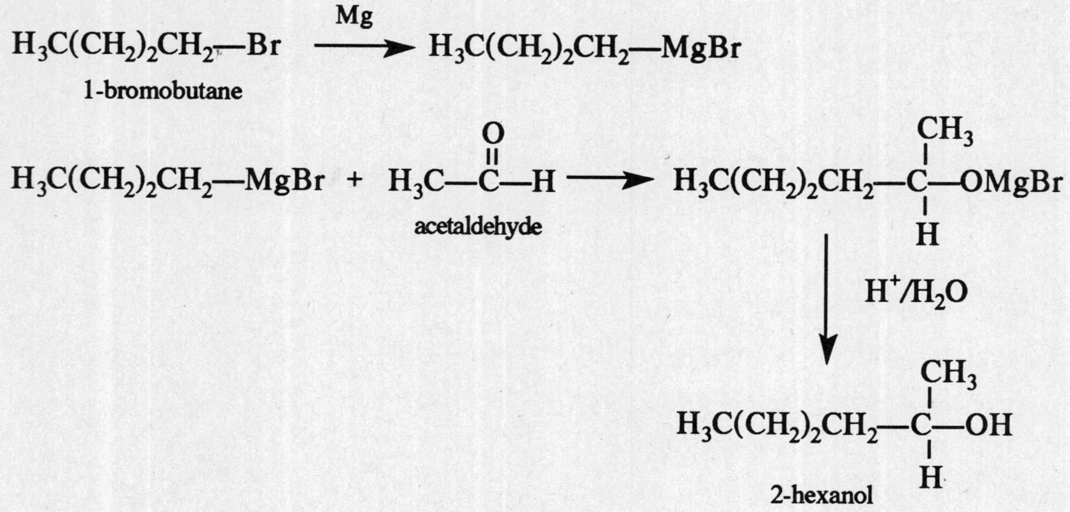

# CHAPTER 27
# POLYMERS: NATURAL AND SYNTHETIC MACROMOLECULES

## INSTRUCTOR'S NOTES

Our goal in writing this chapter was to provide an overview of some aspects of biochemistry and introduce students to the important area of polymer chemistry.

## LIST OF TRANSPARENCIES FOR THIS CHAPTER

| No. | Figure | Page | Subject |
|---|---|---|---|
| 140 | Table 27.1 | 1125 | Ten amino acids |
| 141 | 27.2 | 1126 | Polypeptide chain |
| 142 | 27.4 | 1128 | Helical structure of proteins |
| 143 | 27.5 | 1129 | Myoglobin |
| 144 | --- | 1131 | β - Glucose (structure at top of page) |
| 145 | 27.7(a) | 1132 | Sucrose |
| 146 | --- | 1134 | Five organic bases of nucleic acid and nucleotides (margin) |
| 147 | 27.9 | 1135 | Formations of nucleosides and nucleotides |
| 148 | 27.10 | 1135 | DNA backbone |
| 149 | 27.11 | 1136 | Structure of DNA |
| 150 | 2712 | 1137 | DNA double helix |

## SUGGESTED DEMONSTRATIONS

Molecular models in the area of biochemistry would be very helpful. Beginning students have a difficult time visualizing molecules in three dimensions and mirror image molecules give them particular difficulties.

Several reactions have been photographed for this chapter. In Figure 27.6 (page 1130)

$H_2SO_4$ is used as a dehydrating agent to convert a carbohydrate to carbon and water.

Cross-linked polyvinyl alcohol (Figure 27.14, page 1139) is a student favorite. About 40 grams of PVA are dissolved by heating in a liter of water. (Leave several hours for this process.) Solid sodium borate is stirred into a small quantity of this solution until cross-linking occurs to give the slimy material. To make the demonstration even more graphic, the slime can be colored with food coloring.

Directions for the preparation of a number of different polymers are given in Volume 1 of Shakhashiri's book of chemical demonstrations. The phenol-formaldehyde polymer in Figure 27.15 was synthesized according to experiment 3.3 (page 219). It is a dramatic reaction, since the polymer forms quite suddenly.

Nylon-66 is quite simple to synthesize (Figure 27.16). With care our students once were able to pull a continuous filament that stretched from the front of a large lecture room to the back.

The following demonstration experiments have been obtained from the database published by Project SERAPHIM at the University of Wisconsin (see the introduction to this *Manual*).

- Sperling, L.H. "On the Cross-Linked Structure of Rubber" *J. Chem. Educ.* **1982**, *59*, 651.
- Bramwell, F.B.; Spinner, M.L. "Phosphorescence" *ibid.* **1977**, *54*, 167.
- Schmulbach, C.D.; Ahmed, I.Y. "Demonstration Tests" (one involving the thermodynamics of a rubber band) *ibid.* **1976**, *53*, 775.
- Dole, M. *ibid.* **1977**, *54*, 754. ("Lecture Table Demonstration of Entropy," the classical "thermodynamics of a rubber band" experiment. See also Laswick, D.H. *ibid.* **1972**, *49*, 469.)
- Bieber, T.I. "Improving the Nylon Rope Trick" *ibid.* **1979**, *56*, 409.
- Rodriguez, F. "Lecture Demonstrations of Polymer Structure using Polarized Light" *ibid.* **1969**, *46*, 456.

# SOLUTIONS TO SELECTED STUDY QUESTIONS

## STEREOCHEMISTRY AND AMINO ACIDS

27-2. The molecule $CH_2ClBr$ does NOT have four different groups attached to the central atom.

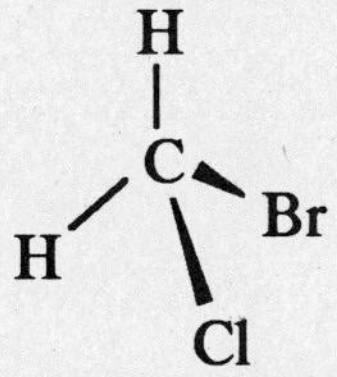

As manipulation of models would show, one cannot arrange the groups around the C atom to generate a molecule that is a nonsuperimposable mirror image of the structure above.

27-4. *Glycine* is not a chiral molecule. The C atom adjacent to the carboxylic acid group (—COOH) has two H atoms attached. Hence, it is not a chiral center.

## PROTEINS

27-6. Tripeptide from glycine, cysteine, and histidine

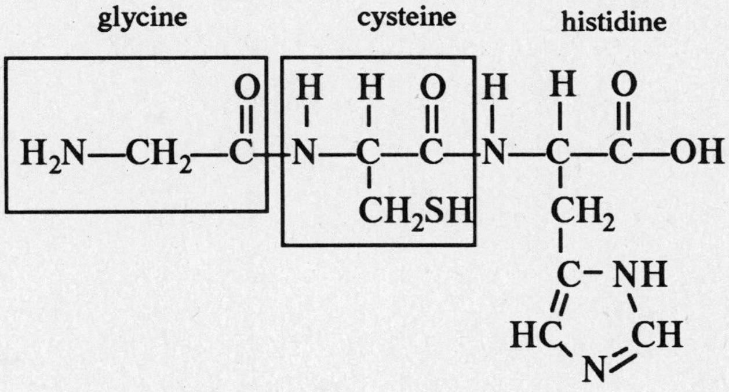

27-8. The α-helix (Figure 27.4) differs from the β-pleated sheet (Figure 13.7) at least in the positions of the hydrogen bonds. In the β-pleated sheet the H-bonds are between the peptide chains, whereas the H-bonds are within the chain in the α-helix and lead to the helical structure.

27-10. The interaction between the serine and aspartic acid units is due to hydrogen bonding, which may be represented as

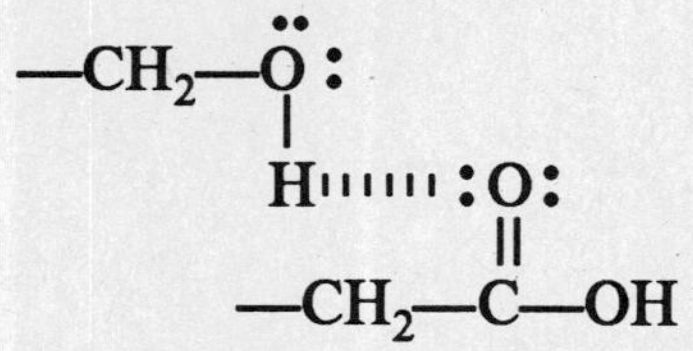


## CARBOHYDRATES

27-12. Chiral center: carbons 1, 2, 3, 4

Carbon 5 is NOT a chiral center.

27-14. Humans lack the ability to degrade the β-1,4 bonds in cellulose.

## NUCLEIC ACIDS

27-16. Reaction of β-D-ribose with adenine

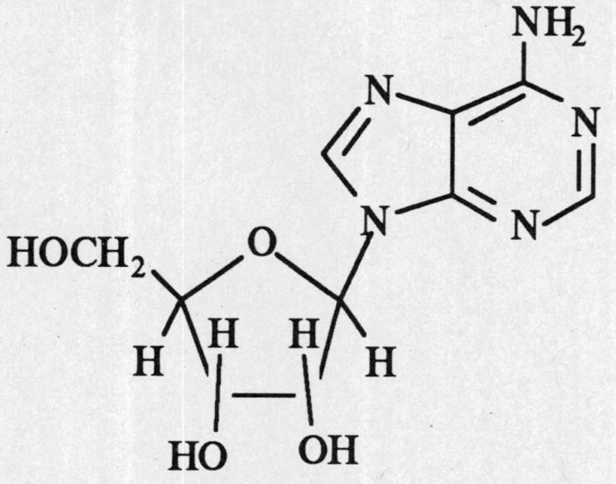


27-18. Synthesized RNA: Base 1 = uracil; Base 2 = adenine; and Base 3 = guanine

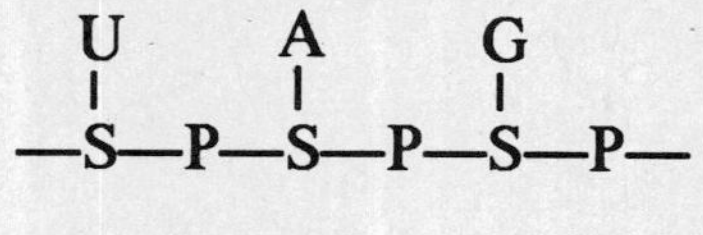


## SYNTHETIC POLYMERS

27-20. Nylon, Dacron, Orlon: clothes, fibers

Polyethylene: packaging, wrapping materials

Polyurethanes: “stretchy” fabrics

Neoprene rubber: tires
Polyvinyl chloride: household plumbing

27-22. 8.0 g polystyrene • 1.0 g styrene/1.0 g polystyrene • 1.0 mol styrene/104 g styrene
= 0.077 mol styrene

27-24. Copolymers are composed of two types of monomers. Examples include nylon, saran, and dacron.

27-26. Disaccharide formation: see pages 1131 and 1132
Polysaccharide formation: see page 1133
Forming nucleosides: see page 1135
Nylon formation: see page 1145

27-28. Structure of Saran

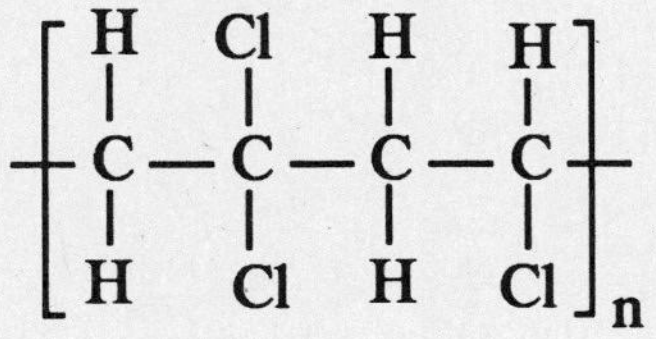

27-30. $2.74 \times 10^6$ lb nylon • 454 g nylon / 1.00 lb nylon • 1 mol nylon / 226 g nylon
= $5.50 \times 10^6$ mol nylon

$5.50 \times 10^6$ mol nylon • 1 mol adipic acid / 1 mol nylon • 146.1 g/ mol
= $8.04 \times 10^8$ g adipic acid

$5.50 \times 10^6$ mol nylon • 1 mol $(CH_2)_6N_2H_4$ / 1 mol nylon • 116.2 g/ mol
= $6.40 \times 10^8$ g diamine